Laboratory Manual for

Fundamentals of HVACR

Fourth Edition

Carter Stanfield
Athens Technical College

David Skaves
Maine Maritime Academy

Pearson

8 2023

 Pearson

ISBN 10:   0-13-684063-9

ISBN 13: 978-0-13-684063-3

INTRODUCTION

This Laboratory Manual is designed to accompany the *Fundamentals of HVACR*, fourth edition. The laboratory exercises are grouped by similar topics into twelve sections. Exercise number designations are related to specific units appearing in the textbook. As an example, the first lab exercise *1.1 HVACR Jobs* is supported by information presented in Unit 1 of the textbook. Each title reflects what is to be covered during the lab exercise. As an example, the exercise title, *Setting a Thermostatic Expansion Valve*, clearly identifies the general subject of the exercise. The Table of Contents contains a listing of each exercise number and title in sequence. This provides a simple reference for locating specific exercises, which best match your individualized program curriculum.

Laboratory worksheets are 3-hole punched and can be readily detached from the manual. This allows them to be separated and inserted into course folders or notebooks and sequenced in a manner most suitable for your program curriculum. The student name, exercise date, and instructor's approval are located in the top right corner on every worksheet. In this manner, student work can be organized, collected, and easily assessed.

Refer to textbook Unit 1

Lab 1.1 HVACR Jobs

Exercise 2 Unit 1

Lab 1.2 HVACR Organizations

Lab 3.1 Safety Data Sheets (SDS)

Refer to textbook Unit 3

CONTENTS

SECTION 5 HVACR ELECTRICAL SYSTEMS AND COMPONENTS

SECTION 7 HEATING SYSTEMS

LAB 1.1 HVACR JOBS

LABORATORY OBJECTIVE

You will learn about job opportunities in the HVACR field.

FUNDAMENTALS OF HVACR TEXT REFERENCE

Unit 1 Introduction to Heating, Ventilation, Air Conditioning, and Refrigeration

REQUIRED TOOLS, EQUIPMENT, AND MATERIALS

Computer with Internet access

PROCEDURE

Do online research to learn about the job prospects for the field of Air Conditioning Technology. Here are a few suggested web sites:

1. AHRI – ahrinet.org
2. Department of Labor – https://www.bls.gov/ooh/home.htm
3. HVAC Excellence – http://HVACexcellence.org/
4. Interstate Renewable Energy Council – https://www.hvaccareermap.org/

Based on your research, list three different HVACR jobs.

Based on your research, list three employment advantages of the HVACR field.

Based on your research, what type of preparation is required for the HVACR field?

LAB 2.1 HVACR ORGANIZATIONS

LABORATORY OBJECTIVE
You will learn about professional organizations in the HVACR field.

FUNDAMENTALS OF HVACR **TEXT REFERENCE**
Unit 1 Introduction to Heating, Ventilation, Air Conditioning, and Refrigeration

REQUIRED TOOLS, EQUIPMENT, AND MATERIALS
Computer with Internet access

PROCEDURE
Do online research to learn about HVACR trade organizations. Here are some suggested web sites:

ACCA www.acca.org

AHRI www.ahrinet.org

ASHRAE www.ashrae.org

CARE www.escogroup.org/care

HVAC EXCELLENCE www.escogroup.org/hvac

NATE www.natex.org

PAHRA www.pahrahvacr.org

PHCC www.phccweb.org

RSES www.RSES.org
   Register for FREE subscription to RSES Journal at www.rses.org/studentjournal.aspx

SKILLS USA www.skillsusa.org

WOMEN IN HVACR www.womeninhvacr.org

Give the purpose and/or mission of each organization.

How do you become a member of each organization?

Where is each organization located?

LAB 3.1 SAFETY DATA SHEETS (SDS)

LABORATORY OBJECTIVE
The student will demonstrate an understanding of the information provided on Safety Data Sheets for one of the refrigerant types to be used in the refrigeration laboratory.

***FUNDAMENTALS OF HVACR* TEXT REFERENCE**
Unit 3 Safety

LABORATORY NOTES
This lab exercise should always be the first one performed at the beginning of the course. Students should be introduced to Safety Data Sheets on the first day of lab.

Safety Data Sheets are required by law and have important information listed in specific areas so that they are easily read by emergency personnel.

You should read Safety Data Sheets on any material before you use it so you know how to use it properly and safely as well as knowing what to do if there is an accident involving the material.

REQUIRED TOOLS, EQUIPMENT, AND MATERIALS
Refrigerant Material Safety Data Sheet

SAFETY REQUIREMENTS
None

PROCEDURE

Step 1. Review a copy of a refrigerant SDS supplied by the Lab Instructor and answer the following questions:

    A. What are the First Aid Measures for inhalation?

    B. What are the First Aid Measures for refrigerant exposure to the eyes?

    C. What are the First Aid Measures for refrigerant exposure to the skin?

4

D. What are the First Aid Measures for refrigerant ingestion?

E. What type of protective clothing is worn for the eyes and face?

F. What type of protective clothing is worn for the hands, arms, and body?

QUESTIONS

To get help in answering some of the following questions, refer to the *Fundamentals of HVACR* text Unit 3.

(Circle the letter that indicates the correct answer.)

1. If an area on a SDS does not apply to a product:
   A. it should be left blank.
   B. it should be filed in with the refrigerant type.
   C. there should be a line through it.
   D. it should be marked as non-applicable.
2. The section of an SDS that provides the properties of the material, such as boiling point, vapor pressure, etc. is:
   A. Section I.
   B. Section II.
   C. Section III.
   D. Section IV.
3. If you do not follow the instructions for proper use and handling of a material as listed on a SDS:
   A. you could be injured.
   B. a customer could be injured.
   C. you could be fired.
   D. All of the above are correct.
4. Section I of the SDS provides:
   A. the manufacturer's name and address.
   B. the fire and explosion hazard data.
   C. hazardous ingredients / identity information.
   D. All of the above.
5. The HMIS information tells health care workers a relative number according to how significantly the material will affect health, how reactive it is, and its flammability.
   A. True.
   B. False.

LAB 3.2 HVACR PPE

LABORATORY OBJECTIVE
You will learn how and when to use the types of personal protective equipment (PPE) used in HVACR.

FUNDAMENTALS OF HVACR TEXT REFERENCE
Unit 3 Safety

LABORATORY NOTES
We will identify the types of PPE used in the HVACR field, discuss when it should be worn, and demonstrate its use.

REQUIRED TOOLS, EQUIPMENT, AND MATERIALS
Shop safety manual
Safety glasses
General mechanics gloves
Welding/Brazing dark safety glasses
Leather gloves for brazing
Arc-blast rated gloves
Respirator
Self-contained breathing apparatus SCBA
Examples of proper and improper apparel
Examples of proper and improper footwear

PROCEDURE
The instructor will discuss and demonstrate the use of HVACR PPE. You will be prepared to discuss and demonstrate proper PPE, including:

- Appropriate shop clothing
- Appropriate shop footwear
- When safety glasses should be worn
- When welding glasses are required
- When to use general shop gloves, welding gloves, or arc blast gloves
- When to use a respirator
- When to use SCBA equipment
- How to find out the correct PPE for any type of job

LAB 3.3 LOCATE SAFETY EQUIPMENT

LABORATORY OBJECTIVE

You will learn the location of safety equipment in the shop.

FUNDAMENTALS OF HVACR TEXT REFERENCE

Unit 3 Safety

LABORATORY NOTES

We will use the shop safety manual to locate and identify all the shop safety equipment.

REQUIRED TOOLS, EQUIPMENT, AND MATERIALS

Shop safety manual
Safety glasses
SDS Book
First aid kit
Electrical lock out kit
Fire extinguishers
Eye wash station
Shop vent fan
Main electric switchgear
Main gas valve shutoff
Emergency exit plan

PROCEDURE

Be prepared to show the instructor the following shop safety equipment:

- SDS book
- First aid kit
- Electrical lock out kit
- Fire extinguishers
- Eye wash stations
- Shop vent fan
- Main electric switchgear
- Main gas valve shutoff
- Emergency exit plan

LAB 3.4 LADDER SAFETY

LABORATORY OBJECTIVE
You will demonstrate how to safety set up and climb both step ladders and extension ladders.

FUNDAMENTALS OF HVACR TEXT REFERENCE
Unit 3 Safety

LABORATORY NOTES
The instructor will demonstrate the correct way to set up a step ladder and an extension ladder. They will then discuss and demonstrate the correct way to climb and work on a ladder. You will demonstrate your ability to safely set up both a step ladder and an extension ladder. You will then demonstrate your ability to safely climb a ladder.

REQUIRED TOOLS, EQUIPMENT, AND MATERIALS
Shop safety manual

Safety glasses

Six-foot fiberglass step ladder

Extension ladder

SAFETY REQUIREMENTS
  A. Safety glasses
  B. Ladder should be fiberglass or wood—NOT aluminum

PROCEDURE
The instructor will demonstrate how to set up and climb a typical extension ladder and a typical extension ladder. You will then demonstrate your ability to:

- Safely set up a six-foot step ladder.
- Safely climb a six-foot step ladder.
- Safely set up an extension ladder.
- Safely climb an extension ladder.
- Discuss what NOT to do on a step ladder.
- Discuss what NOT to do on an extension ladder.

LAB 4.1 USING HAND TOOLS

LABORATORY OBJECTIVE

You will demonstrate your ability to safely perform the following tasks:

- use screwdrivers and nut drivers to secure air conditioning unit panels
- tighten and loosen an assembly using wrenches
- change a hacksaw blade
- cut a piece of metal with a hack saw
- change a drill bit in a power drill
- drill a hole in a piece of metal

FUNDAMENTALS OF HVACR TEXT REFERENCE

Unit 4 Hand and Power Tools

LABORATORY NOTES

You will perform a variety of manipulative tasks using hand tools. These include removing and replacing panels on units, disassembling and reassembling a bolted assembly, changing a hack saw blade, cutting metal with a hack saw, and drilling a hole in a piece of metal.

REQUIRED TOOLS, EQUIPMENT, AND MATERIALS

Safety glasses

Flat blade screwdriver

Phillips screwdriver

¼" nut driver

5/16" nut driver

OR 6-in-1 with all the above

Adjustable jaw wrench

Air conditioning units with panels

Metal for cutting and drilling

Hack saw

Power drill

Extension cord

Drill bits

Fixed wrenches

Assembly to tighten

SAFETY REQUIREMENTS
    A. Safety glasses
    B. Gloves
    C. All power tools MUST be grounded
    D. Never hold items in your hand while drilling into them

PROCEDURE

Safety Glasses are required for all tasks!

Panels

Determine which tool you need: flat blade, Phillips, ¼" nut driver, or 5/16" nut driver. If the fasteners can use more than one type, the order of preference is nut driver, Phillips, then flat blade.

Bolted Assembly

Disassemble and reassemble the bolted assembly assigned by the instructor. Care should be taken to use tools appropriately. Make certain that wrenches fit properly. Pliers are NOT a substitute for a wrench where a wrench is needed!

Hack Saw Blade

To change the hacksaw blade: loosen the knurled wheel on the handle end. The blade should now remove easily. Pay attention to the direction of the teeth when replacing the blade; they should face forward. Many blades have an arrow that should point towards the front of the saw, away from the handle. After replacing the blade, tighten the knurled wheel by turning it clockwise.

Hack Saw

Mount the metal to be cut securely in a vise. If it is round, it should go in the jaws designed for holding tubing and round objects. Place the blade on the metal and use approximately one forward stroke per second with a moderate down pressure.

Drill Bit

Make sure the drill is unplugged! The chuck is the part of the drill that holds the drill bit. Two types of chucks are used: keyed chucks and hand chucks. Keyed chucks have holes in the base of the chuck to insert the chuck key. If the drill uses a chuck key, use the chuck key to loosen the chuck. If it is a hand chuck, loosen by hand. Insert the drill bit and tighten. The bit should be in the center of the chuck.

Drilling

NEVER USE A POWER DRILL WITHOUT SAFETY GLASSES!

Mount the metal to be drilled in a vise. Making a dimple in the metal where the hole will be helps keep the drill bit from wandering when the hole is started. Use a punch to make a dimple in the metal where the hole will be drilled by striking the punch with a hammer. Place the point of the drill bit in the dimple before stating the drill. Start the drill using a moderate amount of pressure. Forcing the bit to drill too quickly will only dull the bit and slow down the entire process.

LAB 5.1 FASTENER IDENTIFICATION

LABORATORY OBJECTIVE

You will identify the following types of fasteners and describe their use:

- sheet metal screws
- wood screws
- nuts and bolts
- nails

FUNDAMENTALS OF HVACR TEXT REFERENCE

Unit 5 Fasteners

LABORATORY NOTES

You will receive an assortment of fasteners. You will identify each fastener, what it is used for, and the tool(s) required to use it.

REQUIRED TOOLS, EQUIPMENT, AND MATERIALS

Safety glasses

Assorted fasteners

PROCEDURE

Examine the assortment of fasteners provided by the instructor; identify the type of fastener, what the fastener is used for, and what tools are used with the fastener. Complete the chart on the next page as you identify the fasteners.

#	FASTENER NAME	TOOL	USE
1			
2			
3			
4			
5			
6			
7			
8			
9			
10			
11			
12			
13			
14			
15			
16			
17			
18			
19			
20			

LAB 6.1 TAKING MEASUREMENTS

LABORATORY OBJECTIVE
You will take measurements of the following:

- Area of a square filter
- Are of a round duct
- Volume of a cylinder

FUNDAMENTALS OF HVACR TEXT REFERENCE
Unit 6 Measurements

LABORATORY NOTES
You will be using measuring equipment to calculate the areas of a square filter, a round duct, and the volume of a cylinder.

REQUIRED TOOLS, EQUIPMENT, AND MATERIALS
Yard stick
Meter stick
Square air filter
Round duct
Small cylinder
Measuring cup
One gallon jug

PROCEDURE

Measuring the Area of a Square Filter (English Units)
Examine a yard stick and note how it is divided into fractions of an inch. Measure the length and the width of a square air filter to the nearest ¼ inch.

Length _____ Width_____
Area = L X W =_____in^2

Area in ft^2 = in^2 ÷ 144 in^2/ft^2 = _____ft^2

Measuring the Area of a Square Filter (SI Units)
Using a meter stick, measure the length and the width of the same square air filter to the nearest centimeter.

Length_____ Width_____

Area = L X W =_____cm^2

Now convert your answer from cm^2 to in^2

(cm^2 x 0.155 in^2/cm^2) _____in^2

Measuring the Area of a round duct (English Units)

Measure the diameter of a round duct to the nearest ¼ inch.

Diameter_____ Area $=\dfrac{\pi d^2}{4}$ = _____in^2

Area in ft^2 = $\dfrac{in^2}{144\dfrac{in^2}{ft^2}}$ = _____ft^2

Measuring the Volume of a Cylinder (English Units)
Measure the diameter and the length of a cylinder to the nearest ¼ inch.

Diameter_____ Length _____

Area $=\dfrac{\pi d^2}{4}$ = _____in^2

Volume = Area x Length _____in^3

Volume in ft^3 = $\dfrac{in^3}{1728\dfrac{in^3}{ft^3}}$ = _____ft^3

Measuring Liquids
Fill a measuring cup with water to the level of one cup. Pour it into the gallon jug container. Record the total number of cups required to fill the container.

Total cups = _____

Now begin filling the cylinder that had previously been measured using one cup of water at a time. Record the total number of cups required to fill the cylinder.

Total cups = _____

Now calculate how many gallons of water the cylinder can hold.

(# of cups to fill the cylinder) ÷ (# of cups to fill the one gallon jug) = _____ gallon(s).

Calculated Value vs Measured Value
You had previously calculated the volume of the cylinder.
One ft^3 is equal to 7.48 gallons.
Calculate the cylinder volume in gallons: (Volume in ft^3) ÷ 7.48 gallons/ ft^3 = _____ gallon(s).

How does the calculated volume in gallons compare to the measured volume?

LAB 7.1 BALLOON—GAS VOLUME VS. TEMPERATURE

LABORATORY OBJECTIVE

The purpose of this lab is to demonstrate the change in gas volume accompanying a change in temperature.

FUNDAMENTALS OF HVACR TEXT REFERENCE

Unit 7 Properties of Matter

LABORATORY NOTES

You will compare the volume of a fully inflated mylar balloon at room temperature to the volume of the same balloon after it has been cooled below freezing.

REQUIRED TOOLS, EQUIPMENT, AND MATERIALS

1 ultralow temp freezer, –40°
1 mylar balloon

PROCEDURE

Observe the volume of a fully inflated mylar balloon.

Place a fully inflated mylar balloon in the freezer and wait 5 minutes for the gas in the balloon to cool.

Remove the balloon from the freezer and observe its reaction.

Note: This step must be done quickly because the gas in the balloon heats up quickly.

LAB 7.2 WEIGHING SOLIDS, LIQUIDS, AND GASES

LABORATORY OBJECTIVE
The purpose of this lab is to learn to weigh solids, liquids, and gases using the shop scales.

FUNDAMENTALS OF HVACR TEXT REFERENCE
Unit 6 Measurements
Unit 7 Properties of Matter

LABORATORY NOTES
You will weigh a solid, a liquid, and a gas using both traditional and metric weights. For the liquid and the gas you will use the scale's zero setting to compensate for the weight of the container.

REQUIRED TOOLS, EQUIPMENT, AND MATERIALS
1 brick

1 quart of oil

1 empty oil quart container

1 empty recovery cylinder

Vacuum pump

PROCEDURE
Solids

- Zero the scale with nothing on it.
- Place the brick on the scale.
- Use the scale units key to see the weight in pounds, pounds and ounces, and kilograms.
- Record below.

Item	Pounds	Pounds & Ounces	Kilograms	grams
Brick				

Liquids

- Place an empty container on the scale.
- Zero the scale with the empty oil container on the scale.
- Replace the empty container with the oil you wish to weigh.
- Use the units key to see the weight in pounds, pounds and ounces, and kilograms.
- Record below.

Item	Pounds	Pounds & Ounces	Kilograms	grams
Oil				

Gas

- With the instructor's help, evacuate an empty refrigerant recovery cylinder.
- Zero the scale with the recovery cylinder on it.
- Open the valve and let air enter the cylinder.
- Record the weight of the air below.

Item	Pounds	Pounds & Ounces	Kilograms	grams
Air				

SUMMARY

1. What is the difference between pounds and pounds and ounces?

2. Why is it necessary to zero the scale with the container on it when weighing a liquid or a gas?

3. What does the increase in the scale reading before and after letting air into the cylinder tell you about air?

LAB 7.3 MEASURING DENSITY, SPECIFIC VOLUME, AND SPECIFIC GRAVITY

LABORATORY OBJECTIVE
Demonstrate an understanding of density, specific volume, and specific gravity.

FUNDAMENTALS OF HVACR TEXT REFERENCE
Unit 7 Properties of Matter

LABORATORY NOTES
You will measure the weight and volume of a solid, a liquid, and a gas. Using those measurements you will calculate the density, specific volume, and specific gravity of each.

REQUIRED TOOLS, EQUIPMENT, AND MATERIALS
Standard ruler
Calculator
Metric ruler
Digital scale
Vacuum pump
Bicycle air pump
1 quart of oil
1 empty oil quart container
1 empty recovery cylinder

Procedure is on following pages.

PROCEDURE

Testing Solids—A Brick

Measure the volume and weight of the brick and complete the table below.

Dimension	Inches	Centimeters
Width		
Length		
Height		
Volume (WxLxH)		

Weight in Ounces	Weight in grams

Use the figures from the table and the formulas below to calculate the density, specific volume, and specific gravity of the brick

Brick Characteristics, English Units

Density =_____ ÷ _____ = _____ ounces per cubic inch

      ounces   cubic inches

Specific Volume= _____ ÷ _____ = _____ cubic inches per ounce

      cubic inches   ounces

Specific Gravity = _____ ÷ 0.58 = _____ times as heavy as water

      Density   (Density of water in ounces per cubic inch)

Brick Characteristics, Metric Units

Density =_____ ÷ _____ = ____

      grams ÷ cubic centimeters = grams per cubic centimeter

Specific Volume_____ ÷ _____ = _____

      cubic centimeters   grams   cubic centimeters per gram

Specific Gravity = Density without the units

Note: Since a gram is defined as the weight of 1 cubic centimeter of water, the specific gravity is the same thing as the density when using grams and cubic centimeters.

Testing Liquids—Oil

- Place the empty container on the scale and zero the scale.
- Replace the empty container with the full quart of oil.
- Record the weight of the full quart of oil in both ounces and grams.

Weight in Ounces	Weight in grams

Note: A quart IS a volume measurement, so we already know the volume. However we need to convert the quart to cubic inches. There are approximately 30 quarts in a cubic foot, and there are 1728 cubic inches in a cubic foot. Dividing 1728 by 30, 1 quart would be 57.6 cubic inches. Look on the quart container for the metric volume. You will see 946 milliliters. Since a milliliter is defined as 1 cubic centimeter in volume, 946 cubic centimeters is the metric volume. Use these figures to calculate the Density, Specific Volume and Specific Gravity.

<u>Oil Characteristics, English Units</u>

Density = _____ ÷ _____ = _____ounces per cubic inch
          ounces   cubic inches

Specific Volume = _____ ÷ _____ = _____ cubic inches per ounce
          cubic inches   ounces

Specific Gravity = _____ ÷ 0.58 = _____ times as heavy as water
          Density   (Density of water in ounces per cubic inch)

<u>Oil Characteristics, Metric Units</u>

Density = _____ ÷ _____ = _____ grams per cubic centimeter
          grams   cubic centimeters

Specific Volume = _____ ÷ _____ = _____
          cubic centimeters   grams   cubic centimeters per gram

Specific Gravity = Density

Note: Since a gram is defined as the weight of 1 cubic centimeter of water, the specific gravity is the same thing as the density when using grams and cubic centimeters.

Testing Gases

You will need an instructor's help for this portion of the experiment. Evacuate the recovery cylinder marked "Air Only." Set the scale to read ounces, place the cylinder on a digital scale, and zero the scale. Open the valve on the cylinder and let air in. You should see a small weight increase on the scale. Record this weight. Use the tire pump to increase the pressure in the tank to 20 psig. Record this weight. Use the formulas below to calculate the density, specific volume, and specific gravity. *Note*: the recovery cylinder volume is 0.76 cubic feet.

Gas Density at 0 psig

Density =_____ ÷_____= ounces per cubic foot
         ounces   cubic feet

Specific Volume = _____ ÷ _____ = _____ cubic feet per ounce
                  cubic feet   ounces

Specific Gravity_____ ÷ 1.2 = _____
                  Density   (Density of Air in ounces per cubic foot)

Gas Density at 20 psig

Density =_____ ÷_____= _____ ounces per cubic foot
         ounces   cubic feet

Specific Volume _____ ÷ _____ = _____ cubic feet per ounce
                cubic feet   ounces

Specific Gravity _____ ÷ 1.2 = _____
                  Density   (Density of Air in ounces per cubic foot)

SUMMARY

1. In general, which state has the highest density, solid, liquid, or gas?

2. In general, which state has the lowest density, solid, liquid, or gas?

3. In general, which state has the highest specific volume, solid, liquid, or gas?

4. In general, which state has the lowest specific volume, solid, liquid, or gas?

5. Which two states have the most similar density?

6. Which two states have the most similar specific volume?

7. Which state's density and specific volume is most subject to change?

8. What advantage do metric units have when performing specific gravity measurements?

9. What effect does increased pressure have on the density and specific volume of a gas?

LAB 8.1 ENERGY CONVERSION

LABORATORY OBJECTIVE
You will demonstrate that energy can be converted from one form to another and that electricity generated from a rotating magnetic device is not free.

FUNDAMENTALS OF HVACR TEXT REFERENCE
Unit 8 Types of Energy and Their Properties

LABORATORY NOTES
We will use a hand cranked electric generator, which is connected to several electric light bulbs. The lights will be controlled by switches. You will turn the generator by hand, producing electricity. You will control the lights with switches to demonstrate the effect of electrical loads on the generator.

REQUIRED TOOLS, EQUIPMENT, AND MATERIALS
1 hand-cranked generator

3 electric spot switches

3 lights

PROCEDURE
1. Turn the generator with all the switches off. This is the amount of energy required to overcome the mechanical inefficiencies in the machine.
2. While you are turning the generator, turn on the smallest light.
3. Notice that the generator is now harder to turn.
4. Turn on the other lights, one at a time.
5. Note the change each time another load is added.
6. Now turn the lights off one at a time.
7. Note the change each time another load is removed.

SUMMARY
What types of energy are represented in this experiment?

LAB 9.1 MEASURING TEMPERATURE

LABORATORY OBJECTIVE
You will demonstrate your ability to measure the temperature of refrigerant lines, air, and objects.

FUNDAMENTALS OF HVACR TEXT REFERENCE
Unit 9 Temperature and Thermodynamics

LABORATORY NOTES
You will use a thermocouple bead and a Fluke 16 multimeter, a thermocouple clamp and a Fluke 16 multimeter, and an infrared thermometer to measure several temperatures. You will give the temperature readings in Fahrenheit, Celsius, Rankin, and Kelvin.

REQUIRED TOOLS, EQUIPMENT, AND MATERIALS
Safety glasses
Fluke 16 multimeter
Temperature clamp
Thermocouple bead
Infrared thermometer
Ice
Operating air conditioner
Operating furnace
Shop heater
Shop air conditioner

PROCEDURE
Read the following temperatures using the indicated instrument. Take readings in both Fahrenheit and Celsius, and then convert them to Rankin and Kelvin. Record your readings on the chart below.

Temperature	Using	Fahrenheit	Rankin	Celsius	Kelvin
Ice	Bead				
Ice	Infrared				
Shop	Bead				
Shop Wall	Infrared				
Skin	Bead				
Skin	Infrared				
Cold Air	Bead				

LAB 9.2 BOILING POINT VS. PRESSURE

LABORATORY OBJECTIVE
The purpose of this lab is to demonstrate the effect of pressure on the boiling point of a liquid.

FUNDAMENTALS OF HVACR TEXT REFERENCE
Unit 9 Temperature and Thermodynamics

LABORATORY NOTES
We know that the boiling point of water at atmospheric pressure is 212 degrees Fahrenheit. We will use a vacuum pump to reduce the pressure on a flask of water and observe the effect of reduced pressure. A vacuum pump reduces the pressure of a closed container by removing gas from the container.

REQUIRED TOOLS, EQUIPMENT, AND MATERIALS
Flask with rubber stopper
Vacuum pump
Vacuum gauge that reads in inches of mercury vacuum

SAFETY REQUIREMENTS
None

PROCEDURE
Step 1
Obtain an empty flask and rubber stopper.

1. Fill the flask half full with water.
2. Stopper the top of the flask.
3. Connect the flask to the vacuum pump.
4. Connect the vacuum gauge to the vacuum pump.

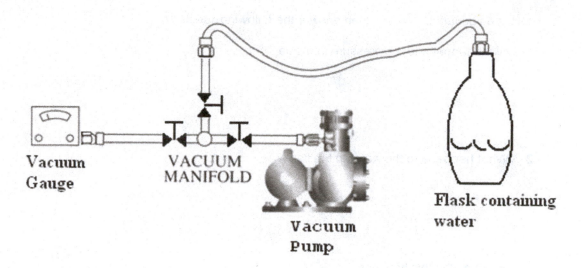

Figure 9-2-1

Step 2

Operate the vacuum pump and then answer the following questions

1. What happens to the pressure in the flask?

2. What happens to the water in the flask?

3. Explain why this happens.

Step 3

Turn off the vacuum pump and then answer the following questions.

4. What happens to the pressure in the flask?

5. Why does this happen?

6. What happens to the water?

7. Why does this happen?

28

LAB 9.3 PRESSURE–TEMPERATURE RELATIONSHIP OF SATURATED MIXTURES

LABORATORY OBJECTIVE

You will demonstrate the relationship of pressure and temperature for saturated gas–liquid mixes.

FUNDAMENTALS OF HVACR TEXT REFERENCE

Unit 9 Temperature and Thermodynamics

LABORATORY NOTES

You will observe the behavior of saturated gas by comparing the pressure of two cylinders containing different weights a saturated mix at the same temperature. Then we will change the temperature and observe the effect on pressure.

You will be using two recovery cylinders with refrigerant. Both cylinders should be the same type of refrigerant, but they should contain different amounts of refrigerant. Both cylinders should be the same temperature and both **should contain some liquid and some vapor.**

REQUIRED TOOLS, EQUIPMENT, AND MATERIALS

Two refrigerant recovery cylinders with the same refrigerant type

Freezer or a bucket of ice

Heat gun or a sink with hot water

Gauge manifold

Infrared thermometer

Digital scale

SAFETY REQUIREMENTS

Wear safety goggles and gloves when working with refrigerants. Liquid refrigerant can cause frostbite when in contact with eyes and skin.

PROCEDURE

Step 1

Measure the temperature, pressure, and weight of the first refrigerant recovery cylinder as follows.

1. Place one of the refrigerant recovery cylinders on top of a digital scale and then connect the refrigerant recovery cylinder to the high side of the gauge manifold as shown in Figure 9-3-1.

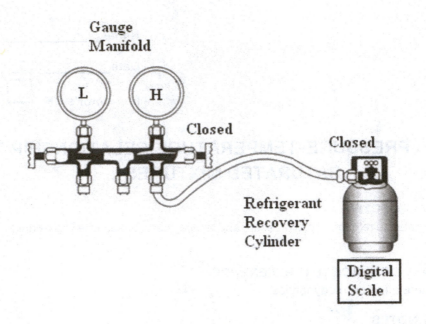

Figure 9-3-1

2. Slowly open the refrigerant recovery cylinder valve as shown in Figure 9-3-2.

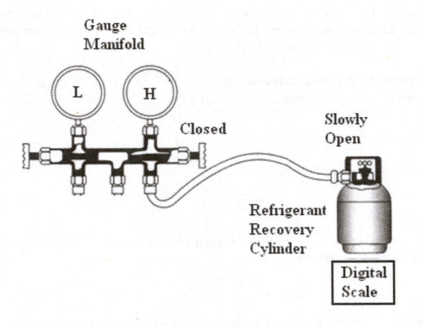

Figure 9-3-2

3. Record the pressure from the reading on the high side gauge.

Recovery Cylinder #1 pressure:

4. Measure the cylinder temperature using the infrared thermometer.

   Recovery Cylinder #1 temperature:

5. Record the cylinder weight from the reading on the digital scale.

   Recovery Cylinder #1 weight:

6. After recording the cylinder pressure and temperature, close the cylinder valve as shown in Figure 9-3-3.

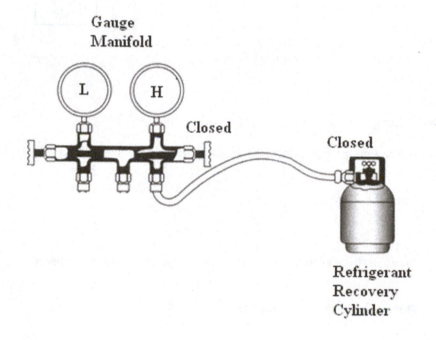

Figure 9-3-3

7. You must now bleed off the pressure remaining in the hose before you disconnect it. To do this, make sure that the cylinder valve is closed. Next take the center hose of the gauge manifold and direct it away from any person including yourself. Slowly open the high side gauge manifold valve as shown in Figure 9-3-4 and the pressure in the line will bleed off. Be careful as the hose may whip around slightly as it drains.

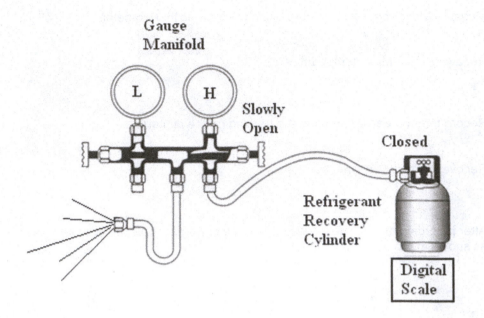

Gauge
Manifold

L   H

Slowly
Open

Closed

Refrigerant
Recovery
Cylinder

Digital
Scale

Figure 9-3-4

Step 2
Repeat step 1 using the second refrigerant recovery cylinder.

1.   Record the pressure from the reading on the high side gauge.

Recovery Cylinder #2 pressure:

2.   Measure the cylinder temperature using the infrared thermometer.

Recovery Cylinder #2 temperature:

3.   Record the cylinder weight from the reading on the digital scale.

Recovery Cylinder #2 weight:

Step 3
Referring to the measurements you recorded in steps 1 and 2, answer the following questions.

1.   How does the pressure of the cylinder that contains the most refrigerant compare
to the cylinder with the least refrigerant?

2. Do you think that this would be different for a cylinder containing only gas without any liquid? Explain your answer.

Step 4

Place one of the cylinders in a bucket of ice or in the freezer.

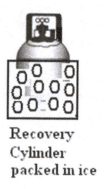

Recovery
Cylinder
packed in ice

Figure 9-3-5

1. Allow the cylinder to cool for 30 minutes and then measure its pressure and temperature as you did in step 1.

2. Record the pressure from the reading on the high side gauge.

    Cooled Recovery Cylinder pressure:

3. Measure the cylinder temperature using the infrared thermometer.

    Cooled Recovery Cylinder temperature:

Step 5

Calculate the expected pressure after cooling the cylinder. For this calculation use the starting pressure, starting temperature, and ending temperature. Refer to the Ideal Gas Laws from Unit 10 in the *Fundamentals of HVACR* text.

(Remember that all pressures and temperatures must be converted to absolute readings BEFORE working the problem.)

SHOW ALL OF YOUR CALCULATIONS IN THE SPACE PROVIDED.

1. How does the actual pressure compare to the pressure predicted by the gas law?

2. Use a saturation chart such as the one found in Unit 10 from the HVACR text and determine the pressure that corresponds to the temperature of the refrigerant in the cylinder.

    Saturated pressure from chart:

3. How does the pressure found from the chart compare to the actual pressure?

4. Which was more accurate in predicting the results, the ideal gas law or using the saturation chart?

Step 6
Place the recovery cylinder into a sink and run warm water over the cylinder.

CAUTION: The refrigerant cylinder temperature should NEVER exceed 125 °F.

Figure 9-3-6

1. Allow the cylinder to cool for 15 minutes and then measure its pressure and temperature as you did in step 1.

2. Record the pressure from the reading on the high side gauge.

   Heated Recovery Cylinder pressure:

3. Measure the cylinder temperature using the infrared thermometer.

   Heated Recovery Cylinder temperature:

Step 7

Calculate the expected pressure after cooling the cylinder. For this calculation use the starting pressure, starting temperature, and ending temperature. Refer to the Ideal Gas Laws from Unit 10 in the *Fundamentals of HVACR* text.

(Remember that all pressures and temperatures must be converted to absolute readings BEFORE working the problem.)

SHOW ALL OF YOUR CALCULATIONS IN THE SPACE PROVIDED BELOW.

1. How does the actual pressure compare to the pressure predicted by the gas law?

2. Use a saturation chart such as the one found in Unit 10 from the HVACR text and determine the pressure that corresponds to the temperature of the refrigerant in the cylinder.

    Saturated pressure from chart:

3. How does the pressure found from the chart compare to the actual pressure?

4. Which was more accurate in predicting the results, the ideal gas law or using the saturation chart?

5. What conclusions can you draw about the behavior of saturated liquid-gas mixtures to temperature changes?

LAB 9.4 SENSIBLE AND LATENT HEAT

LABORATORY OBJECTIVE
You will demonstrate sensible and latent heat properties.

FUNDAMENTALS OF HVACR TEXT REFERENCE
Unit 9 Temperature and Thermodynamics

LABORATORY NOTES
We will demonstrate sensible and latent heat properties by adding heat to water while we monitor the amount of energy used, the temperature of the water, and the weight of the water. Energy input will be determined by measuring the wattage used, temperature change will be measured with a thermocouple, and weight change will be measured with a digital scale.

REQUIRED TOOLS, EQUIPMENT, AND MATERIALS
Temperature sensor with temperature clamp and thermocouple clamp

Electric cook pot

Wattmeter

SAFETY REQUIREMENTS
A. Wear safety goggles and gloves when working with high temperature liquids.
B. The steam that is generated can easily burn your skin. Be careful not to come in contact with the hot steam.

PROCEDURE
Step 1
Obtain an electric cook pot, and digital scale.

1. Place the cook pot on top of the digital scale and pour approximately 3 pounds of water into the pot.
2. Zero the digital scale with the pot and the water.
3. Measure the temperature of the water.

    Water Temperature _____

4. Connect the wattmeter to the cook pot circuit to measure the power that will be used during the heating of the water.

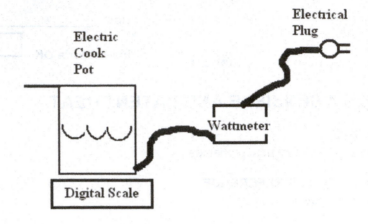

Figure 9-4-1

5. Plug in the cook pot and record the time and wattage.

    Time started:

    Wattage:

Step 2
Monitor the time, temperature, and wattage as follows.

1. When the temperature reaches approximately 150 °F, record the time, the exact temperature and the wattage.

    Time _____

    Temperature _____

    Wattage _____

2. When water begins to boil, record the time, the exact temperature and the wattage.

    Time _____

    Temperature_____

    Wattage _____

38

3. When the water reaches a rolling boil record the time, the exact temperature, and the weight on the scale. Let the water boil for three minutes and then record these values again in part 4.

Time _____

Temperature _____

Weight _____

Wattage _____

4. Record the values after three minutes of boiling.

Time _____

Temperature _____

Weight _____

Wattage _____

Step 3
Calculate the amount of energy in BTUs required to heat the water to 150 °F.

1. Since it takes 1 BTU to heat 1 pound of water 1 °F at atmospheric pressure, you can use the formula:

(weight of the water in pounds) x (1 BTU/lb/°F) x (ending temperature – beginning temperature) = BTU required to heat the water.

SHOW ALL OF YOUR CALCULATIONS IN THE SPACE PROVIDED ON THE NEXT PAGE.

TOTAL AMOUNT OF HEAT REQUIRED (BTU)_____

2. Now calculate the measured amount of electrical heat input. One Watt is equivalent to 3.41 BTU per hour or divide this by sixty minutes in one hour for 0.057 BTU per minute. With this we can convert our measured Wattage into BTU using the formula:

(measured wattage) x (minutes to 150ºF) x (0.057 BTU/min/Watt) = BTU input

SHOW ALL OF YOUR CALCULATIONS IN THE SPACE PROVIDED BELOW.

TOTAL AMOUNT OF HEAT INPUT (BTU) _____

3. How does the amount of heat required to heat the water in BTUs compare to the measured amount of heat input? If they are not equal, then explain why.

Step 4

Calculate the amount of energy in BTUs required to boil the water for three minutes.

1. Since it takes 970 BTU to boil 1 pound of water at atmospheric pressure, you can use the formula:

   (pounds of water beginning – pounds of water ending for the three minute period) x (970 BTU/lb) = BTU required to boil the water for three minutes.

SHOW ALL OF YOUR CALCULATIONS IN THE SPACE PROVIDED BELOW.

TOTAL AMOUNT OF HEAT REQUIRED (BTU) _____

2. Now calculate the measured amount of electrical heat input. One Watt is equivalent to 3.41 BTU per hour or divide this by sixty minutes in one hour for 0.057 BTU per minute. With this we can convert our measured Wattage into BTU using the formula:

   (measured wattage) x (three minutes) x (0.057 BTU/min/Watt) = BTU input

SHOW ALL OF YOUR CALCULATIONS IN THE SPACE PROVIDED ON THE NEXT PAGE.

TOTAL AMOUNT OF HEAT INPUT (BTU) _____

3.   How does the amount of heat required to boil the water in BTUs compare to the measured amount of heat input? If they are not equal, then explain why.

4.   What happened to the temperature of the water as it was boiling?

5.   What do you think the temperature of the steam would be?

6.   Does it take more energy to heat the water or to boil the water?

LAB 9.5 DETERMINING REFRIGERANT CONDITION

LABORATORY OBJECTIVE

The purpose of this lab is to demonstrate your ability to determine if refrigerant is saturated, superheated, or subcooled.

FUNDAMENTALS OF HVACR TEXT REFERENCE

Unit 9 Temperature and Thermodynamics

LABORATORY NOTES

You will measure the pressure and temperature of several refrigeration system components. You will determine if the refrigerant in the component is saturated, superheated, or subcooled by comparing the pressure and temperature to a saturation chart for that particular refrigerant.

REQUIRED TOOLS, EQUIPMENT, AND MATERIALS

Instrumental refrigerant trainer

Pressure-temperature chart

SAFETY REQUIREMENTS

None

PROCEDURE

Step 1

Identify the components on the refrigeration trainer.

Step 2

Line up the components and then start the refrigeration trainer using the trainer's instruction manual if necessary.

Step 3

Allow the refrigerant trainer to run until the temperatures and pressures have begun to stabilize and then complete the chart below. You will need to identify for each situation whether the refrigerant is subcooled, saturated, or superheated. Not all refrigeration trainers are instrumented the same, so if there is no reading available on the refrigerant trainer for the specific condition, then leave that space blank.

LOCATION	PRESSURE	TEMPERATURE FROM CHART	ACTUAL TEMPERATURE	CONDITION
Compressor IN				
Compressor OUT				
Condenser IN				
Condenser OUT				
Evaporator IN				
Evaporator OUT				

LAB 9.6 MEASURING WATER SOURCE HEAT PUMP SYSTEM CAPACITY

LABORATORY OBJECTIVE

You will demonstrate your ability to determine the system capacity of an operating water source heat pump by measuring its water flow and temperature rise.

FUNDAMENTALS OF HVACR TEXT REFERENCE

Unit 9 Temperature and Thermodynamics

LABORATORY NOTES

In this lab we will calculate the capacity of an operating water source heat pump system by measuring the water flow rate and the water temperature difference. A BTU is defined as a 1 degree temperature rise in 1 pound of water at atmospheric pressure. We can calculate the actual capacity accurately by multiplying the water flow in pounds per hour times the temperature rise in degrees Fahrenheit.

REQUIRED TOOLS, EQUIPMENT, AND MATERIALS

Temperature sensor with temperature clamp and thermocouple clamp

Water source heat pump

1 gallon bucket

Timer or watch that can measure seconds

SAFETY REQUIREMENTS

Always familiarize yourself with the equipment and operating manuals prior to starting up any system.

PROCEDURE

Step 1

Trace out the system and make sure that you understand the operation of the heat pump.

Start the water loop through the heat pump before starting the unit.
1. After water flow has been established, you may start the heat pump and run it in the cooling mode for 15 minutes.
2. Measure the flow rate of the water. This can be done by allowing the water leaving the heat pump to flow into the one gallon bucket. Time the duration in seconds from the time the bucket is initially empty until it is full. Repeat this procedure for three different sets of readings.

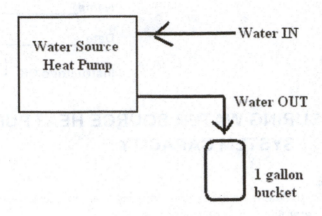

Figure 9-6-1

    a.   First measured time in seconds:

    b.   Second measured time in seconds:

    c.   Third measured time in seconds:

3.   The average time in seconds is equal to the three readings added together and then divided by three.

    Average time in seconds = (Time 1 + Time 2 + Time 3) / 3

    Average time in seconds =

4.   To determine how many gallons per minute is flowing first convert the seconds to minutes. Take the average time in seconds and divide that value by 60 (this is because there are 60 seconds in one minute).

    Gallons per minute = Total seconds / 60

    Gallons per minute (GPM) =

5.   Convert gallons per minute (GPM) to pounds per minute by multiplying GPM by 8.34 (this is because there are 8.34 pounds of water in one gallon).

    Pounds per minute = GPM x 8.34

    Pounds per minute =

6. Measure the temperature of the water going in to the heat pump and the temperature of the water leaving the heat pump.

   Temperature of the water IN:

   Temperature of the water OUT:

7. Calculate the temperature difference that is equal to the temperature OUT minus the temperature IN ($°F_{OUT} - °F_{IN}$).

   Temperature Difference (ΔT) =

8. Calculate the system capacity. To do this multiply pounds per minute by the temperature difference. This will then be multiplied by the specific heat of water at atmospheric pressure which is simply 1 (one BTU for every °F for every lb of water). The calculated value will be in BTU per minute.

   (lb/min x ($°F_{OUT} - °F_{IN}$) x 1 BTU/lb/°F) = BTU/min

   System capacity in BTU/min =

9. Calculate the value for tons of refrigeration.

   Remember that one ton of refrigeration is the equivalent of melting one ton (2,000 lbs) of ice in 24 hours. Remember that the latent heat of fusion (ice to water) is 144 BTU/lb. 2,000 lbs of ice multiplied by 144 BTU/lb is equal to 288,000 BTU for one ton in 24 hours. Divide this by 24 hours per day and you have 12,000 BTU/hr. Divide this by 60 minutes per hour and you have 200 BTU/min.

   Therefore a ton of system capacity is equal to:

   1 ton = 288,000 BTU/day

   1 ton = 12,000 BTU/hr

   1 ton = 200 BTU/min

   To find capacity in tons, divide the system capacity in BTU/min by 200 BTU/min–ton.

   System capacity in tons = (BTU/min) / 200 BTU/min–ton

   System capacity in tons =

10. How does the system capacity that you calculated compare to the nameplate rating of the heat pump? Is the heat pump operating at full capacity?

Step 2
Reverse the heat pump so that now it is in the heating mode and run it in this mode for 15 minutes before recording your next measurements.

1. Measure the flow rate of the water. Repeat this procedure for three different sets of readings.
   a) First measured time in seconds:
   b) Second measured time in seconds:
   c) Third measured time in seconds:

2. The average time in seconds is equal to the three readings added together and then divided by three.

   Average time in seconds = (Time 1 + Time 2 + Time 3) / 3

   Average time in seconds =

3. To determine how many gallons per minute is flowing first convert the seconds to minutes. Take the average time in seconds and divide that value by 60 (this is because there are 60 seconds in one minute).

   Gallons per minute = Total seconds/60

   Gallons per minute (GPM) =

4. Convert gallons per minute (GPM) to pounds per minute by multiplying GPM by 8.34 (this is because there are 8.34 pounds of water in one gallon).

   Pounds per minute = GPM x 8.34

   Pounds per minute =

5. Measure the temperature of the water going in to the heat pump and the temperature of the water leaving the heat pump.

   Temperature of the water IN:

   Temperature of the water OUT:

6. Calculate the temperature difference that is equal to the temperature OUT minus the temperature IN ($°F_{OUT} - °F_{IN}$).

   (*Note*: The temperature OUT minus the temperature IN will have a negative value. This is because in the heating mode the heat pump is absorbing heat from the water rather than rejecting heat. You do not need to use a negative value because you are just calculating the temperature difference so you may ignore the negative sign for this experiment.)

   Temperature Difference (ΔT) =

7. Calculate the system capacity. To do this, multiply pounds per minute by the temperature difference. This will then be multiplied by the specific heat of water at atmospheric pressure which is simply 1 (one BTU for every °F for every lb of water). The calculated value will be in BTU per minute.

$$(\text{lb/min} \times (°F_{OUT} - °F_{IN}) \times 1 \text{ BTU/lb/°F}) = \text{BTU/min}$$

System capacity in BTU/min =

8. Calculate the value for tons of refrigeration.

Remember that one ton of refrigeration is the equivalent of melting one ton (2,000 lbs) of ice in 24 hours. Remember that the latent heat of fusion (ice to water) is 144 BTU/lb. 2,000 lbs of ice multiplied by 144 BTU/lb is equal to 288,000 BTU for one ton in 24 hours. Divide this by 24 hours per day and you have 12,000 BTU/hr. Divide this by 60 minutes per hour and you have 200 BTU/min.

Therefore a ton of system capacity is equal to:

1 ton = 288,000 BTU/day

1 ton = 12,000 BTU/hr

1 ton = 200 BTU/min

To find capacity in tons, divide the system capacity in BTU/min by 200 BTU/min–ton.

System capacity in tons = (BTU/min) / 200 BTU/min–ton

System capacity in tons =

9. How does the system capacity that you calculated compare to the nameplate rating of the heat pump? Is the heat pump operating at full capacity?

10. How does the system capacity when operating the heat pump in the cooling mode differ from running the heat pump in the heating mode. Explain your answer.

LAB 10.1 QUANTITY–PRESSURE RELATIONSHIP OF IDEAL GASES

LABORATORY OBJECTIVE

The student will demonstrate that the pressure of an ideal gas contained in a cylinder depends upon the amount of gas in the cylinder.

FUNDAMENTALS OF HVACR TEXT REFERENCE

Unit 10 Pressure and Vacuum

LABORATORY NOTES

We will compare the pressure of a cylinder with two different weights of the same gas at the same temperature. Using the same cylinder will insure that the volume remains constant. Temperature will be verified using an infrared thermometer.

REQUIRED TOOLS, EQUIPMENT, AND MATERIALS:

An empty refillable refrigerant recovery cylinder
Vacuum pump
Digital scale
Cylinder of nitrogen gas with regulator
Infrared thermometer
Gauge manifold

SAFETY REQUIREMENTS

    A. Wear safety glasses and gloves whenever handling gas cylinders and regulators.
    B. Familiarize yourself with the proper procedures for operating gas cylinder regulators.

PROCEDURE

Step 1

Locate an empty refillable refrigerant recovery cylinder, a nitrogen cylinder with regulator, a vacuum pump, a gauge manifold, and a digital refrigerant charging scale.

    1. Place the refrigerant recovery cylinder on the digital scale and connect the vacuum pump using the gauge manifold. Also connect the gauge manifold to the nitrogen cylinder but keep the nitrogen cylinder valves closed as shown in Figure 10-1-1.

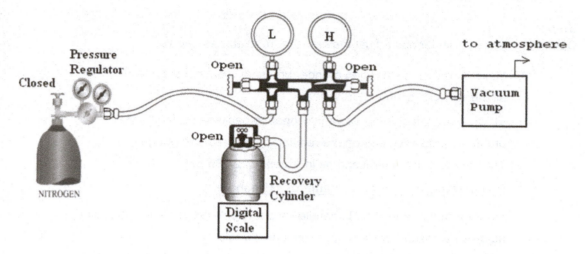

Figure 10-1-1

2. Run the vacuum pump to evacuate the cylinder and gauge manifold hoses to insure that the cylinder and hoses are empty and free from all gas. Close the high side valve (H) on the gauge manifold before shutting of the vacuum pump or else air will leak back into the lines and cylinder.

3. Close the low side valve (L) on the gauge manifold.

4. Zero the digital scale.

5. The valve arrangement should now be as shown in Figure 10-1-2.

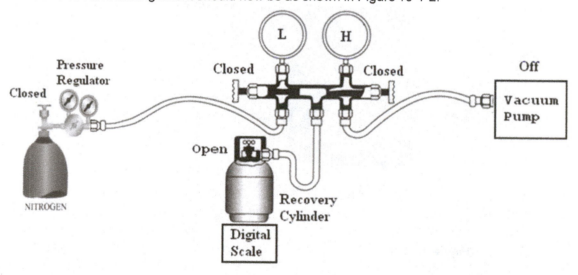

Figure 10-1-2

Step 2

Open the nitrogen cylinder and adjust the pressure regulator as follows:

    A. Make sure that the nitrogen cylinder pressure regulator is turned all the way out (counterclockwise).

    B. Slowly open the cylinder valve fully open to backseat it. The tank pressure should register on the regulator high pressure gauge. The pressure in the tank can be in excess of 2,000 psi. <u>Do not stand in front of the regulator "T" handle</u>.

    C. Slowly turn the regulator "T" handle inward (clockwise) until the regulator adjusted pressure reaches approximately 50 psig.

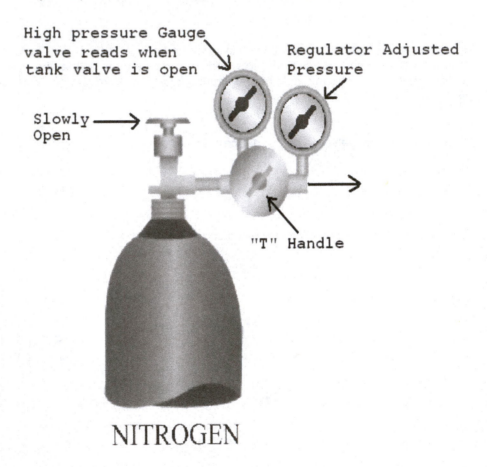

Figure 10-1-3

Step 3

Slowly open the low side valve on the gauge manifold (L) until you see a weight increase on the digital scale. Then close the gauge manifold valve and record your readings.

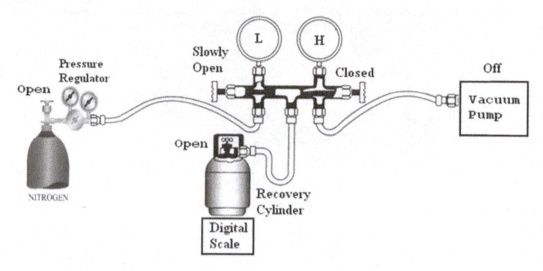

Figure 10-1-4

Gas Pressure:

Gas Temperature:

Gas Weight:

Step 4

Once again slowly open the low side valve on the gauge manifold (L) and add more nitrogen until you see a weight increase on the digital scale. Then close the gauge manifold valve and record your readings.

Gas Pressure:

Gas Temperature:

Gas Weight:

Step 5

After recording the second set of readings you can prepare to disconnect the gauge manifold as follows:

A. Close the shut off valve on the nitrogen cylinder.

B. Disconnect the vacuum pump from the gauge manifold.

C. Open the low side valve on the gauge manifold (L).

D. Slowly open the high side valve on the gauge manifold to bleed the gas pressure from the recovery cylinder and all hoses.

Once the pressure has been bled, you may back all the way off on the "T" handle (counterclockwise) for the pressure regulator on the Nitrogen cylinder.

After all of the pressure has been bled off, the hoses may be disconnected.

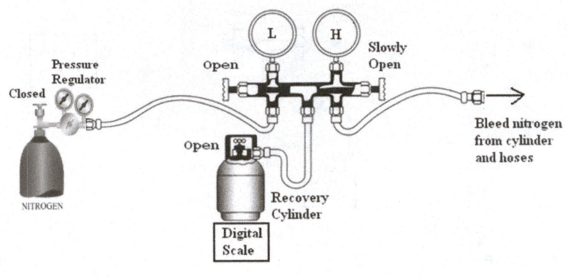

Figure 10-1-5

QUESTIONS

1. How does the pressure of the cylinder change when the amount of nitrogen changes?

2. What conclusions can you draw about the behavior of an ideal gas?

LAB 10.2 GAS TEMPERATURE AND VOLUME AT CONSTANT PRESSURE

LABORATORY OBJECTIVE
The purpose of this lab is to demonstrate the effect of temperature change on the volume of a gas that is at constant pressure.

FUNDAMENTALS OF HVACR TEXT REFERENCE
Unit 10 Pressure and Vacuum

LABORATORY NOTES
The gas will be contained in a mylar balloon. The balloon's volume can change because the balloon can expand and contract. The pressure inside the balloon will stay the same as will the atmospheric pressure surrounding the balloon. We will change the temperature of the gas and observe the volume change by observing the size of the balloon.

REQUIRED TOOLS, EQUIPMENT, AND MATERIALS
Mylar balloon
Freezer
Hair dryer

SAFETY REQUIREMENTS
    A. Wear safety glasses and gloves whenever handling gas cylinders and regulators.
    B. Familiarize yourself with the proper procedures for operating gas cylinder regulators.

PROCEDURE
Step 1
Inflate the mylar balloon at room temperature until it is full, but not taut.

Step 2
Place the balloon in the freezer for five minutes.

Step 3
Remove the balloon and compare its shape and size to its original shape and size.

Step 4
Heat the balloon with a hair dryer.

Step 5
Compare its shape and size to its original shape and size.

QUESTIONS

1. What happened to the balloon when the temperature dropped?

2. What happened to the balloon when the temperature rose?

3. What does this tell us about the volume of a gas compared to its temperature?

LAB 10.3 GAS PRESSURE–TEMPERATURE RELATIONSHIP AT A CONSTANT VOLUME

LABORATORY OBJECTIVE
The purpose of this lab is to demonstrate the effect of gas temperature changes on gas pressure when the volume remains constant.

FUNDAMENTALS OF HVACR TEXT REFERENCE
Unit 10 Pressure and Vacuum

LABORATORY NOTES
The gas will be contained in a recovery cylinder. Since the cylinder is a fixed volume, the gas volume will not change. We will observe the pressure and temperature of the cylinder filled with nitrogen at room temperature. Then we will cool the cylinder in ice and observe the effect of reduced temperature on the gas pressure. Next, we will heat the cylinder and observe the effect of increased temperature on gas pressure. We will use the ideal gas law to verify our results.

REQUIRED TOOLS, EQUIPMENT, AND MATERIALS
Refrigerant recovery cylinder containing
nitrogen
Mop bucket
Ice
Infrared thermometer
Gauge manifold

SAFETY REQUIREMENTS
Wear safety glasses and gloves whenever handling gas cylinders and regulators.

PROCEDURE
Step 1
Obtain a refrigerant recovery cylinder containing nitrogen gas.

1. Measure and record the pressure and temperature of the cylinder.

   Pressure:

   Temperature:

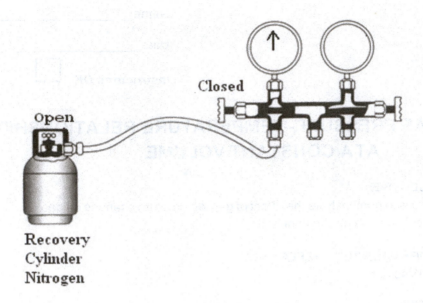

Closed

Open

Recovery
Cylinder
Nitrogen

Figure 10-3-1

Step 2

Place the recovery cylinder into the mop bucket and pack it with ice.

A. Wait 15 minutes and then record the pressure and temperature of the cylinder.

Pressure:

Temperature:

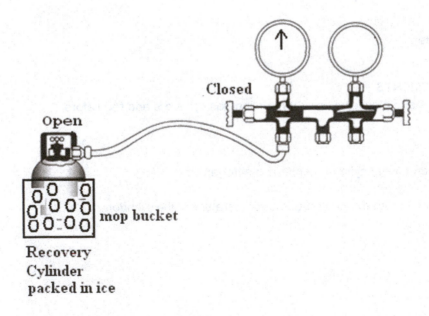

Closed

Open

mop bucket

Recovery
Cylinder
packed in ice

Figure 10-3-2

Step 3

Calculate the expected pressure after cooling the cylinder. For this calculation use the starting pressure, starting temperature, and ending temperature. Refer to the Ideal Gas Laws from Unit 10 in the *Fundamentalsof HVACR* text.

(Remember that all pressures and temperatures must be converted to absolute readings BEFORE working the problem.)

SHOW ALL OF YOUR CALCULATIONS IN THE SPACE PROVIDED BELOW.

Calculated Pressure:

Measured Pressure:

A.   What happened to the pressure in the cylinder when the temperature dropped?

B.   How do your measured results compare to your calculated results?

Step 4

Place the recovery cylinder into a sink and run warm water over the cylinder.

CAUTION: The refrigerant cylinder temperature should NEVER exceed 125 °F.

A.   Record the new pressure and temperature of the cylinder.

Pressure:

Temperature:

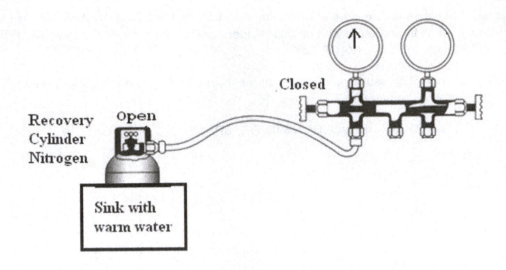

Figure 10-3-3

Step 5

Calculate the expected pressure after warming the cylinder. For this calculation use the starting pressure, starting temperature, and ending temperature. Refer to the Ideal Gas Laws from Unit 10 in the *Fundamentalsof HVACR* text.

(Remember that all pressures and temperatures must be converted to absolute readings BEFORE working the problem.)

SHOW ALL OF YOUR CALCULATIONS IN THE SPACE PROVIDED BELOW.

Calculated Pressure:

Measured Pressure:

60

A. What happened to the pressure in the cylinder when the temperature increased?

B. How do your measured results compare to your calculated results?

C. Based upon the results from cooling and then heating the cylinder, what conclusions can you draw about the pressure of a gas compared to its temperature at constant volume?

LAB 10.4 GAS PRESSURE-TEMPERATURE-VOLUME RELATIONSHIP

LABORATORY OBJECTIVE
The purpose of this lab is to demonstrate the combined effect of gas volume and pressure changes on gas temperature.

FUNDAMENTALS OF HVACR TEXT REFERENCE
Unit 10 Pressure and Vacuum

LABORATORY NOTES
First, we will observe the temperature change in a gas that is being compressed. During compression the volume will be reduced and the pressure will be increased. Second, we will observe the temperature change in a gas as it is expanded. During expansion the volume will increase and the pressure will decrease.

REQUIRED TOOLS, EQUIPMENT, AND MATERIALS
Empty refrigerant recovery cylinder

Bicycle pump
Temperature sensor with thermocouple probe
Infrared thermometer
Gauge manifold

SAFETY REQUIREMENTS
Wear safety glasses and gloves whenever handling gas cylinders and regulators.

PROCEDURE
Step 1
Obtain an empty refrigerant recovery cylinder.

1. Measure and record the pressure and temperature of the cylinder.
   (The pressure should be atmospheric)

   Pressure:

   Temperature:

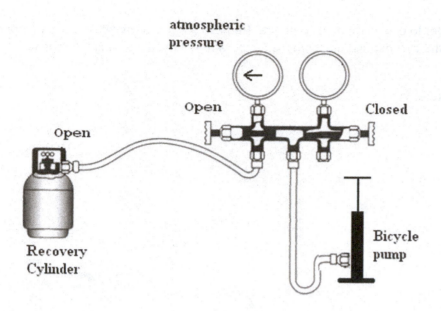

Figure 10-4-1

Step 2

Use the bicycle pump to raise the pressure of the cylinder to 40 psig then close the valve on the cylinder and record the pressure and temperature of the cylinder.

Pressure:

Temperature:

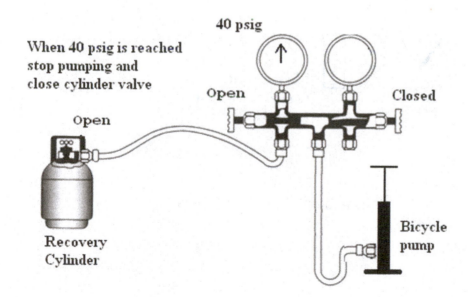

Figure 10-4-2

Step 3

Wait for the cylinder to cool to room temperature. Remove the gauge manifold and then open the gas valve on the cylinder and measure the temperature of the air leaving the cylinder (the air is expanding).

Temperature:

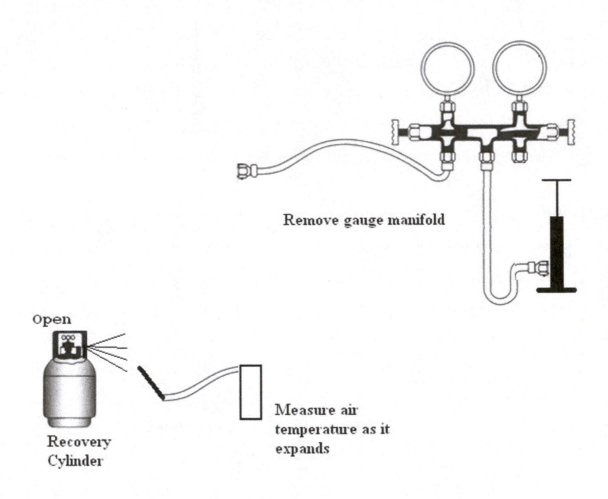

Remove gauge manifold

Open

Recovery Cylinder

Measure air temperature as it expands

Figure 10-4-3

QUESTIONS

1. What happened to the temperature of the air when it was compressed with the bicycle pump?

2. Why did this happen?

3. What happened to the temperature of the air when it was expanded?

4. Why did this happen?

5. How could these physical relationships of pressure-volume-temperature be useful in an air conditioning system?

LAB 10.5 IDEAL GAS LAWS

LABORATORY OBJECTIVE
The purpose of this lab is to demonstrate the effect of gas temperature changes on gas pressure when the volume remains constant.

FUNDAMENTALS OF HVACR TEXT REFERENCE
Unit 10 Pressure and Vacuum

LABORATORY NOTES
The gas will be contained in a recovery cylinder. Since the cylinder is a fixed volume, the gas volume will not change. We will observe the pressure and temperature of the cylinder filled with nitrogen at room temperature. Then we will cool the cylinder in ice and observe the effect of reduced temperature on the gas pressure. Next, we will heat the cylinder and observe the effect of increased temperature on gas pressure. We will use the ideal gas law to verify our results.

REQUIRED TOOLS, EQUIPMENT, AND MATERIALS
Safety glasses
Refrigerant recovery cylinder containing nitrogen infrared thermometer
Refrigerant gauges
Mop bucket
Ice

PROCEDURE

1. Measure and record the temperature of the cylinder.
2. Measure and record the pressure of cylinder.
3. Use the mop bucket to pack the cylinder in ice.
4. Wait 15 minutes for the cylinder to cool off.
5. Measure and record the new cylinder temperature and pressure.
6. Calculate the expected pressure using the starting pressure, starting temperature, and ending temperature. Remember that all pressures and temperatures must be converted to absolute readings BEFORE working the problem. Review the Ideal Gas Laws from Unit 10 in the *Fundamentals of HVACR* text if you need help.
7. Place the cylinder in the sink and run warm water over the cylinder.
8. Measure and record the new temperature and pressure.
9. Calculate the expected pressure using the original starting pressure, original starting temperature, and ending temperature.

SUMMARY

1. What happened to the pressure in the cylinder when the temperature dropped?

2. How did your results compare to the calculated results?

3. What happened to the pressure in the cylinder when the temperature rose?

4. How did your results compare to the calculated results?

5. What does this tell us about the pressure of a gas compared to its temperature?

LAB 11.1 Types of Refrigeration Systems

LABORATORY OBJECTIVE
The purpose of this lab is to demonstrate your ability to distinguish between mechanical vapor compression, absorption, evaporative cooling, and thermoelectric refrigeration systems.

FUNDAMENTALS OF HVACR TEXT REFERENCE
Unit 11 Types of Refrigeration Systems

REQUIRED TOOLS, EQUIPMENT, AND MATERIALS
A representative example of each type of system is not required although this would be helpful if available.

PROCEDURE
Use the Figures 11-1-1 through 11-1-4 provided along with any representative examples to complete the Tables 11-1-1 through 11-1-4.

PART 1
Complete table 11-1-1 by identifying the conditions of the refrigerant labeled A through J for the mechanical vapor compression cycle (Figure 11-1-1) as: saturated vapor, saturated liquid, saturated mixture, subcooled liquid, superheated vapor.

FIGURE 11-1-1

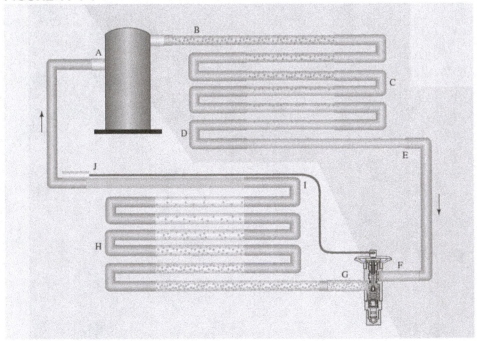

TABLE 11-1-1

A	
B	
C	
D	
E	
F	
G	
H	
I	
J	

Instructor Check_____

PART 2

Complete table 11-1-2 by identifying the components labeled A through I for the absorption cycle (Figure 11-1-2).

FIGURE11-1-2

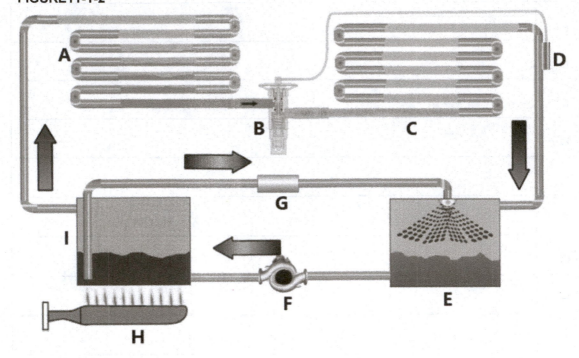

TABLE 11-1-2

A	
B	
C	
D	
E	
F	
G	
H	
I	

Instructor Check_____

PART 3

Complete table 11-1-3 by identifying the components labeled A through L for the evaporative cooler (Figure 11-1-3).

FIGURE 11-1-3

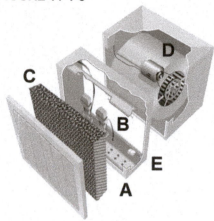

TABLE 11-1-3

A	
B	
C	
D	
E	

Instructor Check_____

PART 4
Complete table 11-1-4 identifying the components labeled A through K for the rooftop packaged unit with evaporative pre-cooling.

FIGURE 11-1-4

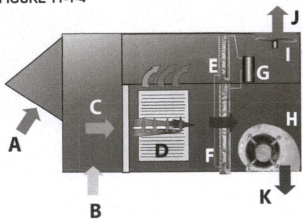

TABLE 11-1-4

A	
B	
C	
D	
E	
F	
G	
H	
I	
J	
K	

Instructor Check_____

PART 5

Complete table 11-1-5 by identifying the components labeled A through F for the thermoelectric cooling module (Figure 11-1-5).

FIGURE 11-1-5

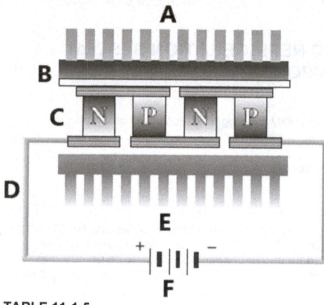

TABLE 11-1-5

A	
B	
C	
D	
E	
F	
G	
H	
I	

Instructor Check_____

LAB 12.1 IDENTIFYING REFRIGERATION SYSTEM COMPONENTS

LABORATORY OBJECTIVE
The purpose of this lab is to demonstrate your ability to recognize each of the four main refrigeration system components on refrigeration systems.

FUNDAMENTALS OF HVACR TEXT REFERENCE
Unit 12 The Refrigeration Cycle

LABORATORY NOTES
You will be assigned four typical refrigeration systems. The units should be labeled SYSTEM 1, SYSTEM 2, SYSTEM 3, and SYSTEM 4. You will locate the four major refrigeration system components in each system and complete a data sheet for each unit. You will be prepared to show the instructor each of these components and discuss the basic characteristics of each of these components.

REQUIRED TOOLS, EQUIPMENT, AND MATERIALS
Packaged refrigeration unit
Split system
Small commercial refrigeration unit
Any other unit of the instructor's choosing
Six-in-one screwdriver for removing system panels

SAFETY REQUIREMENTS
Be careful of sharp edges when removing sheet metal panels.

PROCEDURE
Examine each unit and complete the appropriate data sheet. Remove any panels necessary to gain access to the unit components.

REFRIGERATION SYSTEM COMPONENTS				
	System 1	**System 2**	**System 3**	**System 4**
Type of System	Packaged Split System	Packaged Split System	Packaged Split System	Packaged Split System
Compressor	Open Hermetic Semi-hermetic	Open Hermetic Semi-hermetic	Open Hermetic Semi-hermetic	Open Hermetic Semi-hermetic
Condenser	Air Cooled Water Cooled Evaporative	Air Cooled Water Cooled Evaporative	Air Cooled Water Cooled Evaporative	Air Cooled Water Cooled Evaporative
Evaporator	Cools Air Cools Water	Cools Air Cools Water	Cools Air Cools Water	Cools Air Cools Water
Expansion Device	Orifice Capillary Tube Expansion Valve	Orifice Capillary Tube Expansion Valve	Orifice Capillary Tube Expansion Valve	Orifice Capillary Tube Expansion Valve

Instructor Check_____

Datasheet System 2

REFRIGERATION SYSTEM COMPONENTS				
	System 1	**System 2**	**System 3**	**System 4**
Type of System	Packaged Split System	Packaged Split System	Packaged Split System	Packaged Split System
Compressor	Open Hermetic Semi-hermetic	Open Hermetic Semi-hermetic	Open Hermetic Semi-hermetic	Open Hermetic Semi-hermetic
Condenser	Air Cooled Water Cooled Evaporative	Air Cooled Water Cooled Evaporative	Air Cooled Water Cooled Evaporative	Air Cooled Water Cooled Evaporative
Evaporator	Cools Air Cools Water	Cools Air Cools Water	Cools Air Cools Water	Cools Air Cools Water
Expansion Device	Orifice Capillary Tube Expansion Valve	Orifice Capillary Tube Expansion Valve	Orifice Capillary Tube Expansion Valve	Orifice Capillary Tube Expansion Valve
Refrigerant Storage	Receiver Accumulator Neither	Receiver Accumulator Neither	Receiver Accumulator Neither	Receiver Accumulator Neither

Instructor Check_____

Datasheet System 3

REFRIGERATION SYSTEM COMPONENTS				
	System 1	**System 2**	**System 3**	**System 4**
Type of System	Packaged Split System	Packaged Split System	Packaged Split System	Packaged Split System
Compressor	Open Hermetic Semi-hermetic	Open Hermetic Semi-hermetic	Open Hermetic Semi-hermetic	Open Hermetic S emi-hermetic
Condenser	Air Cooled Water Cooled Evaporative	Air Cooled Water Cooled Evaporative	Air Cooled Water Cooled Evaporative	Air Cooled Water Cooled Evaporative
Evaporator	Cools Air Cools Water	Cools Air Cools Water	Cools Air Cools Water	Cools Air Cools Water
Expansion Device	Orifice Capillary Tube Expansion Valve	Orifice Capillary Tube Expansion Valve	Orifice Capillary Tube Expansion Valve	Orifice Capillary Tube Expansion Valve
Refrigerant Storage	Receiver Accumulator Neither	Receiver Accumulator Neither	Receiver Accumulator Neither	Receiver Accumulator Neither

Instructor Check_____

Datasheet System 4

REFRIGERATION SYSTEM COMPONENTS				
	System 1	**System 2**	**System 3**	**System 4**
Type of System	Packaged Split System	Packaged Split System	Packaged Split System	Packaged Split System
Compressor	Open Hermetic Semi-hermetic	Open Hermetic Semi-hermetic	Open Hermetic Semi-hermetic	Open Hermetic Semi-hermetic
Condenser	Air Cooled Water Cooled Evaporative	Air Cooled Water Cooled Evaporative	Air Cooled Water Cooled Evaporative	Air Cooled Water Cooled Evaporative
Evaporator	Cools Air Cools Water	Cools Air Cools Water	Cools Air Cools Water	Cools Air Cools Water
Expansion Device	Orifice Capillary Tube Expansion Valve	Orifice Capillary Tube Expansion Valve	Orifice Capillary Tube Expansion Valve	Orifice Capillary Tube Expansion Valve
Refrigerant Storage	Receiver Accumulator Neither	Receiver Accumulator Neither	Receiver Accumulator Neither	Receiver Accumulator Neither

Instructor Check_____

LAB 12.2 TRAINER REFRIGERATION SYSTEM CHARACTERISTICS

LABORATORY OBJECTIVE
The purpose of this lab is to demonstrate your ability to discuss the function of the four main refrigeration system components and describe the refrigerant characteristics throughout the system on the refrigeration trainer.

FUNDAMENTALS OF HVACR TEXT REFERENCE
Unit 12 The Refrigeration Cycle

LABORATORY NOTES
You will be assigned a refrigeration trainer. You should be prepared to show the instructor each of the four main refrigeration components in order and discuss their function in the system. You will also discuss the characteristics of the refrigerant as it travels through the system using pressure and temperature measurements to help you determine the condition of the refrigerant throughout the trainer.

REQUIRED TOOLS, EQUIPMENT, AND MATERIALS
Safety glasses
Refrigeration trainer with clear tubes in each coil.
Electronic Thermometer
Temperature clamp

SAFETY REQUIREMENTS
Use caution when touching components. Some components are HOT!

PROCEDURE
Run the trainer and observe its operation. Note the pressure gauges that tell you what the refrigerant pressure is at every point in the system. You can measure the temperature at each location using a Fluke meter. You may also touch components to get a feel for their temperature.

CAUTION: Some components are HOT! Proceed carefully!

If you hold your hand just above a warm component you can tell if it is extremely hot or not. If you can feel radiant heat coming from it, it may be too hot to touch safely. Use the charts below to organize your data. However, the job is not complete until you TELL the instructor how the system works and respond to questions.

Completing the Refrigerant Characteristics Data Table
Record the pressure and temperature at each system location indicated on the chart. Compare the pressure and temperature using a pressure-temperature chart to determine the condition: saturated, superheated, or subcooled. Next, write in the state: gas, liquid, or mixed. Use the pressure, temperature, condition, and state to determine if the refrigerant is at a high heat content or a low heat content.

Completing the Refrigerant Changes Data Table
Look at the data to determine how the refrigerant changed as it went through each component and circle the description that best describes the changes in the conditions through each component.

Trainer Refrigeration System Characteristics
Use the following table to record system operating data for the trainer.

Location	Temperature	Pressure	Condition	State	Heat Content
Compressor in					
Compressor out					
Discharge line					
Condenser in					
Condenser center					
Condenser out					
Liquid line					
Meter device in					
Meter device out					
Evaporator in					
Evaporator center					
Evaporator out					
Suction line					

Trainer Refrigerant Changes Through the Components

Circle the description that best describes the changes that occur through each refrigeration system component.

No change to describe very slight or no significant change in a condition.

Small to describe a measurable change in a condition such as 40°F to 50°F.

Large to describe a very significant change, such as 120°F to 40°F.

Type of Change	Compressor	Condenser	Meter Device	Evaporator
Change in Pressure	No change Small increase Large increase Small decrease Large decrease	No change Small increase Large increase Small decrease Large decrease	No change Small increase Large increase Small decrease Large decrease	No change Small increase Large increase Small decrease Large decrease
Change in Temp	No change Small increase Large increase Small decrease Large decrease	No change Small increase Large increase Small decrease Large decrease	No change Small increase Large increase Small decrease Large decrease	No change Small increase Large increase Small decrease Large decrease
Change in Heat	No change Small increase Large increase Small decrease Large decrease	No change Small increase Large increase Small decrease Large decrease	No change Small increase Large increase Small decrease Large decrease	No change Small increase Large increase Small decrease Large decrease
Change in State	No change Condensation Evaporation	No change Condensation Evaporation	No change Condensation Evaporation	No change Condensation Evaporation
Change in Volume	No change Small increase Large increase Small decrease Large decrease	No change Small increase Large increase Small decrease Large decrease	No change Small increase Large increase Small decrease Large decrease	No change Small increase Large increase Small decrease Large decrease

LAB 12.3 PACKAGED UNIT REFRIGERATION SYSTEM CHARACTERISTICS

LABORATORY OBJECTIVE
The purpose of this lab is to demonstrate your ability to discuss the function of the four main refrigeration system components and describe the refrigerant characteristics throughout the system on a packaged unit.

FUNDAMENTALS OF HVACR TEXT REFERENCE
Unit 12 The Refrigeration Cycle

LABORATORY NOTES
You will be assigned a packaged unit air conditioner. You should be prepared to show the instructor each of the four main refrigeration components in order and discuss their function in the system. You will also discuss the characteristics of the refrigerant as it travels through the system using pressure and temperature measurements to help you determine the condition of the refrigerant throughout the unit.

REQUIRED TOOLS, EQUIPMENT, AND MATERIALS
Safety glasses
Packaged air conditioner.
Electronic Thermometer
Temperature clamp

SAFETY REQUIREMENTS
 A. Use caution when touching components. Some components are HOT!
 B. Be careful around moving fan blades.
 C. Do NOT touch any electrical connections.

PROCEDURE
Run the packaged unit and observe its operation. Note the pressure gauges that tell you what the refrigerant pressure is at every point in the system. You can measure the temperature at each location using a Fluke meter. You may also touch components to get a feel for their temperature.

CAUTION: Some components are HOT! Proceed carefully!

If you hold your hand just above a warm component you can tell if it is extremely hot or not. If you can feel radiant heat coming from it, it may be too hot to touch safely. Use the charts below to organize your data. However, the job is not complete until you TELL the instructor how the system works and respond to questions.

Completing the Refrigerant Characteristics Data Table

Record the pressure and temperature at each system location indicated on the chart. Compare the pressure and temperature using a pressure-temperature chart to determine the condition: saturated, superheated, or subcooled. Next, write in the state: gas, liquid, or mixed. Use the pressure, temperature, condition, and state to determine if the refrigerant is at a high heat content or a low heat content.

Completing the Refrigerant Changes Data Table

Look at the data to determine how the refrigerant changed as it went through each component and circle the description that best describes the changes in the conditions through each component.

Packaged Unit Refrigeration System Characteristics

Use the following table to record system operating data for the trainer.

Location	Temperature	Pressure	Condition	State	Heat Content
Compressor in					
Compressor out					
Discharge line					
Condenser in					
Condenser center					
Condenser out					
Liquid line					
Meter device in					
Meter device out					
Evaporator in					
Evaporator center					
Evaporator out					
Suction line					

Packaged Unit Refrigerant Changes Through the Components

Circle the description that best describes the changes that occur through each refrigeration system component.

No change to describe very slight or no significant change in a condition.

Small to describe a measurable change in a condition, such as 40°F to 50°F.

Large to describe a very significant change, such as 120°F to 40°F.

Type of Change	Compressor	Condenser	Meter Device	Evaporator
Change in Pressure	No change Small increase Large increase Small decrease Large decrease	No change Small increase Large increase Small decrease Large decrease	No change Small increase Large increase Small decrease Large decrease	No change Small increase Large increase Small decrease Large decrease
Change in Temp	No change Small increase Large increase Small decrease Large decrease	No change Small increase Large increase Small decrease Large decrease	No change Small increase Large increase Small decrease Large decrease	No change Small increase Large increase Small decrease Large decrease
Change in Heat	No change Small increase Large increase Small decrease Large decrease	No change Small increase Large increase Small decrease Large decrease	No change Small increase Large increase Small decrease Large decrease	No change Small increase Large increase Small decrease Large decrease
Change in State	No change Condensation Evaporation	No change Condensation Evaporation	No change Condensation Evaporation	No change Condensation Evaporation
Change in Volume	No change Small increase Large increase Small decrease Large decrease	No change Small increase Large increase Small decrease Large decrease	No change Small increase Large increase Small decrease Large decrease	No change Small increase Large increase Small decrease Large decrease

Name_____

Date_____

Instructor's OK ☐

LAB 12.4 SPLIT SYSTEM REFRIGERATION CHARACTERISTICS

LABORATORY OBJECTIVE

The purpose of this lab is to demonstrate your ability to discuss the function of the four main refrigeration system components and describe the refrigerant characteristics throughout the system on a split system air conditioner.

FUNDAMENTALS OF HVACR TEXT REFERENCE

Unit 12 The Refrigeration Cycle

LABORATORY NOTES

You will be assigned a packaged unit air conditioner. You should be prepared to show the instructor each of the four main refrigeration components in order and discuss their function in the system. You will also discuss the characteristics of the refrigerant as it travels through the system using pressure and temperature measurements to help you determine the condition of the refrigerant throughout the unit.

LABORATORY NOTES

You will be assigned a packaged unit air conditioner. You should be prepared to show the instructor each of the four main refrigeration components in order and discuss their function in the system. You will also discuss the characteristics of the refrigerant as it travels through the system using pressure and temperature measurements to help you determine the condition of the refrigerant throughout the unit.

REQUIRED TOOLS, EQUIPMENT, AND MATERIALS

Safety glasses
Split-system air conditioner
Electronic Thermometer
Temperature clamp

SAFETY REQUIREMENTS

    A. Use caution when touching components. Some components are HOT!
    B. Be careful around moving fan blades.
    C. Do NOT touch any electrical connections.

PROCEDURE

Run the split system and observe its operation. Note the pressure gauges that tell you what the refrigerant pressure is at every point in the system. You can measure the temperature at each location using a Fluke meter. You may also touch components to get a feel for their temperature.

CAUTION: Some components are HOT! Proceed carefully!

If you hold your hand just above a warm component you can tell if it is extremely hot or not. If you can feel radiant heat coming from it, it may be too hot to touch safely. Use the charts below to organize your data. However, the job is not complete until you TELL the instructor how the system works and respond to questions.

Completing the Refrigerant Characteristics Data Table

Record the pressure and temperature at each system location indicated on the chart. Compare the pressure and temperature using a pressure-temperature chart to determine the condition: saturated, superheated, or subcooled. Next, write in the state: gas, liquid, or mixed. Use the pressure, temperature, condition, and state to determine if the refrigerant is at a high heat content or a low heat content.

Completing the Refrigerant Changes Data Table

Look at the data to determine how the refrigerant changed as it went through each component and circle the description that best describes the changes in the conditions through each component.

Split System Refrigeration Characteristics

Use the following table to record system operating data for the trainer.

Location	Temperature	Pressure	Condition	State	Heat Content
Compressor in					
Compressor out					
Discharge line					
Condenser in					
Condenser center					
Condenser out					
Liquid line					
Meter device in					
Meter device out					
Evaporator in					
Evaporator center					
Evaporator out					
Suction line					

Split System Refrigerant Changes Through the Components

Circle the description that best describes the changes that occur through each refrigeration system component.

No change to describe very slight or no significant change in a condition.

Small to describe a measurable change in a condition such as 40°F to 50°F.

Large to describe a very significant change, such as 120°F to 40°F.

Type of Change	Compressor	Condenser	Meter Device	Evaporator
Change in Pressure	No change Small increase Large increase Small decrease Large decrease	No change Small increase Large increase Small decrease Large decrease	No change Small increase Large increase Small decrease Large decrease	No change Small increase Large increase Small decrease Large decrease
Change in Temp	No change Small increase Large increase Small decrease Large decrease	No change Small increase Large increase Small decrease Large decrease	No change Small increase Large increase Small decrease Large decrease	No change Small increase Large increase Small decrease Large decrease
Change in Heat	No change Small increase Large increase Small decrease Large decrease	No change Small increase Large increase Small decrease Large decrease	No change Small increase Large increase Small decrease Large decrease	No change Small increase Large increase Small decrease Large decrease
Change in State	No change Condensation Evaporation	No change Condensation Evaporation	No change Condensation Evaporation	No change Condensation Evaporation
Change in Volume	No change Small increase Large increase Small decrease Large decrease	No change Small increase Large increase Small decrease Large decrease	No change Small increase Large increase Small decrease Large decrease	No change Small increase Large increase Small decrease Large decrease

LAB 12.5 BENCH UNIT REFRIGERATION SYSTEM CHARACTERISTICS

LABORATORY OBJECTIVE
The purpose of this lab is to demonstrate your ability to discuss the function of the four main refrigeration system components and describe the refrigerant characteristics throughout the system on a bench unit.

FUNDAMENTALS OF HVACR TEXT REFERENCE
Unit 12 The Refrigeration Cycle

LABORATORY NOTES
You will be assigned a bench unit refrigeration system. You should be prepared to show the instructor each of the four main refrigeration components in order and discuss their function in the system.
You will also discuss the characteristics of the refrigerant as it travels through the system using pressure and temperature measurements to help you determine the condition of the refrigerant throughout the unit.

REQUIRED TOOLS, EQUIPMENT, AND MATERIALS
Safety glasses
Bench-unit refrigeration system
Electronic Thermometer
Temperature clamp

SAFETY REQUIREMENTS
    A. Use caution when touching components. Some components are HOT!
    B. Be careful around moving fan blades.
    C. Do NOT touch any electrical connections.

PROCEDURE
Run the bench unit refrigeration system and observe its operation. Note the pressure gauges that tell you what the refrigerant pressure is at every point in the system. You can measure the temperature at each location using a Fluke meter. You may also touch components to get a feel for their temperature.

CAUTION: Some components are HOT! Proceed carefully!

If you hold your hand just above a warm component you can tell if it is extremely hot or not. If you can feel radiant heat coming from it, it may be too hot to touch safely. Use the charts below to organize your data. However, the job is not complete until you TELL the instructor how the system works and respond to questions.

Completing the Refrigerant Characteristics Data Table

Record the pressure and temperature at each system location indicated on the chart. Compare the pressure and temperature using a pressure-temperature chart to determine the condition: saturated, superheated, or subcooled. Next, write in the state: gas, liquid, or mixed. Use the pressure, temperature, condition, and state to determine if the refrigerant is at a high heat content or a low heat content.

Completing the Refrigerant Changes Data Table

Look at the data to determine how the refrigerant changed as it went through each component and circle the description that best describes the changes in the conditions through each component.

Bench Unit Refrigeration System Characteristics

Use the following table to record system operating data for the trainer.

Location	Temperature	Pressure	Condition	State	Heat Content
Compressor in					
Compressor out					
Discharge line					
Condenser in					
Condenser center					
Condenser out					
Liquid line					
Meter device in					
Meter device out					
Evaporator in					
Evaporator center					
Evaporator out					
Suction line					

Bench Unit Refrigerant Changes Through the Components

Circle the description that best describes the changes that occur through each refrigeration system component.

No change to describe very slight or no significant change in a condition.

Small to describe a measurable change in a condition, such as 40°F to 50°F.

Large to describe a very significant change, such as 120°F to 40°F.

Type of Change	Compressor	Condenser	Meter Device	Evaporator
Change in Pressure	No change Small increase Large increase Small decrease Large decrease	No change Small increase Large increase Small decrease Large decrease	No change Small increase Large increase Small decrease Large decrease	No change Small increase Large increase Small decrease Large decrease
Change in Temp	No change Small increase Large increase Small decrease Large decrease	No change Small increase Large increase Small decrease Large decrease	No change Small increase Large increase Small decrease Large decrease	No change Small increase Large increase Small decrease Large decrease
Change in Heat	No change Small increase Large increase Small decrease Large decrease	No change Small increase Large increase Small decrease Large decrease	No change Small increase Large increase Small decrease Large decrease	No change Small increase Large increase Small decrease Large decrease
Change in State	No change Condensation Evaporation	No change Condensation Evaporation	No change Condensation Evaporation	No change Condensation Evaporation
Change in Volume	No change Small increase Large increase Small decrease Large decrease	No change Small increase Large increase Small decrease Large decrease	No change Small increase Large increase Small decrease Large decrease	No change Small increase Large increase Small decrease Large decrease

LAB 12.6 WATER COOLED SYSTEM CHARACTERISTICS

LABORATORY OBJECTIVE
The purpose of this lab is to demonstrate your ability to discuss the function of the four main refrigeration system components and describe the refrigerant characteristics throughout the system on a water cooled unit.

FUNDAMENTALS OF HVACR TEXT REFERENCE
Unit 12 The Refrigeration Cycle

LABORATORY NOTES
You will be assigned a water cooled unit. You should be prepared to show the instructor each of the four main refrigeration components in order and discuss their function in the system. You will also discuss the characteristics of the refrigerant as it travels through the system using pressure and temperature measurements to help you determine the condition of the refrigerant throughout the unit.

REQUIRED TOOLS, EQUIPMENT, AND MATERIALS
Safety glasses
Split-system air conditioner
Electronic Thermometer
Temperature clamp

SAFETY REQUIREMENTS
    A. Use caution when touching components. Some components are HOT!
    B. Be careful around moving fan blades.
    C. Do NOT touch any electrical connections.

PROCEDURE
Run the system and observe its operation. Note the pressure gauges that tell you what the refrigerant pressure is at every point in the system. You can measure the temperature at each location using a Fluke meter. You may also touch components to get a feel for their temperature.

    CAUTION: Some components are HOT! Proceed carefully!

If you hold your hand just above a warm component you can tell if it is extremely hot or not. If you can feel radiant heat coming from it, it may be too hot to touch safely. Use the charts below to organize your data. However, the job is not complete until you TELL the instructor how the system works and respond to questions.

Completing the Refrigerant Characteristics Data Table

Record the pressure and temperature at each system location indicated on the chart. Compare the pressure and temperature using a pressure-temperature chart to determine the condition: saturated, superheated, or subcooled. Next, write in the state: gas, liquid, or mixed. Use the pressure, temperature, condition, and state to determine if the refrigerant is at a high heat content or a low heat content.

Completing the Refrigerant Changes Data Table

Look at the data to determine how the refrigerant changed as it went through each component and circle the description that best describes the changes in the conditions through each component.

Water Cooled Refrigeration System Characteristics

Use the following table to record system operating data for the trainer.

Location	Temperature	Pressure	Condition	State	Heat Content
Compressor in					
Compressor out					
Discharge line					
Condenser in					
Condenser center					
Condenser out					
Liquid line					
Meter device in					
Meter device out					
Evaporator in					
Evaporator center					
Evaporator out					
Suction line					

Water Cooled System Refrigerant Changes Through the Components

Circle the description that best describes the changes that occur through each refrigeration system component.

No change to describe very slight or no significant change in a condition.

Small to describe a measurable change in a condition, such as 40°F to 50°F.

Large to describe a very significant change, such as 120°F to 40°F.

Type of Change	Compressor	Condenser	Meter Device	Evaporator
Change in Pressure	No change Small increase Large increase Small decrease Large decrease	No change Small increase Large increase Small decrease Large decrease	No change Small increase Large increase Small decrease Large decrease	No change Small increase Large increase Small decrease Large decrease
Change in Temp	No change Small increase Large increase Small decrease Large decrease	No change Small increase Large increase Small decrease Large decrease	No change Small increase Large increase Small decrease Large decrease	No change Small increase Large increase Small decrease Large decrease
Change in Heat	No change Small increase Large increase Small decrease Large decrease	No change Small increase Large increase Small decrease Large decrease	No change Small increase Large increase Small decrease Large decrease	No change Small increase Large increase Small decrease Large decrease
Change in State	No change Condensation Evaporation	No change Condensation Evaporation	No change Condensation Evaporation	No change Condensation Evaporation
Change in Volume	No change Small increase Large increase Small decrease Large decrease	No change Small increase Large increase Small decrease Large decrease	No change Small increase Large increase Small decrease Large decrease	No change Small increase Large increase Small decrease Large decrease

LAB 12.7 REFRIGERANT IDENTIFICATION

LABORATORY OBJECTIVE

Identify the type and amount of refrigerant in four refrigeration systems. You will identify the refrigerant by:
- Name
- Pressure (Very High, High, or Low)
- Ozone Depletion Potential
- Global Warming Potential
- Toxicity
- Flammability
- Chemical Composition
- Formulation

FUNDAMENTALS OF HVACR TEXT REFERENCE

Unit 12 The Refrigeration Cycle

LABORATORY NOTES

You will find the type of refrigerant, the amount of refrigerant, and the system test pressures on the unit data plate of each system. Then you will research the reference material to find the refrigerant characteristics.

REQUIRED TOOLS, EQUIPMENT, AND MATERIALS

Safety glasses
Four refrigeration systems (They should contain different refrigerants.)
Pencil and paper

SAFETY REQUIREMENTS

A. Wear safety glasses
B. Use caution when touching components.
C. Some components are HOT!
D. Be careful around moving fan blades.
E. Do NOT touch any electrical connections.

PROCEDURE FOR EACH UNIT ASSIGNED

1. Locate the nameplate on the unit. The nameplate will be in different places on the units, you simply have to look around.
2. Find and record the unit model number in the chart below.
3. Find and record the refrigerant type in the chart below.
4. Find and record the high side test pressure on the chart below.
5. Find and record the low side test pressure on the chart below.

6. Find a cylinder of the same refrigerant as in the system and read the instructions on the cylinder.
7. Use this information to research this unit's refrigerant characteristics and record them in the chart below.

Refrigerant Characteristics

	Unit 1	*Unit 2*	*Unit 3*	*Unit 4*
Model Number				
Refrigerant Name				
Refrigerant Quantity				
High Side Test Pressure				
Low Side Test Pressure				
Pressure (Very High, High, Medium, Low)				
Ozone Depletion Potential				
Global Warming Potential				
Safety Rating				
Chemical Composition				
Formulation (Compound, zeotrope or azeotrope)				

SUMMARY

1. Were there any refrigerants that did not have either ozone depletion potential or global warming potential?
2. What are the high and low side test pressures used for?
3. Which refrigerants may safely leave the cylinder as either a gas or a liquid?
4. Which refrigerants must leave the cylinder as a liquid only?
5. Which refrigerant had the highest system test pressures?
6. Which refrigerant had the lowest test pressures?
7. What hazards do you need to be aware of when working with these refrigerants?
8. Which of these refrigerants will NOT be used in equipment of the future?
9. Which of these refrigerants WILL be in equipment of the future?

LAB 12.8 REFRIGERANT OIL IDENTIFICATION

LABORATORY OBJECTIVE
Identify different types of refrigeration lubricant and recommend the proper lubricant for four refrigeration systems.

FUNDAMENTALS OF HVACR TEXT REFERENCE
Unit 12 The Refrigeration Cycle

LABORATORY NOTES
You will examine three different types of refrigeration lubricant and recommend where they can be used. You will recommend the correct type of lubricant for four refrigeration systems.

REQUIRED TOOLS, EQUIPMENT, AND MATERIALS
Safety glasses
Three containers of refrigeration lubricant representing three different types of refrigeration lubricant. Four refrigeration systems. They should contain different refrigerants.
Pencil and paper

SAFETY REQUIREMENTS
A. Wear Safety Glasses
B. Use caution when touching components. Some components are HOT!
C. Be careful around moving fan blades.
D. Do NOT touch any electrical connections.

PROCEDURE
Examining Refrigeration Lubricant

1. Identify the type of lubricant in each container.
2. Record the type of lubricant, its viscosity, its recommended evaporator temperature range, and the refrigerants that it is compatible with in the chart below.

	Lubricant 1	Lubricant 2	Lubricant 3
Type of Lubricant			
Viscosity			
Refrigerant Compatibility			
Evaporator Temperature			

For each unit assigned:

1. Locate the nameplate on the unit. The nameplate will be in different places on the units, you simply have to look around.
2. Find and record the unit model number in the chart below.
3. Find and record the refrigerant type in the chart below.
4. Recommend a lubricant that is compatible with this unit

	Unit 1	Unit 2	Unit 3
Type of Lubricant			
Viscosity			
Refrigerant Compatibility			
Evaporator Temperature			

SUMMARY

1. Which lubricant can be used with the widest range of systems?

2. What is the difference in viscosity between lubricants designed for higher evaporator temperatures and lubricants that are designed for use in lower temperature systems?

3. What is the most common viscosity for lubricants used in air conditioning systems?

LAB 13.1 IDENTIFYING COMPRESSORS

LABORATORY OBJECTIVE
You will identify six different compressors and classify them according to their mechanical type, body, cooling method, size, and electrical characteristics.

FUNDAMENTALS OF HVACR TEXT REFERENCE
Unit 13 Compressors

LABORATORY NOTES
The instructor will assign six units. You will examine the compressor in each unit and record it's characteristics in the data table.

REQUIRED TOOLS, EQUIPMENT, AND MATERIALS
Safety glasses
Gloves
Flat blade screwdriver
Phillips screwdriver
¼" nut driver
5/16" nut driver
OR
6-in-1 with all the above
Adjustable jaw wrench
Air conditioning units with compressors

SAFETY REQUIREMENTS
    A. Wear safety glasses
    B. Use caution when touching components. Some components are HOT!
    C. Be careful around moving fan blades.
    D. Do NOT touch any electrical connections.

PROCEDURE
Examine the compressors on the units assigned by the instructor to determine the compressor

    operational type (reciprocating, rotary, etc.)

    body style (hermetic, open, semi-hermetic)

    motor cooling method (air cooled or refrigerant cooled)

    application (low temp, medium temp, high temp)

    size (horsepower and/or tonnage)

    electrical data (phase and voltage)

Use the pictures in the book to identify the operational type and body style, and motor cooling method. Hermetic compressors are in welded steel cans that sit up vertically. Semi hermetic compressors are bolted together and lay horizontally. Open compressors are bolted and have a shaft sticking out. Most hermetic reciprocating compressors are oval shaped. Rotary compressors and scroll compressors are round. The rotaries are smaller with the large suction line entering near the bottom. Scrolls are larger with the large suction line entering midway, closer to the top.

The compressor electrical data comes from the compressor data plate. The compressor size is determined by the model number. Common compressor manufacturer's model number explanations are given in the following pages:

Tecumseh Hermetic Compressors						
AE	A	4	4	40	Y	XA
Family	Generation	Application	Number of Digits in BTU/Hr Capacity	First two digits	Refrigerant	Voltage

Application   1 Low Temp −10°F, Normal Start Torque

                    2 Low Temp −10°F, High Start Torque

                    3 High Temp 45°F, Normal Start Torque

                    4 High Temp 45°F, High Start Torque

                    5 Air Cond 45°F, Normal Start Torque

                    6 Medium Temp 20°F, Normal Start Torque

                    7 Medium Temp 20°F, High Start Torque

                    8 Air Cond   49°F, Normal Start Torque

                    9 Commercial 20°F, Normal Start Torque

                    A Medium/Low 20°F, Normal Start Torque

Refrigerant   A R12

                    B R410A

                    C R407C

                    E R22

                    J R502

                    Y R134a

                    Z R404A/R507

Voltage XA 115 volt 60 hz single phase

                    XB 230 volt 60 hz single phase

                    XD 208-230 volt 60 hz single phase

                    XF 208-230 volt 60 hz three phase

                    XG 460 volt 60 hz three phase

                    XT 200-230 volt 60 hz three phase

                    AB 115 volt 60 hz single phase

Size   The number after the application tells how many digits are in the BTU/hr rating and the next two numbers are the first two digits. In the example above, they compressor rating is 4000 BTU/hr – four digits with the first two being 40.

Bristol Hermetic Compressors								
H	8	2	J	193	A	B	C	A
Application	Refrigerant	Generation	Family	Capacity	Motor	Protect	Electric	Feet

Application
    L Low Temp
    H High Temp
    M Medium Temp 7 Medium Temp 20°F, High Start Torque
    R Multiple Refrigerants
    T Two Capacity High Temp
    V Variable Speed

Refrigerant
    1 R12
    2 R22
    3 R134a
    5 R502 or R402B
    6 R404A
    7 R407C
    8 R410A
    9 R407C, R22, R404A

Motor
    A PSC   G 3 Phase 2 Speed
    B CSR   J Single phase 2 speed
    C RSCR   K 3 phase dual voltage
    D 3 PHASE   L 3 phase wye-delta
    E CSIR   M 3 phase variable speed
    F 3 Phase Part Winding Start

Protection
    B Internal line break overload
    P Pilot Duty Solid State
    R Pilot Duty Solid State 2nd generation
    T Pilot Duty Internal Thermostat

Electrical
    A 115 volt 60 hz single phase
    B 230 volt 60 hz single phase
    C 208-230 volt 60 hz single phase
    D 208-230 volt 60 hz three phase
    E 460 volt 60 hz three phase
    L 200-230 volt 60 hz three phase
    T 208 volts 60 hz
    Y 208-230 volt 60 hz single phase

Size
    The three numbers in the middle of the model number give the nominal
    compressor capacity in BTU/hr. The first two numbers are the first two
    digits in the capacity. The third number is the number of zeros after the
    first two digits. In the example above, they compressor rating is 19000 BTU/hr.

Copeland Semi-Hermetic Compressors								
4	R	R	2	3000	T	S	K	800
Family	Type	Displace	Variant	Horsepower	Motor	Protect	Electric	BOM

Type

A Air Cooled
D Discuss
R Refrigerant Cooled
T Two Stage
W Water Cooled

Motor

C Capacitor Run Capacitor Start
I Induction Run Capacitor Start
E Three phase
F 3 Phase Part Winding Start
T 3 phase Specialized

Protection

A External line break overload
F Internal line break overload
H Pilot Duty Internal Thermostat and External Overload
L Pilot Duty Internal Thermostat and 3 External Overloads
S Internal electronic thermal sensor

Electrical

A 115 volt 60 hz single phase
B 230 volt 60 hz single phase
C 208-230 volt 60 hz single phase
D 460 volt 60 hz three phase
H 208 volts 60 hz single phase
I 208-230 volt 60 hz single phase

Horsepower   The four numbers in the middle of the model number give the nominal compressor capacity in horsepower. The decimal goes in the middle of the four numbers. In the example above, the compressor rating is 30 horsepower.

Copeland Hermetic Compressors									
Z	R	90	K	3	E	T	W	D	551
Family	Application	First 2 digits in BTU/hr capacity	Capacity Multiplier	Variant	Oil	Motor	Protect	Voltage	Misc

Family   C   Reciprocating
            Z   Scroll

Application   B   High/Medium Temp
                 BD   High/Medium digital
                 BH   High/Medium horizontal
                 F   Low Temp with Injection
                 FH   Low Temp Horizontal
                 H   Heat Pump
                 P   Air Condition 410A
                 R   Air Condition
                 RT   Air Condition Even Tandem
                 RU   Air Condition Uneven Tandem
                 S   Medium Temperature

Multiplier   K   1000
                M   10000

Oil   E   Polyol Ester (POE)
         No Letter = Mineral Oil

Motor   T   Three Phase
           P   Single Phase

Protection   F   Internal line break overload
                W   External electronic

Voltage   C   208-230 volt 60 hz
             D   460 volt 60 hz
             5   200-220 volt 60 hz

Capacity   The two numbers are the first two digits of the compressor capacity in BTU/hr.
              They are multiplied by the capacity multiplier. In this example, the capacity is
              90,000 BTU/hr 90 x 1000

LAB 13.1 COMPRESSOR IDENTIFICATION DATA SHEET						
	Unit 1	Unit 2	Unit 3	Unit 4	Unit 5	Unit 6
Type Reciprocating Rotary Scroll						
Body Open Semi-hermetic Hermetic						
Cooling Air Refrigerant						
Application AC High Medium Low						
Size Btuh or Horsepower						
Electrical Volts Phase Amps						

LAB 13.2 SEMI-HERMETIC COMPRESSOR INSPECTION

LABORATORY OBJECTIVE
You will disassemble a semi-hermetic compressor, identify its parts, discuss its operation, and reassemble it.

FUNDAMENTALS OF HVACR TEXT REFERENCE
Unit 13 Compressors

LABORATORY NOTES
The instructor will assign a semi-hermetic compressor for you to disassemble. You will examine the parts and discuss their function with the instructor. You will then reassemble the compressor.

REQUIRED TOOLS, EQUIPMENT, AND MATERIALS
Safety glasses
Gloves
6-in-1
Adjustable jaw wrench
Semi-hermetic compressor
Socket set
Oil drain pan

SAFETY REQUIREMENTS
    A. Wear safety glasses
    B. Wear gloves when handling compressor components

PROCEDURE

1. Obtain semi-hermetic compressor from instructor.

2. Release pressure from crankcase through suction service valve.

3. Drain oil from crankcase of compressor.

4. Remove cylinder head and valve plate.

5. Identify the suction and discharge valves.

6. Mark both end bells and carefully examine their orientation on the compressor.

7. Remove bolts from end bells.

8. Remove end bells.

9. Remove bolts from the bottom plate.

10. Remove bottom plate from the crankcase.

11. Have instructor check compressor after teardown.

12. Examine the parts and discus their function with the instructor.

13. Complete data sheet.

14. Assemble compressor in reverse order of disassembly.

LAB 13.2 SEMI-HERMETIC COMPRESSOR DISASSEMBLY	
Manufacturer	
Model Number	
Number of Pistons	
Crankshaft Type **(crank throw or eccentric)**	
Valve Type **(reed, ring, discuss)**	
Cooling **(Air or Refrigerant)**	
Lubrication **(Splash or Pressure)**	
Application **(AC, High, Medium, Low)**	

LAB 13.3 HERMETIC COMPRESSOR INSPECTION

LABORATORY OBJECTIVE
You will disassemble a hermetic compressor, identify its parts, discuss its operation, and reassemble it.

FUNDAMENTALS OF HVACR TEXT REFERENCE
Unit 13 Compressors

LABORATORY NOTES
The instructor will assign a hermetic compressor for you to disassemble. You will examine the parts and discuss their function with the instructor. You will then reassemble the compressor.

REQUIRED TOOLS, EQUIPMENT, AND MATERIALS
Safety glasses
Gloves
6-in-1
Adjustable jaw wrench
Hermetic compressor
Socket set
Oil drain pan
Electric grinder or Reciprocating saw
Extension cord

SAFETY REQUIREMENTS
  A. Wear safety glasses
  B. Wear gloves
  C. All power tools MUST BE GROUNDED!
  D. Do **NOT** use cutting torch! The oil in the compressor shell and oxygen from the cutting torch can explode!

PROCEDURE

Procedure for cutting open compressor shell

  1. Release pressure from shell through process tube.
  2. Drain oil from crankcase of compressor by inverting compressor and pouring the oil out of the process tube.
  3. Purge the shell with nitrogen to prevent combustion of the oil residue while cutting open the shell.
  4. Cut the shell in two at the weld seam. A reciprocating saw with a metal blade makes the cleanest and safest cut, but does leave metal filings in the shell. A grinder works quickly, but leaves a sharp edge on the metal casing. Do NOT use a cutting torch. The oxygen can mix with the oil in the compressor to create an explosion.
  5. Separate the shell and cut the discharge line at a convenient point.

Procedure for disassembly

1. Remove the compressor-motor assembly from the shell.
2. Remove the compressor head and valve plate.
3. Have an instructor check your disassembly.
4. Examine the parts and discus their function with the instructor.
5. Answer the questions on the following page.
6. Reassemble valve plate and head.
7. Place compressor back in shell.

QUESTIONS

1. What type of lubrication system does this compressor use?

2. Is this compressor motor air-cooled or refrigerant cooled?

3. What is in the shell: low pressure suction gas or high pressure discharge gas?

4. What type of duty was this unit designed for? (high, medium, or low temp)

5. What size compressor is this? (horsepower and/or BTU)

6. What type of valves doe this compressor have?

7. How was the compressor mounted inside the shell?

8. Why is the discharge line shaped the way it is?

9. How does the suction gas reach the cylinders in the compressor?

LAB 13.4 ROTARY COMPRESSOR INSPECTION

LABORATORY OBJECTIVE
You will disassemble a rotary compressor, identify its parts, discuss its operation, and reassemble it.

***FUNDAMENTALS OF HVACR* TEXT REFERENCE**
Unit 13 Compressors

LABORATORY NOTES
The instructor will assign a rotary compressor for you to disassemble. You will examine the parts and discuss their function with the instructor. You will then reassemble the compressor.

REQUIRED TOOLS, EQUIPMENT, AND MATERIALS
Safety glasses
Gloves
6-in-1
Adjustable jaw wrench
Rotary compressor
Socket set
Oil drain pan

SAFETY REQUIREMENTS
   A. Wear safety glasses
   B. Wear gloves when handling compressor parts

PROCEDURE
1. Obtain hermetic rotary compressor from instructor.
2. Remove the hermetic casing from the rotary compressor body.
3. Remove the top plate from the compressor body.
4. Examine the blades of the compressor.
5. Turn rotor and examine action of the blades when compressor is rotated.
6. Replace top plate.
7. Replace compressor hermetic casing.
8. Answer the questions below:

   a. Is this stationary blade or rotating blade compressor?

   b. What is in the shell: low-pressure suction gas or high pressure discharge gas?

   c. How does the suction gas get into the compressor?

   d. How does the discharge gas get from the compressor to the discharge line?

LAB 13.5 SCROLL COMPRESSOR INSPECTION

LABORATORY OBJECTIVE
You will disassemble a scroll compressor, identify its parts, discuss its operation, and reassemble it.

FUNDAMENTALS OF HVACR TEXT REFERENCE
Unit 13 Compressors

LABORATORY NOTES
The instructor will assign a scroll compressor for you to disassemble. You will examine the parts and discuss their function with the instructor. You will then reassemble the compressor.

REQUIRED TOOLS, EQUIPMENT, AND MATERIALS
Safety glasses
Gloves
6-in-1
Adjustable jaw wrench
Scroll compressor Socket set
Oil drain pan

SAFETY REQUIREMENTS
    A. Wear safety glasses
    B. Wear gloves when handling compressor parts

PROCEDURE
1. Obtain hermetic scroll compressor from instructor.
2. Remove the hermetic casing from the scroll compressor body.
3. Remove the top plate from the compressor body.
4. Examine the scrolls of the compressor.
5. Turn rotor and examine action of the scrolls when compressor is rotated.
6. Replace top plate.
7. Replace compressor hermetic casing.
8. Answer the questions below:

    a. What is in the shell: low-pressure suction gas or high pressure discharge gas?

    b. How does the suction gas get into the compressor?

    c. How does the discharge gas get from the compressor to the discharge line?

    d. How many moving parts does this compressor have?

LAB 13.6 DETERMINING COMPRESSION RATIO

LABORATORY OBJECTIVE
You will check the suction and discharge pressures on an operating compressor and determine its compression ratio.

FUNDAMENTALS OF HVACR TEXT REFERENCE
Unit 13 Compressors

LABORATORY NOTES
You will install your gauges on an operating system and use the operating pressures to determine the compressor ratio. You will then block the condenser airflow, retest, and recalculate the compression ratio. Finally, you will block the evaporator airflow, retest, and recalculate the compression ratio.

REQUIRED TOOLS, EQUIPMENT, AND MATERIALS
Safety glasses
Gloves
Two adjustable jaw wrenches
6-in-1
Refrigeration gauges
Refrigeration valve wrench
Calculator
Operating refrigeration system

SAFETY REQUIREMENTS
    A. Wear safety glasses and gloves when accessing refrigeration systems
    B. Use caution around moving fans.
    C. Do NOT touch any electrical connections

PROCEDURE
Operate the system assigned by the instructor until the pressures have stabilized. Measure the suction and discharge pressure and calculate the compression ratio. The compression ratio is found by dividing the absolute discharge pressure by the absolute suction pressure. This is shown in the formula:

$$(\text{Discharge pressure} + 15) \div (\text{Suction Pressure} + 15)$$

Temporarily restrict the airflow across the condenser.

CAUTION: Monitor the high side pressure and remove the restriction if the condenser saturation temperature reaches 140°F.

The high side pressure should increase.

Recheck the system pressures.

Record the new pressures.

Remove the restriction and allow the pressures to return to normal.

Recalculate the compression ratio using the new pressures.

What effect did blocking the condenser airflow have on the compression ratio?

Temporarily restrict the airflow across the evaporator.

The evaporator pressure should drop. Recheck the system pressures.

Record the new pressures.

Remove the restriction and allow the pressures to return to normal.

Recalculate the compression ratio using these new pressures.

What effect did blocking the evaporator airflow have on the compression ratio?

LAB 13.7 COMPRESSOR AUTOPSY

LABORATORY OBJECTIVE
You will disassemble a failed compressor and identify the cause of the compressor failure.

FUNDAMENTALS OF HVACR **TEXT REFERENCE**
Unit 13 Compressors

LABORATORY NOTES
You will disassemble a failed compressor, examine the parts of the compressor, and determine the type of failure. You will then recommend ways to prevent a repeat failure of the replacement compressor.

REQUIRED TOOLS, EQUIPMENT, AND MATERIALS
Safety glasses
Gloves
Two adjustable jaw wrenches
6-in-1
Failed compressor
Socket set

SAFETY REQUIREMENTS
    A. Wear safety glasses
    B. Avoid touching compressor oil - it may be acidic.
    C. Wear gloves when handling compressor parts

PROCEDURE
Release pressure from crankcase through suction service valve.

Drain oil from crankcase of compressor.

Remove cylinder head and inspect the valve plate.

Are there signs of high temperature operation? (dark varnished looking parts)

Are the discharge valves intact?

Remove the valve plate and inspect the suction valves.

Are the suction valves intact?

Mark both end bells and carefully examine their orientation on the compressor.

Remove bolts from end bells.

Remove end bells.

Remove bolts from the bottom plate.

Remove bottom plate from the crankcase.

Inspect the crank and rods.

Are all the parts intact?

Turn the crank and observe the operation of the rods and pistons.

Is there any slop in the fit between the crank, rods, and pistons?

Examine the parts and discus their function with the instructor

Assemble compressor in reverse order of disassembly.

Determine the failure type.

- **Liquid flooding** – broken parts, especially valves, pistons, and rods
- **Overheating** – dark, discolored parts, especially valve plates
- **Oil Failure** – scoring on bearing surfaces or seizure

Recommend ways of preventing a similar repeat failure for the replacement compressor.

LAB 14.1 AIR COOLED CONDENSER PERFORMANCE

LABORATORY OBJECTIVE
You will measure the heat rejection of an air cooled condenser.

***FUNDAMENTALS OF HVACR* TEXT REFERENCE**
Unit 14 Condensers

LABORATORY NOTES
You will measure the airflow, entering air temperature, and leaving air temperature through an air-cooled condenser. Those data will be used to calculate the total condenser heat rejection.

REQUIRED TOOLS, EQUIPMENT, AND MATERIALS
Safety glasses
Gloves
6-in-1
Thermometer
Calculator
Operating unit with air cooled condenser
Rotating vane anemometer
Condenser airflow capture device

SAFETY REQUIREMENTS
   A. Wear safety glasses
   B. Avoid moving fan blades.

PROCEDURE
   1. Start the unit assigned by the instructor.
   2. Allow the unit to operate for at least 10 minutes.
   3. Place the condenser airflow capture device over the condenser discharge.
   4. Measure the temperature of the air entering and leaving the condenser.
   5. Calculate the temperature difference ΔT = Leaving Air Temp − Entering air Temp
   6. Measure the velocity of the air leaving the condenser airflow capture device.
   7. Calculate the air volume CFM = FPM (air velocity) x ft^2 area of airflow capture device.
   8. Calculate the heat rejection. BTUs/hr = ΔT x CFM x 1.08
   9. Complete the chart below.

Air Cooled Condenser Performance	
Entering air temperature	
Leaving air temperature	
Temperature rise	
Air Velocity in Feet per Minute FPM	
Air flow in Cubic Feet per Minute CFM	
Total Heat Rejection in BTUs/hr	

LAB 14.2 WATER COOLED CONDENSER PERFORMANCE

LABORATORY OBJECTIVE
You will measure the heat rejection of a water-cooled condenser.

FUNDAMENTALS OF HVACR TEXT REFERENCE
Unit 14 Condensers

LABORATORY NOTES
You will measure the water flow, entering water temperature, and leaving water temperature through an water cooled condenser. Those data will be used to calculate the total condenser heat rejection.

REQUIRED TOOLS, EQUIPMENT, AND MATERIALS
Safety glasses
Gloves
6-in-1
Thermometer calculator
Operating unit with water-cooled condenser
Water flow indicator

SAFETY REQUIREMENTS
    A. Wear safety glasses
    B. Wear gloves

PROCEDURE
1. Start the unit assigned by the instructor.
2. Allow the unit to operate for at least 10 minutes.
3. Measure the temperature of the water entering and leaving the condenser.
4. Calculate the temperature difference ΔT = Leaving Water Temp – Entering Water Temp
5. Observe the water flow indicator to determine the volume of water moving across the condenser in gallons per minute (GPM).
6. Calculate the total heat rejection BTUs/hr = ΔT x GPM x 8.35 lbs/gallon x 60 min/hr
7. Complete the chart below and calculate the heat rejection.

Water Cooled Condenser Performance	
Entering water temperature	
Leaving water temperature	
Temperature rise	
Water flow in GPM	
Total Heat Rejection	

LAB 15.1 INTERNALLY EQUALIZED THERMOSTATIC EXPANSION VALVES

LABORATORY OBJECTIVE
The student will disassemble an internally equalized thermostatic expansion valve and be able to describe how it operates.

FUNDAMENTALS OF HVACR TEXT REFERENCE
Unit 15 Metering Devices

LABORATORY NOTES:
For this lab exercise there should be one or more internally equalized thermostatic expansion valves available that may be disassembled.

REQUIRED TOOLS, EQUIPMENT, AND MATERIALS
Internally equalized thermostatic expansion valve

SAFETY REQUIREMENTS
None

PROCEDURE

Step 1
Locate an internally equalized thermostatic expansion valve and examine it carefully so that you may complete the following exercise:

A. Record all of the data that can be found on the valve such as refrigerant type, capacity, line size, refrigerant bulb charge, etc.

Thermal
Bulb

Figure 15-1-1

Thermostatic Expansion Valve Data:

B. Disassemble the valve being careful not to lose components while identifying the manner in which the valve came apart so that it may then be put back together again. An internally equalized valve has only an inlet and an outlet. There is no equalizing connection.

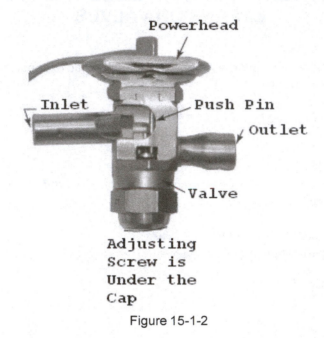

Powerhead

Inlet

Push Pin

Outlet

Valve

Adjusting
Screw is
Under the
Cap

Figure 15-1-2

C. Sketch a cross sectional view of the disassembled valve in the space provided below.

D. Explain how the valve operates using the sketch you provided above.

Step 2

Using the disassembled valve, your sketch, and your description, answer the following questions:

    A. Draw a one-line representation of the diaphragm and the forces acting upon it (bulb pressure, evaporator pressure, and spring pressure).

    B. How is the diaphragm movement transmitted to the valve disk?

    C. Does the spring help to open or to close the valve?

    D. How does the evaporator pressure travel to the underside of the valve diaphragm?

E. Does the evaporator pressure help to open or to close the valve?

F. If the bulb lost its charge would the valve fail open or closed?

G. What type of refrigerant can the valve be used for?

QUESTIONS

(Circle the letter that indicates the correct answer.)

1. A thermostatic expansion valve can be adjusted for:
   A. refrigerant flow through the evaporator.
   B. evaporator refrigerant outlet superheat.
   C. evaporator capacity.
   D. All of the above are correct.

2. An internally equalized thermostatic expansion valve senses pressure.
   A. at the evaporator outlet.
   B. at the compressor inlet.
   C. at the condenser outlet.
   D. at the evaporator inlet.

3. To increase the superheat setting of thermostatic expansion valve you would turn the adjusting screw:
   A. clockwise.
   B. counter clockwise.
   C. It can not be adjusted.
   D. back and forth.

4. If the thermal bulb on a thermostatic expansion valve lost its charge:
   A. the valve would open wide.
   B. the valve would hunt.
   C. the valve would close.
   D. the valve would frost up.

5. The thermal bulb pressure on the top of a thermostatic expansion valve diaphragm is transmitted to the valve by:
   A. the spring.
   B. the evaporator inlet pressure.
   C. the evaporator outlet pressure.
   D. push pins.

6. The thermal bulb force acting on the topside of the diaphragm in a thermostatic expansion valve is balanced by:
   A. spring force only.
   B. evaporator pressure only.
   C. spring force and evaporator pressure.
   D. an orifice plate.

7. Internally equalized thermostatic expansion valves can be used:
   A. on evaporator coils that have a minimal pressure drop.
   B. on evaporator coils that have a large pressure drop.
   C. on beverage coolers only.
   D. A & C are both correct.

8. If an internally equalized thermostatic expansion valve is used on an evaporator coil that has a large pressure drop:
   A. the evaporator coil will flood with refrigerant.
   B. the evaporator coil will freeze up.
   C. the evaporator coil will be starved.
   D. liquid will slug back to the compressor.

LAB 15.2 EXTERNALLY EQUALIZED THERMOSTATIC EXPANSION VALVES

LABORATORY OBJECTIVE
The student will disassemble an externally equalized thermostatic expansion valve and be able to describe how it operates.

FUNDAMENTALS OF HVACR TEXT REFERENCE
Unit 15 Metering Devices

LABORATORY NOTES
For this lab exercise there should be one or more externally equalized thermostatic expansion valves available that may be disassembled.

REQUIRED TOOLS, EQUIPMENT, AND MATERIALS:
Externally equalized thermostatic expansion valve

SAFETY REQUIREMENTS
None

PROCEDURE
Step 1
Locate an externally equalized thermostatic expansion valve and examine it carefully so that you may complete the following exercise:

   A. Record all of the data that can be found on the valve such as refrigerant type, capacity, line size, refrigerant bulb charge, etc.

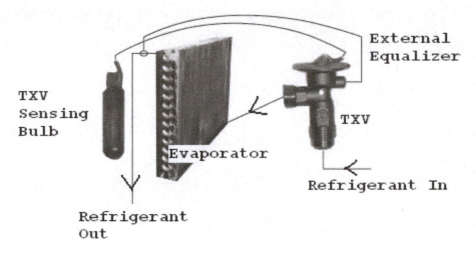

Figure 15-2-1

B. Disassemble the valve being careful not to lose components while identifying the manner in which the valve came apart so that it may then be put back together again. An externally equalized valve has an inlet, an outlet, and an external equalizing connection.

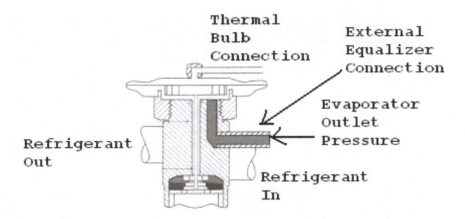

Figure 15-2-2

C. Sketch a cross sectional view of the valve in the space provided below.

D. Explain how the valve operates using the sketch you provided above.

Step 2
Using the disassembled valve, your sketch and your description, answer the following questions:

A. Draw a one-line representation of the diaphragm and the forces acting upon it (bulb pressure, evaporator pressure, and spring pressure).

B. How is the diaphragm movement transmitted to the valve disk?

C. Does the spring help to open or to close the valve?

D. How does the evaporator pressure travel to the underside of the valve diaphragm?

E. Does the evaporator pressure help to open or to close the valve?

F. If the bulb lost its charge would the valve fail open or closed?

G. What type of refrigerant can the valve be used for?

QUESTIONS

(Circle the letter that indicates the correct answer.)

1. A thermostatic expansion valve can be adjusted for:
   A. refrigerant flow through the evaporator.
   B. evaporator refrigerant inlet superheat.
   C. evaporator pressure.
   D. All of the above are correct.

2. An externally equalized thermostatic expansion valve senses pressure.
   A. at the evaporator outlet.
   B. at the compressor inlet.
   C. at the condenser outlet.
   D. at the evaporator inlet.

3. To decrease the superheat setting of thermostatic expansion valve you would turn the adjusting screw:
   A. clockwise.
   B. counter clockwise.
   C. It can not be adjusted.
   D. back and forth.

4. If the thermal bulb on a thermostatic expansion has a liquid charge:
   A. control will always be from the bulb.
   B. control will never be from the bulb.
   C. the valve will have a limited opening.
   D. the valve will open quicker.

5. If the thermal bulb on a thermostatic expansion has a limited liquid (gas) charge:
   A. control will always be from the bulb.
   B. control will never be from the bulb.
   C. the valve will have a limited opening.
   D. the valve will open quicker.

6. If the thermal bulb on a thermostatic expansion has a cross charge:
   A. the fluid in the bulb is the same as the refrigerant in the system.
   B. the fluid in the bulb is different than the refrigerant in the system.
   C. the suction pressure crosses the discharge pressure.
   D. the valve will open quicker.

7. Externally equalized thermostatic expansion valves can be used:
   A. on evaporator coils that have a minimal pressure drop.
   B. on evaporator coils that have a large pressure drop.
   C. on beverage coolers only.
   D. A & C are both correct.

8. If the external equalizing connection on a externally equalized thermostatic expansion valve is capped closed:
   A. the evaporator coil will flood with refrigerant.
   B. the evaporator coil will freeze up.
   C. the evaporator coil will be starved.
   D. liquid will slug back to the compressor.

9. The fluid in a straight charged thermostatic expansion valve bulb:
   A. is the same as the refrigerant in the system.
   B. is different than the refrigerant in the system.
   C. is always liquid.
   D. is always vapor.

Name_____

Date_____

Instructor's OK ☐

LAB 15.3 SETTING A THERMOSTATIC EXPANSION VALVE

LABORATORY OBJECTIVE
The student will demonstrate how to properly set a thermostatic expansion valve to maintain the proper superheat at the evaporator outlet.

FUNDAMENTALS OF HVACR TEXT REFERENCE
Unit 15 Metering Devices

LABORATORY NOTES
For this lab exercise there needs to be an operating refrigeration system that has a thermostatic expansion valve. If there are no thermometers or pressure gauges currently on the unit then you must instrument the refrigeration system using a gauge manifold to measure the evaporator pressure and a thermometer to read the evaporator refrigerant temperature.

If you are unable to place probes directly into the refrigerant stream flow, you can use evaporator coil surface temperatures, however you must adjust for the heat transfer difference through the coil.

REQUIRED TOOLS, EQUIPMENT, AND MATERIALS
Operating refrigeration unit with thermostatic expansion
valve gauge manifold & temperature sensor

SAFETY REQUIREMENTS
   A. Always read the equipment manual to become familiarized with the refrigeration system and its accessory components prior to start up.
   B. Wear safety goggles and gloves when working with refrigerants. Liquid refrigerant can cause frostbite when in contact with eyes and skin.
   C. Use low loss hose fittings, or wrap cloth around hose fittings before removing the fittings from a pressurized system or cylinder. Inspect all fittings before attaching hoses.

PROCEDURE
Step 1
Familiarize yourself with the major components in the refrigeration system including the condenser, compressor, evaporator, and metering device.

   A. The thermostatic expansion (TEV) supplies the evaporator with enough refrigerant for any and all load conditions. It is <u>NOT</u> a temperature, suction pressure, humidity, or operating control.

   B. The thermostatic expansion valve is adjusted to control the refrigerant superheat at the outlet of the evaporator.

   C. Many thermostatic expansion valves come pre-set for 10°F of superheat.

   D. Thermostatic expansion valves should not be adjusted unless the evaporator conditions (temperature & pressure) can be measured.

Step 2
Determine the proper setting for the TEV as follows:

    A. The flow of the refrigerant through the TEV is controlled by three pressures.

        a) The evaporator pressure

        b) The spring pressure acting on the bottom of the diaphragm

        c) The bulb pressure opposing these two pressures and acting on the top of the diaphragm

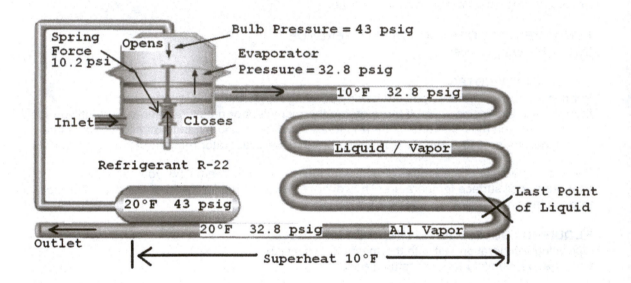

Figure 15-3-1

    B. In Figure 15-3-1, these three pressures are illustrated. When the three forces are balanced and the valve is in equilibrium as shown, then there should be ten degrees of superheat at the evaporator outlet.

    C. The saturation temperature for R-22 at 32.8 psig is 10°F. If the refrigerant pressure is 32.8 psig and the temperature is 20°F, then there would be what is considered as 10°F of superheat.

    D. If the superheat temperature decreases, there will be a corresponding decrease in bulb pressure as the TEV would be allowing too much refrigerant to flow. The balance of the three pressures would be disrupted and the valve would begin to close until the pressures arrived at a new equilibrium point.

    E. To determine the superheat value you will need to measure both the pressure and the temperature at the evaporator outlet.

Assuming we are using R-22 and the evaporator pressure is 32.8 psig as shown in Unit 9, Figure 9-21 of the *Fundamentals of HVACR* text (or Table 49-1-1 in Lab 49.1) we can look up the saturated refrigerant temperature.

From the chart at 32.8 psig, for R-22 the saturated temperature would be 10°F. You would then subtract this saturated temperature from the actual measured temperature from the thermometer to determine the total degrees of superheat.

Step 3

    A. If there are no pressure gauges currently on the unit then you must connect a gauge manifold to read the evaporator pressure. To help guide you, refer to the procedures in Lab 47.1 *Basic Refrigeration System Startup*, Steps 4, 5, & 6.

    B. If you are unable to place temperature probes directly into the refrigerant stream flow, you can use evaporator coil surface temperatures however you must adjust for the heat transfer difference through the coil.

Step 4

    A. Start the refrigeration system in the normal cooling mode and allow the system to stabilize and then record the evaporator pressure and temperature below.

Measured _____ psig   Measured _____ °F

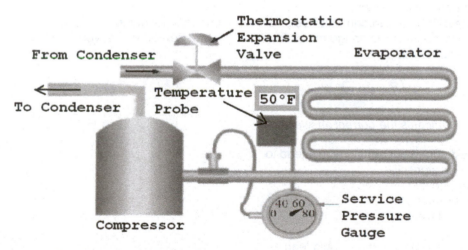

To Measure Superheat:
1. Find Suction Pressure
2. Find Matching Saturation Temperature
3. Read Temperature Leaving Evaporator
4. Superheat = Temp. Leaving - Saturation Temp.

Figure 15-3-2

B. From the saturated pressure-temperature chart determine the saturated temperature at the measured pressure reading.

C. Calculate the superheat.

Superheat = Measured Temp. – Saturation Temp.

D. Determine what adjustments need to be made, if any, and turn the adjusting screw in small increments only. This will change the spring tension that in turn will change the superheat setting.

E. On many valves the adjustment is clockwise to increase the superheat and counterclockwise to decrease the superheat. Valve instructions should be checked to be sure of correct adjustments.

F. Allow the system to run and then recheck the superheat once again before making any further changes. It may take some time before the system stabilizes and the final adjustment is complete.

Step 5

A. Continue making any necessary adjustments until the thermostatic expansion valve superheat is correctly set.
B. Allow the system to run and stabilize prior to shutting down and then carefully disconnect the gauge manifold and any instrumentation that you attached to the unit.

QUESTIONS

(Circle the letter that indicates the correct answer.)

1. Too high of a TEV superheat setting would lead to:
   A. a flooded evaporator coil.
   B. liquid slugging the compressor.
   C. a starved evaporator coil.
   D. A & B are both correct.

2. Too low of a TEV superheat setting would lead to:
   A. a flooded evaporator coil.
   B. liquid slugging the compressor.
   C. a starved evaporator coil.
   D. A & B are both correct.

3. Ice forming on the outside of a thermostatic expansion valve:
   A. indicates blockage.
   B. indicates normal operation.
   C. can crack the diaphragm.
   D. should be chipped off with an ice pick.

4. Raising the temperature of a fluid above its saturation temperature is:
   A. superheating.
   B. subcooling.
   C. sublimation.
   D. saturation absorption.

5. If the sensing bulb for a thermostatic expansion valve becomes unattached from the evaporator coil:
   A. the valve will begin to close and starve the evaporator.
   B. the valve will begin to open and starve the evaporator.
   C. the valve will begin to close and flood the evaporator.
   D. the valve will begin to open and flood the evaporator.

6. If the diaphragm on a thermostatic expansion valve fails the valve.
   A. the valve will begin to close and starve the evaporator.
   B. the valve will begin to open and starve the evaporator.
   C. the valve will begin to close and flood the evaporator.
   D. the valve will begin to open and flood the evaporator.

7. If the evaporator coil is cold at the inlet but is warm about half way down then this could indicate:
   A. a high superheat.
   B. a low superheat.
   C. the proper amount of superheat.
   D. floodback.

LAB 15.4 TYPES OF METERING DEVICES

LABORATORY OBJECTIVE
You will identify different metering devices and discuss their application.

FUNDAMENTALS OF HVACR TEXT REFERENCE
Unit 15 Metering Devices

LABORATORY NOTES
You will examine units with different metering devices, including capillary tube, orifice, and thermostatic expansion valve. You will identify the type of metering device in each unit and discuss the application in that unit.

REQUIRED TOOLS, EQUIPMENT, AND MATERIALS
Safety glasses
Gloves
6-in-1
Adjustable jaw wrench
Refrigeration systems with different metering devices

SAFETY REQUIREMENTS
A. Wear safety glasses
B. Units should be off during inspection.
C. Turn off power to units before inspection.
D. Use caution around panels—sharp edges can cut you.

PROCEDURE
1. Remove panels necessary to locate the metering device in each of the units assigned by the instructor.
2. Identify the type of metering device (cap tube, orifice, TEV).
3. Determine the number of refrigerant circuits.
4. Identify the distribution device for coils with more than one circuit.
5. Locate where the refrigerant from the metering device feeds the evaporator.

LAB 15.5 ADJUSTING AUTOMATIC EXPANSION VALVES

LABORATORY OBJECTIVE
You will adjust an automatic expansion valve to maintain a specified evaporator pressure.

FUNDAMENTALS OF HVACR TEXT REFERENCE
Unit 15 Metering Devices

LABORATORY NOTES
You will adjust the automatic expansion valve on the air dryer to the pressures specified in the lab.

REQUIRED TOOLS, EQUIPMENT, AND MATERIALS
Safety glasses
Gloves
6-in-1
Adjustable jaw wrench
Refrigerated Air Drier with AEV

SAFETY REQUIREMENTS
  A. Wear safety glasses
  B. Avoid moving fan blades.
  C. Do NOT touch any electrical connections.

PROCEDURE
1. Remove panels necessary to locate the automatic expansion valve.
2. Operate the system and record the suction pressure.
3. Turn the valve adjustment stem in clockwise until the suction pressure is 5 psig higher than the original reading.
4. Let the system operate for 5 minutes.
5. Turn the valve-adjusting stem out counterclockwise to set the pressure back to the original setting.
6. Explain how the adjustment on an automatic expansion valve works.

LAB 15.6 TEV INSTALLATION

LABORATORY OBJECTIVE
You will properly install a thermostatic expansion valve and measure the superheat.

FUNDAMENTALS OF HVACR TEXT REFERENCE
Unit 15 Metering Devices

LABORATORY NOTES
You will pump down the refrigerant into the high side of the system and remove the existing metering device. You will then install a TEV and evacuate the refrigerant lines and coil. After the lines and coil are evacuated, you will open the king valve, allowing the refrigerant to flow back into the low side. Finally, you will operate the system and measure the superheat at the expansion valve-sensing bulb.

REQUIRED TOOLS, EQUIPMENT, AND MATERIALS
Safety glasses
Gloves
Two adjustable jaw wrenches
6-in-1
Refrigeration gauges
Refrigeration valve wrench
Temperature tester
Refrigeration system
TEV
Vacuum pump
Vacuum gauge
Refrigerant

SAFETY REQUIREMENTS
    A. Wear safety glasses and gloves when accessing refrigeration systems.
    B. Use caution around moving fans.
    C. Do NOT touch any electrical connections

PROCEDURE
1. Install refrigeration gauges on the system.

2. The compound gauge should be connected to the suction service valve.

3. The high side gauge should be connected to the king valve.

4. Purge the gauges.

5. Front seat liquid service valve.

6. Operate the system to pump down refrigerant into high side.

7. The compound gauge should pull down to 0 psig or into a vacuum.

8. The high side gauge may pull down to 0 psig or remain high, depending on the type of king valve the system has.

9. Clean the metering device connection by wiping with a clean cloth.

10. Loosen the flare connections to the existing metering device.

11. Be sure to use two properly fitting wrenches.

12. Do NOT try to loosen the flare connection with a single wrench. The tubing will twist, damaging the coil.

13. Remove the existing metering device.

14. Install thermostatic expansion valve.

15. Tighten using two wrenches, being careful not to twist the tubing.

16. Secure the TEV bulb to the suction line and insulate it.

17. Pressurize the lines and coil with nitrogen and check for leaks.

18. The king valve should REMAIN FRONTSEATED. Do NOT move it.

19. Evacuate low side of system (high side still has refrigerant in it).

20. Backseat crack the king valve to let the refrigerant back into the system.

21. Start system and adjust charge as necessary.

21. Measure and record system pressures.

22. Measure and record suction line temperature, liquid line temperature, superheat, and subcooling.

23. Have instructor check unit.

LAB 15.6 TEV INSTALLATION	
Evaporator Pressure	
Evaporator Saturation Temp	
Bulb Temperature	
Superheat (Bulb Temp –Evap Saturation)	
Condenser Pressure	
Condenser Saturation	
Liquid Line Temperature	
Subcooling Cond saturation-liquid line	

LAB 15.7 ADJUSTING VALVE SUPERHEAT

LABORATORY OBJECTIVE
You will measure the superheat of thermostatic expansion valve system and adjust it to meet manufacturer's specifications.

FUNDAMENTALS OF HVACR TEXT REFERENCE
Unit 15 Metering Devices

LABORATORY NOTES
You will use the suction pressure, suction line temperature, and PT chart to determine the operating superheat of an expansion valve system. You will then adjust the valve counterclockwise and then clockwise to observe the effects of valve adjustment. Finally, you will adjust the valve to manufacturer specification.

REQUIRED TOOLS, EQUIPMENT, AND MATERIALS
Safety glasses
Gloves
Two adjustable jaw wrenches
6-in-1
Refrigeration gauges
Refrigeration valve wrench
Temperature tester
Operating refrigeration system with TEV
Refrigerant

SAFETY REQUIREMENTS
  A. Wear safety glasses and gloves when accessing refrigeration systems.
  B. Use caution around moving fans.
  C. Do NOT touch any electrical connections

PROCEDURE
  1. Operate the system assigned by the Instructor for 15 minutes.
  2. Measure the superheat of the expansion valve and record all pertinent data on the chart below.
  3. Turn the adjusting stem 1 full turn counter-clockwise. Let system operate or 15 minutes.
  4. Measure the superheat and record all pertinent data.
  5. Turn adjusting stem 2 full turns clockwise. Let system operate for 15 minutes.
  6. Measure the superheat and record all pertinent data.
  7. Adjust T.E.V. for normal operating superheat. Let operate for 15 minutes.
  8. Measure superheat and record all pertinent data.

9. Explain how superheat adjustment works.

10. Summarize the effect of bulb temperature on system operation.

LAB 15.7 ADJUSTING VALVE SUPERHEAT				
	Condenser Pressure	Evaporator Pressure	Saturation Temp	Bulb Temp	Superheat (Bulb –Saturation)
Starting Setting					
1 Turn CCW					
2 Turns CW					
Normal Operation					

LAB 15.8 CAPILLARY TUBE ASSEMBLY & OPERATION

LABORATORY OBJECTIVE
You will assemble, leak test, evacuate, and charge a capillary tube refrigeration system.

***FUNDAMENTALS OF HVACR* TEXT REFERENCE**
Unit 15 Metering Devices

LABORATORY NOTES
You will pump down the system, remove the existing metering device, and install a capillary tube. You will then evacuate the system, open the liquid service valve, and operate the system. You will record all operating data including system pressures, system saturation temperatures, superheat, and subcooling.

REQUIRED TOOLS, EQUIPMENT, AND MATERIALS
Safety glasses
Gloves
Two adjustable jaw wrenches
6-in-1
Refrigeration gauges
Refrigeration valve wrench
Temperature tester
Operating refrigeration system
Refrigerant
Capillary tube
Vacuum pump
Torch
Brazing material

SAFETY REQUIREMENTS
    A. Wear safety glasses and gloves when accessing refrigeration systems.
    B. Use caution around moving fans.
    C. Do NOT touch any electrical connections

PROCEDURE
1. Install your gauges with the high side on the liquid service valve.
2. Make certain to purge your gauges.
3. Front seat liquid service valve and pump down refrigerant into high side.
4. Clean the connection.
5. Remove existing metering device.
6. Install capillary tube.
7. Pressure test system for leaks using nitrogen.
8. Evacuate low side of system (high side still has refrigerant in it).

9. Back seat crack the liquid service valve to let the refrigerant into the system.
10. Start and test system.
11. Adjust charge as necessary.
12. Record system pressures, suction line temperature, liquid line temperature, superheat, and subcooling.
13. Have instructor check unit.

LAB 15.8 DATA	
Evaporator Pressure	
Evaporator Saturation Temperature	
Suction Line Temperature	
Superheat **Suction temp – Evaporator Saturation Temp**	
Condenser Pressure	
Condenser Saturation Temperature	
Liquid Line Temperature	
Subcooling **Condenser Saturation temp – Liquid Temp**	

LAB 16.1 INSTALLING CONDENSATE DRAIN

LABORATORY OBJECTIVE
You will install an evaporator condensate drain and demonstrate that it drains properly.

FUNDAMENTALS OF HVACR TEXT REFERENCE
Unit 16 Evaporators

LABORATORY NOTES
You will install a condensate drain on the evaporator assigned by the instructor. After installing the drain, you will pour water into the drain pan to demonstrate that the drain works properly.

REQUIRED TOOLS, EQUIPMENT, AND MATERIALS
Safety glasses
Gloves
6-in-1
Air conditioning evaporator
PVC
PVC fittings
PVC primer and solvent
PVC shear

SAFETY REQUIREMENTS
A. Wear safety glasses and gloves
B. Avoid contact with PVC cleaner and solvent.
C. Avoid breathing fumes form PVC cleaner and solvent.
D. Turn on shop ventilation fan

PROCEDURE
1. Install the condensate drain for the unit assigned by the instructor.
2. Tubing should be cut with a tubing shear.
3. If a hacksaw is used, the chips must be cleaned out before assembling the drain.
4. Joints should be primed before applying the solvent.
5. The drain line must not run up hill.
6. The drain line should not interfere with panel removal or filter replacement.
7. The drain should have a trap.
8. Verify that the drain works by pouring water into the evaporator drain pan.

LAB 17.1 SOLENOID VALVES

LABORATORY OBJECTIVE
The student will disassemble a solenoid valve and be able describe how it operates.

***FUNDAMENTALS OF HVACR* TEXT REFERENCE**
Unit 17 Special Refrigeration Components

LABORATORY NOTES
For this lab exercise there should be one or more solenoid valves available that may be disassembled.

Solenoid valves can be used for many applications in a refrigeration system. They can be used to regulate flow to the evaporator and as king valves. They can also be used for the defrost cycle as well as other refrigeration applications.

REQUIRED TOOLS, EQUIPMENT, AND MATERIALS
Solenoid valves
Multimeter

SAFETY REQUIREMENTS
None

PROCEDURE
Step 1
Locate a solenoid valve and examine it carefully so that you may complete the following exercise:

  A. Record all of the data that can be found on the valve such as volts, cycles, wattage, manufacturer, etc.

Figure 17-1-1

Solenoid Valve Data:

B. Disassemble the valve being careful not to lose components while identifying the manner in which the valve came apart so that it may then be put back together again.

Figure 17-1-2

C. Sketch a cross sectional view of the disassembled valve in the space provided below.

D. Explain how the valve operates using the sketch you provided above.

Step 2
Measure the coil resistance as follows:

A. The coil removed from the solenoid valve will have an identification number for spare parts purposes. Notice the part identification number on the coil shown in the illustration below.

Figure 17-1-3

B. Using the coil part number and the manufacturer's data, you should be able to determine the normal resistance of the coil.

C. Familiarize yourself with a multimeter and zero it in

(Refer to Fundamentals of HVACR text Unit 33 *Electrical Measuring & Test Instruments*).

D. Connect the one of leads of the multimeter to each wire attached to the coil.

E. Start with the resistance reading on the highest scale.

F. Record the resistance reading_____Ohms.

G. Compare this reading to the manufacturer's data to see if the coil is satisfactory.

Step 3
Re-assemble the solenoid valve, making sure that <u>none of the parts are lost and each is</u> <u>in its proper order.</u>

QUESTIONS

1. A common failure with solenoid valves is that the coil eventually burns out. Regarding the valve you disassembled, if the coil burned out, would the valve fail open or closed?

2. If the measured resistance of the coil is infinite, what would this indicate?

3. If the measured resistance of the coil is zero, what would this indicate?

(Circle the letter that indicates the correct answer.)

4. If the coil on the solenoid supplying refrigerant to an evaporator burned out:
   A. The refrigerated space would warm up.
   B. The refrigerated space would cool down.
   C. The compressor would shut down on high pressure.
   D. The compressor motor would overload.

5. If a solenoid valve is energized.
   A. it will always be closed.
   B. then the oil must be burned out.
   C. it may feel warm to the touch.
   D. then the circuit is faulty.

6. If the measured resistance of a solenoid coil is infinite:
   A. then the coil is satisfactory.
   B. the coil may have a short.
   C. the coil may have an open.
   D. the coil will run hot.

7. If the measured resistance of a solenoid coil is infinite:
   A. then the coil is satisfactory.
   B. the coil may have a short.
   C. the coil may have an open.
   D. the coil will run hot.

8. Many refrigerant solenoid valves often fail in the:
   A. neutral position.
   B. reverse flow position.
   C. open position.
   D. closed position.

9. When the coil of a solenoid valve is energized:
   A. it will spin.
   B. it will act as a magnet.
   C. its polarity will be reversed.
   D. it will slowly rotate.

10. An energized solenoid valve:
   A. will act as a magnet.
   B. may make a slight humming sound.
   C. may feel warm to the touch.
   D. All of the above are correct.

11. Many solenoid valves have a pilot orifice to assist in valve lift:
   A. to speed up the action of the valve.
   B. to reverse the flow through the valve.
   C. for manual operation if the coil burns out.
   D. to allow for smaller size coils to be used.

12. Before removing a coil from a solenoid valve:
   A. make sure that you have a spare.
   B. twist the ends of the wires together.
   C. tap it gently with a wrench.
   D. de-energize the circuit.

LAB 17.2 SETTING AN EVAPORATOR PRESSURE REGULATOR

LABORATORY OBJECTIVE
The student will demonstrate how to properly set an evaporator pressure regulator that is controlling the temperature of an evaporator coil for a refrigerated space.

FUNDAMENTALS OF HVACR TEXT REFERENCE
Unit 17 Special Refrigeration Components

LABORATORY NOTES
For this lab exercise there needs to be an operating refrigeration system that has an evaporator pressure regulator on the outlet of an evaporator for the space being cooled. If there are no thermometers or pressure gauges currently on the unit then you must instrument the refrigeration system using a gauge manifold to measure the evaporator pressure and a thermometer to read the evaporator refrigerant temperature.

If you are unable to place probes directly into the refrigerant stream flow, you can use evaporator coil surface temperatures, however you must adjust for the heat transfer difference through the coil.

Multiple evaporator systems are commonly found in supermarkets. A single compressor may be used to control a number of different case or fixture temperatures.

Without an evaporator pressure regulator (EPR) all of these spaces would have a common evaporator pressure and temperature. The refrigerant would need to be cold enough for the low temperature boxes but this would make it undesirable for the warmer boxes.

The refrigerant in the evaporator coil should be approximately 15°F lower than the space being cooled. If the refrigerant temperature is excessively low, then this will tend to rob the food of its moisture and dry it out. This is particularly applicable to fruits and vegetables. An EPR will elevate the evaporator pressure and thus the refrigerant temperature to bring it on more in line with the box temperature.

REQUIRED TOOLS, EQUIPMENT, AND MATERIALS
Operating refrigeration unit with evaporator pressure regulator
Gauge manifold and temperature sensor

SAFETY REQUIREMENTS
A. Always read the equipment manual to become familiarized with the refrigeration system and its accessory components prior to start up.
B. Wear safety goggles and gloves when working with refrigerants. Liquid refrigerant can cause frostbite when in contact with eyes and skin.
C. Use low loss hose fittings, or wrap cloth around hose fittings before removing the fittings from a pressurized system or cylinder. Inspect all fittings before attaching hoses.

PROCEDURE
Step 1
Familiarize yourself with the major components in the refrigeration system including the condenser, compressor, evaporator, and metering device.

    A. Evaporator pressure regulators (EPRs) are placed at the outlets of the suction lines for the warmer temperature evaporators. These are adjusted to maintain the desired evaporator pressure and thereby controlling evaporator temperature.

    B. A check valve is installed at the outlet of the suction line for the coldest evaporator coil. This prevents migration of the refrigerant from the higher temperature coils to the low temperature coil.

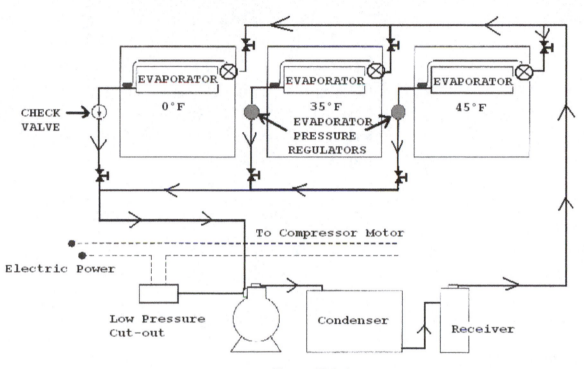

Figure 17-2-1

Step 2
Determine the proper setting for the EPR as follows:

    A. You should be able to determine the proper evaporator pressure based upon the desired space temperature. As an Example, use the 45°F controlled space temperature in Figure 17-2-1. The approximate refrigerant temperature should be 45°F − 15°F = 30°F.

    B. You should also consult any operating manual parameters in regard to the expected temperature difference between the space and the evaporator as recommended by the Manufacturer. We will be using a 15°F temperature differential for this Example which is most common, however different types of systems may require other settings.

C. Inside the evaporator coil the condition of the refrigerant will always be saturated as long as there is liquid and vapor present together. Assuming we are using R-22 then from the P-T chart, Figure 17-2-2, we can look up the saturated refrigerant temperature. At 30°F, for R-22 the EPR set pressure would be approximately 55 psig. For another type of refrigerant such as R-134a the expected evaporator pressure would be somewhat lower at approximately 26 psig.

VAPOR PRESSURE, PSIG									
Temp (°F)	11	12	22	113	114	500	502	134a	123
−50	28.9	15.4	6.2	—	27.1	12.8	0.2	18.7	29.2
−45	28.7	13.3	2.7	—	26.6	10.3	1.9	16.9	29.0
−40	28.4	11.0	0.5	—	26.0	7.6	4.1	14.8	28.9
−35	28.1	8.4	2.6	—	25.4	4.6	6.5	12.5	28.7
−30	27.8	5.5	4.9	29.3	24.6	1.2	9.2	9.8	28.4
−25	27.4	2.3	7.4	29.2	23.8	1.2	12.1	6.9	28.1
−20	27.0	0.6	10.1	29.1	22.9	3.2	15.3	3.7	27.8
−15	26.5	2.4	13.2	28.9	21.8	5.4	18.8	0.1	27.4
−10	26.0	4.5	16.5	28.7	20.6	7.8	22.6	1.9	27.0
−5	25.4	6.7	20.0	28.5	19.3	10.4	26.7	4.1	26.5
0	24.7	9.1	23.9	28.2	17.8	13.3	31.1	6.5	25.9
5	23.9	11.8	28.2	27.9	16.2	16.4	35.9	9.1	25.3
10	23.1	14.6	32.8	27.6	14.4	19.7	41.0	11.9	24.6
15	22.1	17.7	37.7	27.2	12.4	23.3	46.5	15.0	23.7
20	21.1	21.0	43.0	26.8	10.2	27.2	52.5	18.4	22.8
25	19.9	24.6	48.7	26.3	7.8	31.5	58.8	22.1	21.8
30	18.6	28.4	54.9	25.8	5.2	36.0	65.6	26.0	20.7
35	17.2	32.5	61.5	25.2	2.3	40.8	72.8	30.3	19.5
40	15.6	36.9	68.5	24.5	0.4	46.0	80.5	35.0	18.1
45	13.9	41.6	76.0	23.8	2.0	51.6	88.7	40.0	16.6

Figure 17-2-2

Step 3

A. If there are no pressure gauges currently on the unit then you must connect a gauge manifold to read the evaporator pressure.

B. If you are unable to place temperature probes directly into the refrigerant stream flow, you can use evaporator coil surface temperatures, however you must adjust for the heat transfer difference through the coil.

Step 4

    A. Start the refrigeration system in the normal cooling mode and allow the system to stabilize and then record the evaporator pressure and temperature below.

    _____ psig   _____ °F

    B. From your readings, determine what adjustments need to be made if any and set the evaporator pressure regulator by turning the adjusting screw with a screwdriver or Allen wrench dependent on EPR type.

    C. Allow the system to run and stabilize prior to shutting down and then carefully disconnect the gauge manifold and any instrumentation that you attached to the unit.

QUESTIONS

(Circle the letter that indicates the correct answer.)

1. The temperature in an evaporator coil is directly related to:
    A. the temperature of the space to be cooled.
    B. the temperature of the condenser.
    C. the pressure in the evaporator.
    D. the compressor pressure.

2. If an evaporator pressure regulator fails shut.
    A. the space will freeze up.
    B. the system will still operate as normal.
    C. the compressor will run continuously.
    D. the space will warm up.

3. An evaporator pressure regulator:
    A. should be used on the warmer evaporators.
    B. should be used on the colder evaporators.
    C. is a safety device.
    D. unloads the compressor at high space temperatures.

4. On a multiple box system, a check valve is usually installed at the outlet of the:
    A. coldest evaporator.
    B. warmest evaporator.
    C. compressor.
    D. condenser.

LAB 17.3 IDENTIFYING ACCESSORIES

LABORATORY OBJECTIVE
You will correctly identify refrigeration accessories on refrigeration systems and explain their function.

FUNDAMENTALS OF HVACR TEXT REFERENCE
Unit 17 Refrigeration Accessories

LABORATORY NOTES
You will locate the refrigeration accessories on systems assigned by the instructor. You will then point them out to the instructor and describe their function.

REQUIRED TOOLS, EQUIPMENT, AND MATERIALS
Safety glasses
Gloves
Systems with refrigeration accessories

SAFETY REQUIREMENTS
    A. Wear safety glasses
    B. Use caution around moving fans.
    C. Do NOT touch any electrical connections

PROCEDURE

Locate the following accessories on systems in the shop.

LAB 17.3 IDENTIFYING ACCESSORIES	
Refrigeration Accessory	
Liquid Line Filter Drier	
Suction Line Filter Drier	
Muffler	
Sight Glass	
Moisture Indicator	
Receiver	

Accumulator	
Crankcase Heater	
Vibration Eliminator	
Solenoid Valve	
Heat Exchanger	
Crankcase Pressure Regulator	

LAB 17.4 INSTALLING ACCESSORIES

LABORATORY OBJECTIVE
You will demonstrate your ability to install a filter drier on a refrigeration system.

FUNDAMENTALS OF HVACR TEXT REFERENCE
Unit 17 Refrigeration Accessories

LABORATORY NOTES
You will front-seat the king valve or liquid line service valve, operate the system, and pump down the refrigerant into the high side of the system. You will then install a liquid line filter-drier in the liquid line. After installing the drier, yow will leak test the joints with nitrogen, evacuate the lines and coil, and open the king valve to let the system refrigerant into the rest of the system.

REQUIRED TOOLS, EQUIPMENT, AND MATERIALS
Safety glasses
Gloves
Two adjustable jaw wrenches
6-in-1
Refrigeration gauges
Refrigeration valve wrench
Refrigeration system with king valve or liquid line service valve
Liquid line filter drier
Vacuum pump

SAFETY REQUIREMENTS
A. Wear safety glasses and gloves when accessing refrigeration systems.
B. Use caution around moving fans.
C. Do NOT touch any electrical connections

PROCEDURE
1. Follow the procedures given in Lab 26.7 to pump down the system.

2. Use two wrenches to loosen the flare nuts on the existing filter-drier.

3. One wrench holds the filter to keep it from turning while the other turns the flare nut.

4. Remove the existing filter.

5. Remove the protective caps from the new filter.

Note: Do NOT remove the caps until just before installing the filter. If the caps are removed ahead of time and the filter is allowed to sit open in the air, it will absorb moisture from the air, reducing its ability to absorb moisture from the system.

6. Install the new filter with the arrow pointing towards the metering device.
7. Tighten the flare nuts using two wrenches.
8. Leak test with nitrogen and soap bubbles.
9. Release nitrogen and evacuate the lines and coil.
10. Open the king valve or liquid line valve to let refrigerant into the system.
11. Operate the system and check system pressures

LAB 18.1 IDENTIFYING REFRIGERANTS AND REFRIGERANT CHARACTERISTICS

LABORATORY OBJECTIVE
The purpose of this lab is to demonstrate your ability to identify the type of refrigerant contained in each of four different refrigeration systems.

FUNDAMENTALS OF HVACR TEXT REFERENCE
Unit 18 Refrigerants and Their Properties

LABORATORY NOTES
You will find the type of refrigerant, the amount of refrigerant, and the system test pressures on the unit data plate of each system. Then you will research the reference material to find the refrigerant characteristics.

You will identify the refrigerant by:

- Name
- Pressure (Very High, High, or Low)
- Ozone Depletion Potential
- Global Warming Potential
- Toxicity
- Flammability
- Chemical Composition
- Formulation

REQUIRED TOOLS, EQUIPMENT, AND MATERIALS
Four refrigeration systems containing different refrigerants
Refrigerant cylinders
Refrigerant MSDS

SAFETY REQUIREMENTS
Caution!!! Do not open the valves on the refrigeration cylinders to avoid possible injury due to skin contact with refrigerant. Also it is illegal to intentionally vent refrigerants comprised of CFC's and HCFC's.

PROCEDURE
Step 1
Locate the nameplate on the first unit. The nameplate will be **in different locations for each unit you inspect.**

    A.   Find and record the unit model number on the provided Refrigerant Characteristics Chart for the first refrigeration system.

B. Find and record the refrigerant type on the provided Refrigerant Characteristics Chart for the first refrigeration system.

C. Find and record the high side test pressure the provided Refrigerant Characteristics Chart for the first refrigeration system.

D. Find and record the low side test pressure on the provided Refrigerant Characteristics Chart for the first refrigeration system.

E. Locate a cylinder of the same refrigerant type and read the instructions on the cylinder.

F. Locate the MSDS for this refrigerant type and read the information regarding the refrigerant properties.

G. Use the information from the refrigerant cylinder, MSDS and Fundamentals of HVACR text Unit 18 to complete filling out the provided Refrigerant Characteristic Chart for the first refrigeration system.

Step 2

Repeat the process from Step 1 for the other three refrigeration systems completing the provided **Refrigerant Characteristic** Chart for each one. When finished have your Instructor check the information that you recorded and then answer the questions at the end of this laboratory exercise.

Refrigerant Characteristics Chart

	SYSTEM #1	SYSTEM #2	SYSTEM #3	SYSTEM #4
Model Number				
Refrigerant Name				
Refrigerant Quantity				

High Side Test Pressure				
Low Side Test Pressure				
Pressure (Very High, High, Low)				
Ozone Depletion Potential				
Global Warming Potential				
Safety Rating				
Chemical Composition				
Formulation				

QUESTIONS

1. Were there any refrigerants that did not have either an ozone depletion potential or a global warming potential?

2. What are the high and low side test pressures used for?

158

3. Which refrigerants may safely leave the cylinder as either a gas or a liquid?

4. Which refrigerants must leave the cylinder as a liquid only?

5. Which refrigerant had the highest system test pressures?

6. Which refrigerant had the lowest system test pressures?

7. What hazards do you need to be aware of when working with these refrigerants?

8. Which of these refrigerants will NOT be used in equipment of the future?

9. Which of these refrigerants WILL be used in equipment of the future?

LAB 18.2 IDENTIFYING REFRIGERANT LUBRICANT CHARACTERISTICS

LABORATORY OBJECTIVE
The purpose of this lab is to demonstrate your ability to identify three different types of refrigerant lubricant and to recommend the proper lubricant for four refrigeration systems.

FUNDAMENTALS OF HVACR TEXT REFERENCE
Unit 18 Refrigerants and Their Properties

REQUIRED TOOLS, EQUIPMENT, AND MATERIALS
Four refrigeration systems containing different refrigerants
Three containers of different refrigeration lubricant

SAFETY REQUIREMENTS
Caution!!! Do not open the lubricant containers and spill lubricant or allow it to contact your skin.

PROCEDURE
Step 1
Locate the first container of refrigerant lubricant.

    A. Identify the type of lubricant in the first container.
    B. Record in the Lubricant Type Chart provided:
- Type of lubricant
- Lubricant viscosity
- Recommended evaporator temperature range
- Type of refrigerants that the lubricant is compatible

    C. Repeat this process for the other lubricant types.

Lubricant Type Chart

	BRAND NAME	TYPE (MINERAL, PAG, POE, AB)	VISCOSITY	EVAPORATOR TEMPERATURE	REFRIGERANT COMPATABILITY
Container #1					
Container #2					
Container #3					

Step 2

Locate the nameplate on the first refrigeration unit. The nameplate will be **in different locations for each unit you inspect.**

A. Find and record the unit model number on the provided System Refrigerant Lubricant Chart for the first refrigeration system.
B. Find and record the refrigerant type on the provided System Refrigerant Lubricant Chart for the first refrigeration system.
C. Recommend a lubricant that is compatible with this unit.

Step 3

Repeat the process from Step 2 for the other three refrigeration systems completing the provided System **Refrigerant Lubricant Characteristic** Chart for each one. When finished have your Instructor check the information that you recorded and then answer the questions at the end of this laboratory exercise.

System Refrigerant Lubricant Characteristics Chart

	SYSTEM #1	SYSTEM #2	SYSTEM #3	SYSTEM #4
Model Number				
Refrigerant Name				
Recommended Lubricant				

QUESTIONS

1. Which lubricant can be used with the widest range of systems?

2. What is the difference in viscosity between lubricants designed for higher evaporator temperatures and lubricants that are designed for use in lower temperature systems?

3. What is the most common viscosity for lubricants used in air conditioning systems?

162

Name_____

Date_____

Instructor's OK ☐

LAB 19.1 DESCRIBE THE REFRIGERATION SYSTEM CHARACTERISTICS USING A REFRIGERATION TRAINER

LABORATORY OBJECTIVE: The purpose of this lab is to demonstrate your ability to discuss the function of the four main refrigeration system components and describe the refrigerant characteristics throughout the system on the refrigeration trainer.

FUNDAMENTALS OF HVACR TEXT REFERENCE
Unit 19 Plotting the Refrigeration Cycle

LABORATORY NOTES You will be assigned a refrigeration trainer. You should be prepared to show the instructor each of the four main refrigeration components in order and discuss their function in the system. You will also discuss the characteristics of the refrigerant as it travels through the system using pressure and temperature measurements to help you determine the condition of the refrigerant throughout the trainer.

REQUIRED TOOLS, EQUIPMENT, AND MATERIALS
Instrumented refrigerant trainer
Pressure temperature chart
Temperature sensor with temperature clamp and thermocouple clamp

SAFETY REQUIREMENTS
None

PROCEDURE
Step 1
Identify the components on the refrigeration trainer.

Locate the compressor, condenser, metering device, and evaporator.

Draw a sketch of the trainer.

Label the components.

Draw arrows to indicate the direction of refrigerant flow.

Identify high pressure and low pressure circuits.

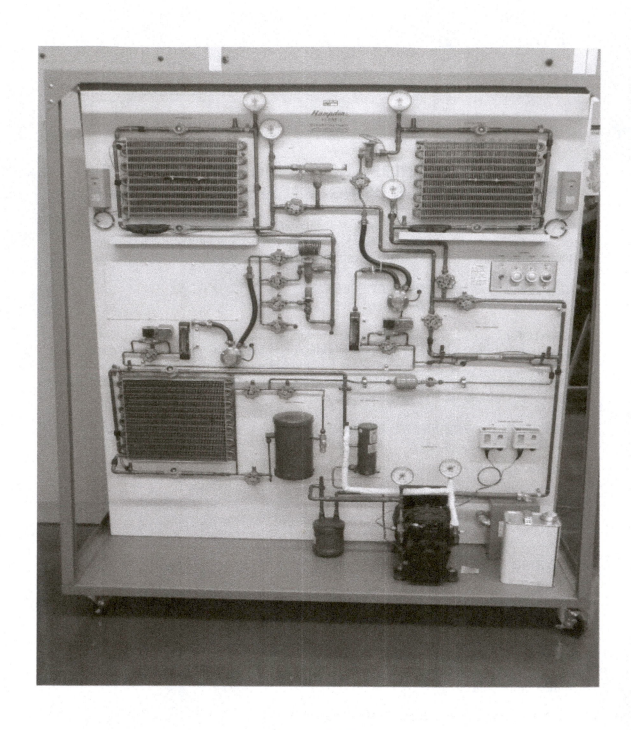

Figure 19-1-1

DRAW SKETCH OF REFRIGERATION TRAINER IN THE SPACE PROVIDED

INSTRUCTOR CHECK BEFORE PROCEEDING TO NEXT STEP _____

Step 2

Line up the components and then start the refrigeration trainer using the trainer's instruction manual if necessary. If you are not sure how to start the trainer, then check with your Instructor.

Step 3

Run the trainer and observe its operation. Note the pressure gauges that tell you what the refrigerant pressure is at every point in the system.

    A. If there are no installed temperature sensors then you can measure the temperature at each location using a portable temperature sensor.

    B. You may also touch components to get a feel for their temperature.
        CAUTION: Some components are HOT! Proceed carefully!

    C. If you hold your hand just above a warm component you can tell if it is extremely hot or not. If you can feel radiant heat coming from it, it may be too hot to touch safely.

Step 4

Use the provided chart to record and organize your data. However, the job is not complete until you TELL the instructor how the system works and respond to questions.

    A. Record the pressure and temperature at each system location indicated on the chart.

    B. Compare the pressure and temperature using a pressure-temperature chart as found in Appendix A of the HVACR text, to determine the condition. These are saturated, superheated, or subcooled.

    C. Write in the state: gas, liquid, or mixed.

    D. Use the pressure, temperature, condition, and state to look up the heat content on an enthalpy chart such as found in Unit 21 of the Fundamentals of HVACR text.

LOCATION	TEMPERATURE	PRESSURE	CONDITION	STATE	HEAT CONTENT
Compressor in					
Compressor out					
Discharge line					
Condenser in					
Condenser center					
Condenser out					
Liquid line					
Metering device in					
Metering device out					
Evaporator in					
Evaporator center					
Evaporator out					
Suction line					

Step 5

Use the following terms to describe the changes in the refrigerant through each of the four major components and fill in the chart provided.

Stable – to describe no significant change in a condition

Small increase – to describe an increase of 30% or less

Large increase – to describe an increase of 200% or more

Small decrease – to describe a decrease of 30% or less

Large decrease – to describe a decrease of 200% or more

TYPE OF CHANGE	COMPRES-SOR	CONDENSER	METERING DEVICE	EVAPORA-TOR
Change in pressure				
Change in temperature				
Change in heat				
Change in state				
Change in volume				

LAB 20.1 DISPOSABLE REFRIGERANT CYLINDERS

LABORATORY OBJECTIVE
The student will demonstrate an understanding of disposable refrigerant cylinder types and handling procedures.

FUNDAMENTALS OF HVACR TEXT REFERENCE
Unit 20 Refrigeration Safety
Unit 26 Refrigerant Management and the EPA

LABORATORY NOTES
For this lab exercise there should be a number of different disposable refrigerant cylinders for the students to inspect. Empty cylinders are acceptable as the student will not be removing any refrigerant from the cylinder. It would be preferable to also have cylinders of different sizes.

REQUIRED TOOLS, EQUIPMENT, AND MATERIALS
Disposable refrigerant cylinders of different sizes and types

SAFETY REQUIREMENTS
A. Wear safety glasses and gloves handling the refrigerant cylinders
B. **Caution!!!** Do not open the valves on the refrigeration cylinders to avoid possible injury due to skin contact with refrigerant. Also, it is illegal to intentionally vent refrigerants comprised of CFC's and HCFCs.

PROCEDURE
Step 1
In 1990 the Congress of the United States passed a series of amendments to the Clean Air Act that greatly affected the refrigeration and air conditioning industry. The Act establishes a set of standards and requirements for the use and disposal of certain common refrigerants containing chlorine.

Review the Information on the Clean Air Act from the *Fundamentals of HVACR* text Unit 26 and answer the following:

A. Can you knowingly release CFC's, HCFCs, or HFC's while repairing appliances?

B. When did it become illegal to vent CFC's and HCFCs?

C. Do you need to be certified to service, maintain, or dispose of appliances containing refrigerants? If yes, when did this become mandatory?

D. How much can you or your company be fined per day for violating Section 608 of the Federal Clean Air Act?

E. How much is the bounty for turning someone in?

Step 2
Locate a <u>disposable</u> refrigerant cylinder and examine it carefully so that you may complete the following exercise.

A. The refrigerant type, chemical designation of the refrigerant, cylinder color code, refrigerant boiling temperature, normal discharge (head) pressure, normal suction pressure, and latent heat value.

B. Refrigerant toxicity, flammability, corrosive tendency, and rated refrigerant weight when cylinder is full.

C. What is the maximum temperature that this cylinder can be exposed to?

D. How often must this cylinder be checked (DOT regulation)?

E. Can you remove both liquid and vapor refrigerant from this cylinder? If yes, then describe how you would remove vapor and how you would remove liquid refrigerant.

F. Can you re-use this cylinder in an emergency?

G. What type of protection against bursting due to excessive pressure does the cylinder have?

H. Draw a rough sketch of the protection device in the space provided below.

QUESTIONS

(Circle the letter that indicates the correct answer.)

1. A disposable cylinder:
   A. can be refilled <u>only</u> with the proper type of refrigerant.
   B. can never be refilled.
   C. is always color coded yellow and gray.
   D. can only be refilled by a certified technician.

2. Cylinders over 4.5 inches in diameter and over 12 inches long:
   A. will always be disposable cylinder.
   B. will always be recovery cylinders.
   C. cannot be used for refrigerant purposes.
   D. must have some type of pressure relief device.

3. When transporting a refrigerant cylinder:
   A. it should be properly secured in an upright position.
   B. it should be properly secured laying down.
   C. it should be properly secured in an inverted position.
   D. it does not need to be secured if it is nearly empty.

4. Refrigerant cylinder temperatures should not exceed:
   A. 92.5°F.
   B. 99.8°F.
   C. minus 25°F.
   D. 125°F.

5. The cardboard boxes that contain new refrigerant cylinders:
   A. can be used to prop the bottle upright when charging.
   B. may contain important safety information.
   C. must be opened and discarded before the cylinder is used.
   D. can be used to hold recovered refrigerant.

6. To charge a liquid using a disposable cylinder:
   A. it must be heated.
   B. it must be cooled.
   C. it must be shaken.
   D. it must be inverted.

7. The burst disk on a disposable refrigerant cylinder:
   A. will re-seat after the pressure is reduced.
   B. will not re-seat once opened–losing all refrigerant from the cylinder.
   C. will only drain the liquid refrigerant.
   D. acts like a relief valve.

LAB 20.2 REFILLABLE REFRIGERANT RECOVERY CYLINDERS

LABORATORY OBJECTIVE
The student will demonstrate an understanding of refillable refrigerant recovery cylinder types and handling procedures.

FUNDAMENTALS OF HVACR TEXT REFERENCE
Unit 20 Refrigeration Safety

LABORATORY NOTES
For this lab exercise there should be a refillable refrigerant recovery cylinder for the students to inspect. Empty cylinders are acceptable as the student will not be removing any refrigerant from the cylinder.

REQUIRED TOOLS, EQUIPMENT, AND MATERIALS:
Refillable refrigerant recovery cylinder

SAFETY REQUIREMENTS

Wear safety glasses and gloves handling the refrigerant cylinders

Caution!!! Do not open the valves on the refrigeration cylinders to avoid possible injury due to skin contact with refrigerant. Also, it is illegal to intentionally vent refrigerants comprised of CFC's and HCFCs.

PROCEDURE
Step 1
Locate a refillable refrigerant recovery cylinder and examine it carefully so that you may complete the following exercise.

A. Sketch a cross sectional view of the cylinder including fittings and valves in the space provided below.

B.   Explain how can you remove either liquid or vapor from this cylinder.

C.   What is the cylinder weight empty (T.W.–Tare Weight)?

D.   What is the maximum refrigerant weight that the cylinder can hold?

E.   What is the weight of a full cylinder including the metal?

F.   What type of protection against bursting due to excessive pressure does the cylinder have?

G.   How often must this cylinder be inspected?

H.   What type of refrigerant can be stored in this cylinder?

QUESTIONS

(Circle the letter that indicates the correct answer.)

1. A refillable refrigerant recovery cylinder:
    A.   can be refilled <u>only</u> with the proper type of refrigerant.
    B.   has an automatic fill valve that insures proper fill level.
    C.   is always color coded according to the refrigerant.
    D.   is covered by EPA cylinder specifications.

2. A refillable refrigerant recovery cylinder must meet DOT approval:
    A.   True.
    B.   False.

LAB 20.3 REFRIGERANT CYLINDERS

LABORATORY OBJECTIVE

You will examine two non-refillable refrigerant cylinders and a refrigerant recovery cylinder and identify the important information found on the cylinders.

FUNDAMENTALS OF HVACR TEXT REFERENCE

Unit 20 Refrigeration Safety
Unit 26 Refrigerant Management and the EPA

LABORATORY NOTES

You will examine two non-refillable refrigerant cylinders and identify the type of refrigerant, the amount of refrigerant the cylinder holds when new, the position of the cylinder when removing refrigerant from it, and the potential fine for refilling it. You will then examine a ref refrigerant recovery cylinder and identify important data found on the cylinder; including, tare weight, water capacity, service pressure, the type of refrigerant in the cylinder, the amount of refrigerant currently in the cylinder, and the maximum amount of refrigerant the cylinder can hold.

REQUIRED TOOLS, EQUIPMENT, AND MATERIALS

Safety glasses
One R-22 non-refillable refrigerant cylinder
One R-410A non-refillable refrigerant cylinder
Refrigerant recovery cylinder
Digital scale

SAFETY REQUIREMENTS

Wear safety glasses and gloves handling the refrigerant cylinders

Caution!!! Do not open the valves on the refrigeration cylinders to avoid possible injury due to skin contact with refrigerant. Also, it is illegal to intentionally vent refrigerants comprised of CFC's and HCFCs.

PROCEDURE

Non-refillable Cylinders

Examine the two non-refillable cylinders and complete the data sheet.

Non-refillable Refrigerant Cylinders		
	Cylinder 1	Cylinder 2
Cylinder color		
Type of refrigerant in cylinder		
Amount of refrigerant in new cylinder		
Position of cylinder during charging		
Fine for refilling and transporting		

Refrigerant Recovery Cylinder

Examine a recovery cylinder and complete the data sheet.

Refrigerant Recovery Cylinder	
Cylinder color	
Cylinder tare weight (TW)	
Cylinder water capacity (WC)	
Cylinder maximum service pressure	
At what pressure does the pressure relief valve open?	
Date when this cylinder must be re-certified	
Type of refrigerant in cylinder	
What color is the handle on the vapor valve?	
What color is the handle on the liquid valve?	
Type of refrigerant in cylinder	
Maximum weight of refrigerant cylinder can safely hold	
Maximum safe gross weight (tare weight + maximum safe fill)	
Weigh the cylinder to determine the current gross cylinder weight	
Weight of refrigerant in cylinder (gross weight – tare weight)	

Show safe fill level calculations below using formula

Safe Fill = Water Capacity x 0.8 x Refrigerant Specific Gravity

Note: Refrigerant specific gravity can be found in Table 26-2 of Unit 26.

LAB 21.1 REFRIGERANT GAUGE MANIFOLDS

LABORATORY OBJECTIVE
The student will demonstrate how to properly use a refrigerant gauge manifold.

FUNDAMENTALS OF HVACR TEXT REFERENCE
Unit 21 Refrigerant System Servicing and Testing Equipment

LABORATORY NOTES
For this lab exercise there should be a typical gauge manifold available for students to inspect.

REQUIRED TOOLS, EQUIPMENT, AND MATERIALS
Gauge manifold

SAFETY REQUIREMENTS
Wear safety glasses

PROCEDURE
Step 1
Locate a refrigerant gauge manifold and examine it carefully so that you may complete the following exercise.

A. Identify the two gauges that are on the manifold including the minimum and maximum readings.

B. What is meant by compound gauge?

C. What is the range of the pressure scale on the compound gauge?

D. List three operations that can be performed with a gauge manifold.

E. What do the colored scales in the center of the gauge indicate?

F. Sketch a cross sectional view of a gauge manifold, labeling all parts and connections in the space provided below.

QUESTIONS

(Circle the letter that indicates the correct answer.)

1. When not in use:
   A. gauge manifolds should be stored in a plastic wrap cloth.
   B. gauge manifolds should be pressurized with an inert gas.
   C. the ports and charging lines should be capped.
   D. disassembled and inspected after each use.

2. Gauge manifolds:
   A. have a small adjustment screw that allows the gauge to be calibrated.
   B. can never be calibrated.
   C. are oiled every week.
   D. must be returned to the factory for recalibration.

3. Gauge manifolds may be used with:
   A. the type of refrigerant for which they were designed.
   B. any refrigerant.
   C. any fluid.
   D. checking water pressure.

4. A compound gauge measures both pressure and vacuum:
   A. True.
   B. False.

LAB 21.2 REFRIGERATION EQUIPMENT FAMILIARIZATION

LABORATORY OBJECTIVE
You will demonstrate how to use a vacuum pump, vacuum gauge, digital scale, and recovery unit.

FUNDAMENTALS OF HVACR TEXT REFERENCE
Unit 21 Refrigerant System Servicing and Testing Equipment

Unit 26 Refrigerant Management and the EPA

Unit 28 Refrigerant System Evacuation

Unit 29 Refrigerant System Charging

LABORATORY NOTES
You will use a vacuum pump to pull a vacuum on your gauges and check the vacuum with a vacuum gauge.

You will weigh a set of screws with a digital scale, zero the scale, and then weigh the screws removed.

You will use a recovery unit to transfer refrigerant from one recovery cylinder to another.

REQUIRED TOOLS, EQUIPMENT, AND MATERIALS
Safety glasses and gloves

Infrared thermometer

Refrigeration gauges with extra refrigeration hoses

Vacuum pump and vacuum gauge

Refrigerant recovery unit

2 Refrigerant recovery cylinders with the same refrigerant

Digital scale

Assortment of small weights (can be a bucket of screws)

SAFETY REQUIREMENTS
A. Wear safety glasses and gloves handling the refrigerant cylinders

B. It is illegal to intentionally vent refrigerants comprised of CFC's and HCFCs.

PROCEDURE

PART A: Vacuum Pump and Vacuum Gauge
1. Tighten the low side hose (blue) and the high side hose (red) on the hanger so they will not leak.
2. Connect the middle hose to the vacuum gauge.
3. Connect another hose from the vacuum gauge to the vacuum pump.
4. *(Depending upon the vacuum gauge used, a separate tee and hose may be required.)*

5. Close the manifold gauge hand wheels by turning both of them clockwise.
6. Open the ballast valve and isolation valves on the vacuum pump.
7. Turn on the vacuum gauge.
8. Start the vacuum pump. It will gurgle loudly at first, then start to get quieter.
9. Close the ballast valve on the vacuum pump.
10. The vacuum gauge reading should drop to less than 100 microns.
11. Open the manifold gauge hand wheels.
12. The pump will get louder again for a minute or so and then quiet down.
13. The vacuum gauge reading will go up and then fall back down.
14. Close the vacuum pump isolation valve and turn off the vacuum pump and vacuum gauge.

(Always close the isolation valve BEFORE turning off the vacuum pump.)

PART B: Digital Scale

1. Turn on the scale and zero it by pushing the "zero" button.
2. Place the bucket of screws on the scale.
3. Push the "Units" button to see the weight displayed in pounds, pounds and ounces, and kilograms.
4. Zero the scale with the screws still on it.
5. Remove some of the screws – the weight of the screws you removed should be displayed on the scale.

PART C: Vapor Refrigerant Recovery

1. Connect the vapor valve of one of the recovery cylinders to the inlet of the recovery unit.
2. Connect the outlet of the recovery unit to the vapor valve of the other recovery cylinder.
3. Purge the hoses and the recovery unit.
   (If the recovery unit was last used on the same type of refrigerant it should not need purging.)
4. Place the recovery cylinder connected to the recovery unit outlet valve on the scale.
5. Zero the scale.
6. Open the valves on the recovery cylinders.
7. Set the recovery unit valves to recover vapor.
8. Turn on the recovery unit.
9. Use infrared thermometer to check the cylinder temperatures periodically as the unit operates.
10. The cylinder connected to the recovery unit inlet should drop in temperature while the cylinder connected to the recovery unit outlet should increase in temperature.
11. The scale should show that the cylinder on the scale is increasing in weight.
12. Turn off the recovery unit and close the recovery unit valves.

PART D: Liquid Refrigerant Recovery

1. Turn the cylinder connected to the recovery unit inlet valve upside down.
2. Position the recovery unit valves to equalize the pressures in the recovery unit.
3. *(This can be done on many units by setting the valves to the purge position.)*
4. Set the valves on the recovery unit to recover liquid.
5. Open the valves on the recovery unit.
6. Start the recovery unit.
7. The scale should indicate that the cylinder weight is increasing faster.
8. Close the valve on the cylinder connected to the recovery unit inlet valve.
9. The recovery unit suction gauge should drop into a vacuum and the recovery unit should shut off. *(Not all recovery units have auto shutoff.)*
10. Turn off the recovery unit.
11. Set the valves on the recovery unit to purge.
12. Start the recovery unit an operate until it has cleared itself.
13. Close the valves on the recovery unit, and turn it off.
14. Make sure the valves are closed on both cylinders before disconnecting the hoses.

LAB 22.1 IDENTIFYING FITTINGS

LABORATORY OBJECTIVE
You will identify the following types of fittings and their sizes:

- pipe fittings
- sweat fittings
- flare fittings
- PVC fittings

***FUNDAMENTALS OF HVACR* TEXT REFERENCE**
Unit 22 Piping and Tubing

LABORATORY NOTES
You will receive an assortment of fittings. You will identify each fitting and list its name and size.

REQUIRED TOOLS, EQUIPMENT, AND MATERIALS
Safety glasses
Ruler
Assorted fittings

SAFETY REQUIREMENTS
Wear safety glasses when working in the shop.

PROCEDURE
Give the name and size of each fitting in the space provided. Line up the fittings and keep them out so that you may show ALL fittings with proper name and size to the instructor.

#	FITTING NAME	SIZE		#	FITTING NAME	SIZE
1				21		
2				22		
3				23		
4				24		
5				25		
6				26		
7				27		
8				28		
9				29		
10				30		
11				31		
12				32		
13				33		
14				34		
15				35		
16				36		
17				37		
18				38		
19				39		
20				40		

LAB 22.2 FLARING

LABORATORY OBJECTIVE
You will construct a leak-proof 2-foot long flare assembly using 1/4-inch, 3/8-inch, ½-inch, and 5/8-inch copper. You will flare both ends of each piece of copper and join them together with flare reducing unions. You will pressurize the completed assembly with nitrogen and check for leaks.

FUNDAMENTALS OF HVACR TEXT REFERENCE
Unit 22 Piping and Tubing

LABORATORY NOTES
You will build a flared assembly that contains four sizes of tubing connected by flare fittings. With the instructor's help, you will then leak test the assembly using nitrogen and soap bubbles.

REQUIRED TOOLS, EQUIPMENT, AND MATERIALS
Safety glasses
Ruler
Tubing cutter and deburring tool
Flaring tool
Wrenches sized to fit flare nuts
Flare torque wrench
Nitrogen cylinder and regulator
1/4" OD, 3/8" OD, 1/2" OD and 5/8" OD copper tubing and 2 flare nuts for each diameter
Flare reducing unions: 1/4" x 3/8", 3/8" x 1/2", 1/2" x 5/8"
5/8" flare plug

SAFETY REQUIREMENTS
A. Wear safety glasses when working in the shop and when pressure testing flared assembly.
B. When pressure testing the assembly, make sure it is not pointed in another person's direction or towards you. Crack the valve slowly to avoid a sudden pressure surge on your assembly wear safety glasses since your assembly will be under pressure.

PROCEDURE
Step 1 – Prepare Tubing
Cut a 6-inch piece of each size copper tubing: 1/4", 3/8", 1/2", and 5/8". Tip: take your time cutting the tubing to reduce the size of the burr created by the tubing cutter. Deburr both ends of each piece of tubing.

Step 2 – Flaring

Flare one end of the 1/4" tubing and slide a flare nut over the completed flare Tip: If the flare has been done properly it will not catch on the threads of the flare nut but will fill the chamfered area at the bottom of the nut, as in Figure 22-2-1. Slide another flare nut over the tubing facing the opposite direction and then flare the other end. When you have finished it should look like Figure 22-2-2. Repeat for each size of copper tubing.

Step 3 – Assembly

Assemble the pieces together using the flare reducing unions, Figure 22-2-3. Tip: a few drops of refrigeration oil or Nylog™ on the flare surfaces will aid in making the flares leak proof by lubricating the mating surfaces. Thread sealing compound should NOT be used because it can cause problems in the refrigeration system. The preferable tools to use when tightening the flare fittings are wrenches sized to fit the flare nuts. Adjustable wrenches may also be used when used correctly. Tip: NEVER use pliers, especially not locking pliers. They can squeeze and distort the soft brass flare nuts, making them out of round and causing leaks.

Step 4 – Leak Test

Connect the flare nut to a 14" flare union and connect a 1/4" refrigeration hose between the flare union and the outlet of the regulator on a nitrogen cylinder. Pressurize the assembly to 100 psig by adjusting the nitrogen regulator, Figure 22-2-4. Use an ultrasonic detector or soap bubbles to check the assembly joints for leaks. Mark any leaking joints, release the nitrogen pressure, and repair the leaking joints. Retest until the assembly does not leak.

Figure 22-2-1 Completed flare   Figure 22-2-2 Flared section

Figure 22-2-3 Flaring project

Figure 22-2-4 Nitrogen cylinder and regulator

Regulator

Relief Valve

LAB 22.3 SWAGING

LABORATORY OBJECTIVE
You will make a total of 60 swage joints: 15 each of 1/4-inch, 3/8-inch, 1/2-inch, and 5/8-inch tubing.

FUNDAMENTALS OF HVACR TEXT REFERENCE
Unit 22 Piping and Tubing

LABORATORY NOTES
You will make swage joints to be used for the soldering and brazing labs.

REQUIRED TOOLS, EQUIPMENT, AND MATERIALS
Safety glasses
Ruler
Copper tubing: 1/4" OD, 3/8" OD, 1/2" OD, and 5/8" **OD**
Hammer
Tubing cutter and deburring tool
Flaring block
Swaging tool(s)

SAFETY REQUIREMENTS
A. Wear safety glasses
B. Wear gloves
C. Tubing edges can be sharp and cut you

PROCEDURE

Step 1 – Tubing Preparation
Cut 15 pieces of copper tubing 3" long for the following tubing sizes: 1/4", 3/8", 1/2", and 5/8".
Tip: Hold the loose end of the roll on the bench and roll out the tubing on the bench to form a long, straight section. It is far easier to straighten the tubing in one long section. If you cut the tubing while it is still rolled you will not be able to straighten the short 4" sections.

Step 2 – Swaging
When possible, use more than one type of swaging tool in order to learn how they work. The swage joint should be as long as the diameter of the tubing. The tubing should fit snugly into the swage joint. Loose joints will make soldering a painful experience. Time spent making good joints will save time in the long run. Save these joints because you will use them in the soldering section coming up in Lab 23.3 and the brazing in Lab 23.4.

LAB 23.1 LIGHTING AIR ACETYLENE TORCH

LABORATORY OBJECTIVE
You will demonstrate your ability to safely adjust, light, and shut down an air acetylene torch.

FUNDAMENTALS OF HVACR TEXT REFERENCE
Unit 23 Soldering and Brazing

LABORATORY NOTES
The instructor will demonstrate how to safely adjust, light, and shut off an air acetylene torch. You will then demonstrate your ability to safely adjust, light, and shut off an air acetylene torch.

REQUIRED TOOLS, EQUIPMENT, AND MATERIALS
Safety glasses and gloves
Air acetylene torch and torch key
Sparker

SAFETY REQUIREMENTS

A. Safety glasses and gloves should always be worn when using a torch.

B. One hundred percent cotton or leather clothing is the best material to wear while brazing, soldering, or welding.

C. Shirts should have long sleeves and work gloves must be worn (cloth, leather palm, or all leather).

D. When using Acetylene, torch pressure should be approximately 5 psig and the cylinder valve should be open no more than one and 1 1/2 turns.

E. NEVER use a butane lighter to light the torch. Only igniters made specifically for lighting a torch should be used.

PROCEDURE
Adjusting
1. Put on your safety glasses!
2. The instructor will demonstrate the proper way to handle and light an air acetylene torch.
3. You will demonstrate your ability for the instructor using the following procedure.
4. The regulator should be adjusted all the out counterclockwise.
5. The torch handle valve should be closed.
6. Place the torch key on the valve stem and open ½ turn counterclockwise.
7. The tank pressure gauge should show the cylinder fill level.
8. Adjust the regulator by turning it in clockwise approximately 2 ½ turns.

Lighting

1. Before lighting the torch, practice sparking the striker.
2. Open the torch handle, hold the striker in front of the torch, and spark it.

Shutting Torch Off

1. Close the torch handle clockwise to extinguish flame.
2. When finished with the torch, close the valve stem clockwise.
3. Open the torch handle to bleed trapped acetylene out of hose and regulator.
4. Put torch back in designated area.

LAB 23.2 OXY-ACETYLENE TORCH SAFETY

LABORATORY OBJECTIVE
You will demonstrate your ability to safely adjust, light, and shut down an oxy-acetylene torch.

***FUNDAMENTALS OF HVACR* TEXT REFERENCE**
Unit 23 Soldering and Brazing

LABORATORY NOTES
The instructor will demonstrate how to safely adjust, light, and shut off an oxy-acetylene torch. You will then demonstrate your ability to safely adjust, light, and shut off an oxy-acetylene torch.

REQUIRED TOOLS, EQUIPMENT, AND MATERIALS
Safety glasses and gloves
Oxy-acetylene torch and torch key
Sparker

SAFETY REQUIREMENTS

READ AND ABSORB THESE RULES CAREFULLY. FAILURE TO DO SO IS PUNISHABLE BY DEATH!!!

Blow out cylinder valve before attaching regulators.
Why do we have this rule? It is simply to get the dust or combustible dirt out of the cylinder valve. Is dust a combustible? Have you ever read of a grain elevator explosion? Almost every case has been pinned down to dust. Combustion requires a fuel, an oxygen source, and a temperature source to cause ignition. The better the combustible material is mixed with oxygen, the better the chance of combustion. In the pure oxygen environment of the oxygen regulator, dust is downright explosive!

Release the adjusting screw on regulator before opening cylinder valve.
Why? Oxygen in the cylinder is compressed to pressures in excess of 2000 psig. When the adjusting screw is released, the seat of the regulator is in contact with the nozzle with sufficient pressure to hold 2000 psig. The oxygen travels only a short distance through the regulator this way. The regulator is not subjected to a sudden rush of pressure and less likely to rupture. If the pressure is allowed to blast into the regulator nozzle, any dust or oil can be easily ignited. This will burn out the diaphragm of the regulator and cause a potentially hazardous situation.

Stand to the side of the regulator when opening the valve.
Where could a man go who stands in front of his regulator when opening the valve? Possibly to the hospital or the morgue. The weakest point of any regulator is the front and back. If one is going to blow, the force will travel forwards and/or backwards. Your chance of surviving a regulator blowout is much greater if you are not standing directly in front of or behind the regulator.

Open the cylinder valve slowly.

There are two reasons for this one. The first is that if we open the cylinder valve slowly, the heat made from the gas travel is very small. This reduces the possibility of an explosion. The second reason is to reduce shock on the regulator. Slow opening of the valve will increase regulator life.

Purge oxygen and fuel lines individually before lighting torch.

Are acetylene and oxygen an explosive mixture? They sure are, especially when they are confined in an area as they are in a torch. Have you ever seen a blown-up torch? What causes these blown up torches? Analyzing the cause of the explosion, you know that fuel, oxygen, and ignition temperatures must be present at the point of the explosion. A mixture of acetylene and oxygen can be a potentially fatal brew. By purging the lines, you prevent mixture of the two gasses other than at the tip.

Light the acetylene first.

Here we go to a short study of the burning rate of acetylene. Acetylene emits from the tip orifice at the correct velocity using this method: open acetylene valve only and light. Adjust the flow between these two points: 1) when the flame flares out in a fan and is luminously bright and 2) when the flame leaves the tip. The flame-burning rate should be set with the acetylene valve only. After adjusting the burn rate, open the oxygen valve and adjust to the desired flame.

Do not use oxygen as a substitute for air.

How many people have you seen use oxygen to blow dirt off of their clothes? Clothes are a fuel; oxygen greatly increases the chance of combustion. A lighted cigarette could burn you up faster than if you had poured gasoline over your body and lit it.

Do not use oil on regulators or torch fittings.

Why shouldn't we use oil to lubricate this equipment? An oil-oxygen mixture can explode. Every gauge made to be used with oxygen has this information printed on the side of it. Oxygen cylinders have as much as 2200 psig pressure on them. When the oxygen is released into the regulator under pressure, it travels faster than the speed of sound. This generates heat due to the friction of the rapidly moving compressed gas. If the smallest amount of combustible material is present, combustion may ensue. Even the oil from your skin is enough to cause an explosion.

Do not use acetylene at pressures above 15 psig.

Acetylene gas is composed of two atoms of hydrogen and two molecules of carbon $C2H2$. The two chemicals do not have a great affinity for each other at pressures above 15 psig, making acetylene very unstable at high pressures. When molecules break down into their constituent elements, the energy holding them together is released. When acetylene breaks down under pressure, it releases enough heat to kindle a chain reaction of chemical breakdown, giving off large quantities of heat. In other words, acetylene can explode at pressures in excess of 15 psig WITHOUT oxygen. This is a very dangerous situation and should be avoided.

Keep the flame away from combustibles.

This rule is usually taken for granted, but it is imperative to make sure that you will not ignite something else when lighting and using a torch.

PROCEDURE

The instructor will demonstrate the proper way to adjust an oxy-acetylene torch.

You will demonstrate your ability to safely adjust an oxy-acetylene torch.

Step 1 – Safety Check

Check to make sure the valves on the torch handle are closed clockwise.

Check to see that the regulators on the acetylene and oxygen cylinders are not adjusted in. They should both be out counterclockwise until there is no spring pressure on the T handle.

Step 2 – Open Acetylene Cylinder Valve

Stand behind the acetylene cylinder on the side opposite the regulator so that you are not in a direct line with the regulator. This reduces the possibility of being struck by the regulator should it fail and come flying apart under pressure. Using the tank valve key, turn the acetylene tank valve counterclockwise the minimum amount required to get a reading on the cylinder pressure gauge. This should never be more than two and a half turns. Make sure to leave the key on the valve while you are using the torch.

Step 3 – Open Oxygen Cylinder Valve

Stand behind the oxygen cylinder on the side opposite the regulator so that you are not in a direct line with the regulator and crack open the cylinder valve counterclockwise. Once you see a reading on the oxygen cylinder pressure gauge, open the oxygen cylinder valve all the way counterclockwise. This is done to prevent the high cylinder pressure from placing a strain on the cylinder valve threads.

Step 4 – Adjust Regulators

Torch manufacturers have designed their equipment to operate properly at specific acetylene and oxygen pressures. Always consult the manufacturer's tip chart when determining the correct pressures for adjusting an oxyacetylene torch. Table 23-2-1 shows a typical brazing tip chart. Regulate the acetylene and oxygen pressures by slowly turning the regulators in clockwise and observing the pressure gauges. Brazing seldom requires acetylene pressures greater than 8 psig, and any acetylene pressure adjustment beyond 15 psig is extremely dangerous. Do not exceed the recommended pressure for any given torch and application.

Table 23-2-1 Oxygen and Acetylene Pressures for Different Tip Sizes

TIP	Acetylene Pressure	Oxygen Pressure
000	3–5 psig	3–5 psig
00	3–5 psig	3–5 psig
0	3–5 psig	3–5 psig
2	3–5 psig	3–5 psig
3	3–6 psig	4–7 psig
4	4–7 psig	5–10 psig
5	5–8 psig	6–12 psig
6	6–9 psig	7–14 psig

7	8–10 psig	8–16 psig
8	9–12 psig	10–19 psig
10	12–15 psig	12–24 psig

Step 5 – Lighting the Torch

To prevent potentially dangerous backfires in the torch, acetylene is always ignited first and turned off last.

    A. Crack open only the torch handle acetylene valve counterclockwise.

    B. light the acetylene gas using a spark or flint lighter.

    C. increase the flow of acetylene until most of the smoke in the flame has disappeared.

    D. crack open the oxygen valve counterclockwise.

    E. Slowly increase the oxygen flow until the center cone is well defined as in Figure 23-2-1

Figure 23-2-1

Step 6 – Putting up the Torch

    A. To shut off the torch, close the oxygen torch handle valve clockwise immediately followed by closing the acetylene torch handle valve clockwise. It is important to shut off the oxygen first and acetylene last to avoid a backfire.

    B. Close both the acetylene and oxygen cylinder valves clockwise.

    C. Bleed off the acetylene gas trapped in the acetylene hose by cracking just the acetylene torch handle valve. Close the valve once the regulator shows 0 psig.

    D. Bleed off the oxygen gas trapped in the oxygen hose by cracking just the oxygen torch handle valve. Close the valve once the regulator shows 0 psig.

    E. Back the T handle on each regulator counterclockwise until you no longer feel spring tension. Try not to over-loosen the T handle, as in some cases it can completely unscrew and fall out of the regulator body. If this does occur, be very careful not to cross-thread the threads in the regulator body when you reinstall it.

LAB 23.3 SOLDERING COPPER PIPE

LABORATORY OBJECTIVE
You will make 9 solder joints each of 1/4", 3/8", 1/2", 5/8", and 3/4" copper using the swage joints you made in Lab 22.3. You will cut open the joints to verify the solder penetration and coverage.

FUNDAMENTALS OF HVACR TEXT REFERENCE
Unit 23 Soldering and Brazing

LABORATORY NOTES
For this lab exercise a fitting will be soldered to copper piping. The fitting will then be tested to determine if it is satisfactory.

Soldering takes place below 840 °F and is used for water or condensate drains. Solder is not approved for refrigerant line joints.

Soldering to join copper pipes uses an air acetylene torch, air MAPP torch, or air propane torch. Oxyacetylene is not recommended for soldering due to high flame temperatures.

Table 23-3-1 Common Soldering and Brazing Metal and Fluxes Showing
        Base Metals That Can Be Joined

	Alloy	Flux Type	Base Metal
S O L D E R	95-5 Tin antimony solder[1]	C-Flux	Copper pipe, brass, steel
	95-5 Tin antimony solder	Rosin	Copper pipe, copper wiring, brass
	95-5 Tin antimony solder	Acid	Copper pipe, brass, steel, galvanized sheet metal
	98-2 Tin silver solder	Mineral based flux	Copper pipe, brass, steel
	40-60 Cadmium zinc solder	Specific flux from solder manufacturer	Aluminum
B R A Z I N G	Copper phosphorus silver brazing BCuP	1% to 15% Silver no flux required	Copper pipe
	Copper phosphorus silver brazing BCuP	1% to 15% Silver mineral based flux	Copper pipe to brass, brass, steel
	Copper silver brazing BCuP	45% Silver mineral flux	Copper pipe to steel, brass, steel

[1]The percentages of the materials in the flux are given in the numbers, for example 95% tin, 5% antimony.

REQUIRED TOOLS, EQUIPMENT, AND MATERIALS
    Safety glasses and gloves
    Air acetylene or propane/MAPP torch, torch Key, sparker
    Sand tape, soft solder flux, soft solder

Copper tubing and fittings
Vise, hacksaw, pliers, and flat screwdriver

SAFETY REQUIREMENTS

A. One hundred percent cotton or leather clothing is the best material to wear while brazing, soldering, or welding.

B. Shirts should have long sleeves and work gloves must be worn (cloth, leather palm, or all leather).

C. Safety glasses and gloves should always be worn when soldering.

D. When using Acetylene, torch pressure should be approximately 5 psig and the cylinder valve should be open no more than one and 1 1/2 turns.

E. Lead solder should be avoided.

F. Ventilation fans should be running.

PROCEDURE
Step 1 – Joint Preparation

Use an abrasive sanding cloth or wire brush to remove all contaminants from the surface of the pipe. Do not touch the cleaned surfaces with your hands as oil from your fingers can prevent solder from flowing completely into a joint. Use a brush to apply the flux to the end of the pipe. Apply the flux to approximately 1/16 to 1/8 from the end of the pipe to avoid flux contamination into the system. Do not apply flux to the very end of the pipe.

Figure 23-3-1   Figure 23-3-2

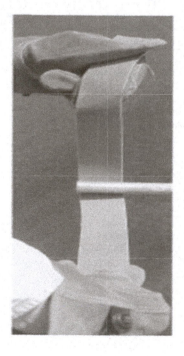

Step 2 – Soldering Procedure

For small tubing, a length of solder approximately equal to the tube diameter is adequate. Use an air fuel torch and begin by heating the pipe. As the pipe begins to be heated, move the torch onto the fitting and pipe. Periodically test the back side of the joint with the tip of the solder. Once the solder begins to melt, remove the torch. Continue adding the solder until the joint is filled. Ideally a small fillet of solder should be left at the joint surface. Pushing more solder into the joint simply results in solder bb's being formed inside the piping.

Figure 23-3-3

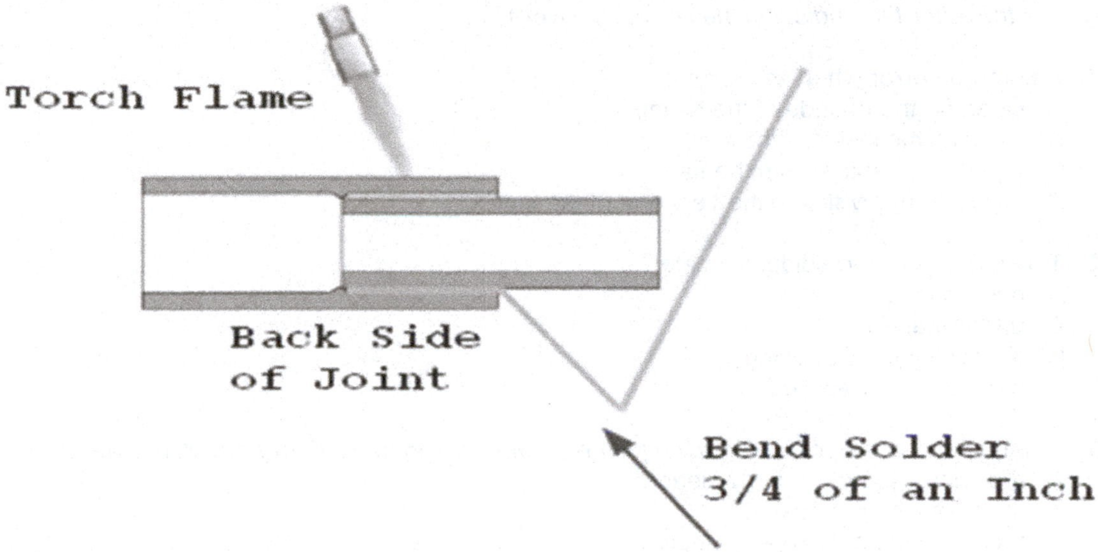

Torch Flame

Back Side
of Joint

Bend Solder
3/4 of an Inch

Step 3 – Number of Joints

You are to solder 3 joints in a horizontal position, 3 joints in a vertical position, and 3 joints in an upside-down position for 1/4", 3/8", 1/2", 5/8", and 3/4" copper. Use the swages that you made in Lab 22.3.

Step 4 – Destructively Test the Soldered Joints

Once cooled, use a hacksaw to slice a 45° angle through the fitting but not the pipe. Cross-cut the tubing just past the joint. Cut completely through the edge of the outer tubing at the diagonal cut. Grab the outer tubing with a pair of pliers and roll off the outer piece. If you cannot grab the outer piece, place a flat bladed screwdriver into the slot and twist to release the pipe fitting from the copper pipe. Use a pair of pliers, peel back the copper fitting to expose the soldered surface. If the surface is smooth and has no large voids, the soldering was successful. Save all your joints for inspection.

Step 5 – Grading

Joints will be graded on the following basis:

1. There should be no solder past the end of the joint inside the tubing.

2. There must be a complete seal around the entire circumference of the joint.

3. There should be few pockets, or bubbles, where voids exist in the solder.

4. The joint should be neat in appearance.

QUESTIONS

(Circle the letter that indicates the correct answer.)

1. A common error when soldering is:
   A. cleaning the outside of the tubing.
   B. cleaning the inside of the joint.
   C. melting the solder with the flame.
   D. failure to apply flux to the very end of the pipe.

2. Tiny bubbles in the solder indicate:
   A. overheating.
   B. underheating.
   C. excessive joint cleaning.
   D. too little solder applied.

3. When solder does not flow into the joint but balls up and falls off, this could indicate that:
   A. the joint was not properly cleaned.
   B. the joint was not fluxed.
   C. the joint was not properly heated.
   D. any of the above could be correct.

4. Soldering flux should be applied:
   A. with a brush to within an 1/8 of an inch from the end of the pipe.
   B. with a brush to the end of the pipe.
   C. with your finger to within an 1/8 of an inch from the end of the pipe.
   D. with your finger to the end of the pipe.

5. The solder:
   A. should be introduced to the back side of the pipe.
   B. should be introduced directly under the flame.
   C. should be wrapped around the joint before heating.
   D. should be introduced after the copper starts to oxidize.

LAB 23.4 COPPER-PHOS BRAZING

LABORATORY OBJECTIVE

The student will demonstrate how to properly braze a copper joint using copper-phos brazing rod.

FUNDAMENTALS OF HVACR TEXT REFERENCE
Unit 23 Soldering and Brazing

LABORATORY NOTES

Using the swages you made earlier in Lab 22.3, you will use copper-phos to braze one copper joint and destructively test it using the method outlined later in this lab. You will then use copper-phos to braze 5 joints of 1/4", 3/8", 1/2", 5/8", and 3/4" copper and assemble them as shown in Figure 23-4-9. You will add a flare nut and flare the 1/4" tubing on one end. The other end will be pinched and brazed shut. The assembly will then be tested for leaks using nitrogen.

Brazing takes place at temperatures above 840 °F and Codes require that refrigerant joints be made using silver brazing alloys. Copper-phos brazing can be done using either an air acetylene torch with a swirl tip or an oxyacetylene torch. Oxyacetylene torches are generally preferred for making joints in copper tubing larger than 1/2 inch in diameter.

REQUIRED TOOLS, EQUIPMENT, AND MATERIALS

Safety glasses and gloves
Air acetylene or oxyacetylene torch, torch key, sparker
Sand tape, copper-phos brazing rod
Five swage joints for 1/4", 3/8", 1/2", 5/8", and 3/4" copper pipe
Vise, hacksaw, and pliers

SAFETY REQUIREMENTS

A. One hundred percent cotton or leather clothing is the best material to wear while brazing, soldering, or welding.

B. Shirts should have long sleeves and work gloves must be worn (cloth, leather palm, or all leather).

C. Safety glasses and gloves should always be worn when brazing.

D. Set acetylene and oxygen pressures according to the manufacturer's instructions.

E. The acetylene cylinder valve should be open no more than one and 1 1/2 turns.

F. Ventilation fans should be running.

PROCEDURE

Step 1 – Preparing Tubing and Joint

Use an abrasive sanding cloth or wire brush to remove all contaminants from the surface of the pipe as shown in Lab 23.3. Do not touch the cleaned surfaces with your hands as oil from your fingers can prevent the brazing alloy from flowing completely into a joint. No flux is necessary when using copper-phos brazing rod.

Step 2 – Torch Setup

Prepare the torch according to the manufacturer's instructions. You can refer to Lab 23.1 for air acetylene torches and Lab 23.2 for oxy-acetylene torches.

Step 3 – Light the Torch

Light the torch according to the manufacturer's instructions. You can refer to Lab 23.1 for air-acetylene torches and Lab 23.2 for oxy-acetylene torches. **NEVER USE A BUTANE LIGHTER TO LIGHT A TORCH!**

The only safe devices to use to light any torch are those specifically designed for that purpose, such as spark lighters or flint lighters. When using a flint lighter hold it slightly off to one side of the torch tip as shown in Figure 23-4-1.

Adjust the torch to achieve a neutral flame, as show in Figure 23-4-2. Excessive pressure will cause he flame to lift off the tip, as in Figure 23-4-3. Too little pressure will cause a small flame, close to the tip, as shown in Figure 23-4-4. Too small a flame can overheat the torch tip.

Figure 23-4-1

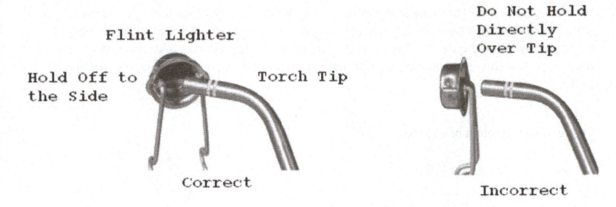

Figure 23-4-2 Normal flame

Figure 23-4-3 Lifting flame, excessive pressure

Figure 23-4-4 Flame too small

Step 4 – Brazing

Heat the pipe first and then the fitting, as shown in Figure 23-4-5. Once the pipe appears shiny, move the torch flame onto the fitting so that it envelops both the fitting and the pipe. Occasionally touch the pipe surface with the tip of the brazing rod as a test of temperature readiness. Continue heating the pipe until the brazing metal begins to flow evenly over the surface.

Figure 23-4-5 Heating sequence

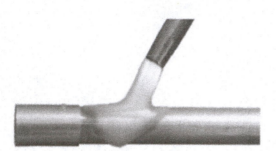

First Heat the Pipe

Then Heat the Fitting

If you watch carefully, you can see a slight change in the fitting's color as the filler metal flows into the joint space. Bringing the torch down on the fitting will help draw the filler metal completely into the joint. Once the joint gap has been filled, continue adding small amounts of filler braze metal until a fillet of metal surrounds the joint.

Step 5 – Destructively Test the Joint

Use a hacksaw to saw off the copper fitting at a point just beyond the depth that the pipe was inserted into the joint, as shown in Figure 23-4-6. Once the fitting has been cut completely apart, clamp the pipe in a vise and cut straight down through the entire joint into the pipe. Rotate the pipe ninety degrees and repeat the process so that there are four cuts, as shown in Figure 23-4-7. Bend each quarter out with a wrench. Using a hammer and an anvil, flatten each of the four corners, as shown in Figure 34-4-8. It is easy to see areas where 100% joint penetration did not occur after the joint pieces are flattened.

Figure 23-4-6

Figure 23-4-7

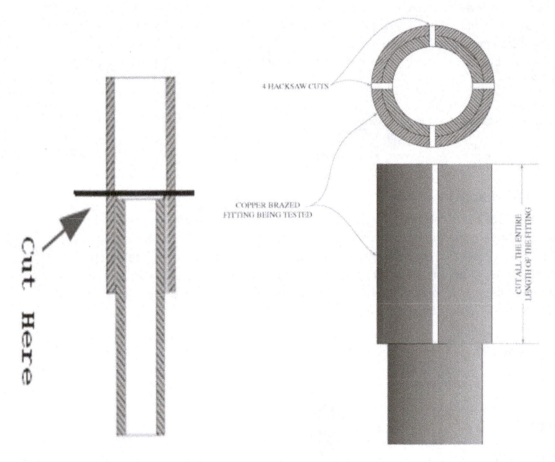

4 HACKSAW CUTS

COPPER BRAZED
FITTING BEING TESTED

CUT ALL THE ENTIRE
LENGTH OF THE FITTING

Cut Here

Figure 23-2-8

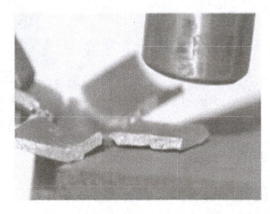

Step 6 – Build Brazed Assembly

Make five assemblies by brazing five joints each, end to end, of 1/4", 3/8", 1/2", 5/8", and 3/4" copper. Flare one end of the 1/4" assembly, slide the other end inside the 3/8" assembly and braze them together. Slide the other end of the 3/8" assembly into one end of the 1/2" assembly and braze them together. Continue until you have one assembly them as shown in Figure 23-4-9.

Figure 23-4-9 Brazing project

Step 7 – Test Brazed Assembly

Connect the flare nut of the brazed assembly to a 14" flare union and connect a 1/4" refrigeration hose between the flare union and the outlet of the regulator on a nitrogen cylinder, Figure 23-4-10. Pressurize the assembly to 100 psig by adjusting the nitrogen regulator. Use an ultrasonic detector or soap bubbles to check the assembly joints for leaks. Mark any leaking joints, release the nitrogen pressure, and repair the leaking joints. Retest until the assembly does not leak.

Figure 23-4-10 Nitrogen regulator

Regulator

Relief Valve

QUESTIONS

(Circle the letter that indicates the correct answer.)

1. Acetylene gas must be stabilized in the cylinder with:
   A. propane.
   B. oxygen.
   C. acetone.
   D. nitrogen.

2. It is illegal to operate a torch with an acetylene pressure:
   A. greater than 15 psi.
   B. less than 15 psi.
   C. greater than 5 psi.
   D. greater than 3 psi.

3. Turning the acetylene pressure up beyond the recommended pressure value for a specific torch will:
   A. make the torch operate better.
   B. cause it to "pop."
   C. not make it work better.
   D. cause it to "crack."

4. It is against OSHA regulations to open any acetylene cylinder more than:
   A. 1 1/2 turns.
   B. 1 turn.
   C. 2 turns.
   D. 2 1/2 turns.

5. When transporting an acetylene cylinder:
   A. lay it on its side.
   B. leave the cylinder "cracked" open.
   C. always remove the regulator.
   D. leave the cylinder cap off.

LAB 23.5 SILVER BRAZING

LABORATORY OBJECTIVE

The student will demonstrate how to properly braze dissimilar metals using high silver content brazing rod.

LABORATORY NOTES

You will braze copper to brass, copper to steel, and steel to brass using high silver content brazing rod. Copper-phos brazing alloy is mainly used for brazing copper to copper. High content silver brazing alloy is used to braze different types of metal together. Typically, this is done when connecting copper refrigerant lines to either steel or brass valves. Unlike copper-phos brazing alloys, high content silver brazing alloy requires brazing flux.

FUNDAMENTALS OF HVACR TEXT REFERENCE

Unit 23 Soldering and Brazing

REQUIRED TOOLS, EQUIPMENT, AND MATERIALS

Safety glasses and gloves

Air acetylene or oxyacetylene torch, torch key, sparker

Sand tape, brazing flux, high silver content brazing alloy

One 5/8" copper swage joint, a short piece of 1/2" black iron pipe, two short pieces of 5/8" diameter brass rod

Vise, hacksaw, and pliers

SAFETY REQUIREMENTS

A. One hundred percent cotton or leather clothing is the best material to wear while brazing, soldering, or welding.

B. Shirts should have long sleeves and work gloves must be worn (cloth, leather palm, or all leather).

C. Safety glasses and gloves should always be worn when brazing.

D. Set acetylene and oxygen pressures according to the manufacturer's instructions.

E. The acetylene cylinder valve should be open no more than one and 1 1/2 turns.

F. Ventilation fans should be running.

G. Use cadmium-free brazing alloy whenever possible.

CADMIUM POISONING SAFETY NOTE!

Some high content silver brazing alloys contain cadmium which can cause serious illness or even death if inhaled in large enough quantities. Alloys containing cadmium can produce cadmium vapors when heated. Steps to avoid cadmium poisoning include:

A. Use cadmium-free brazing alloy whenever possible

B. Always braze in well ventilated areas

C. Avoid overheating the braze joint

PROCEDURE

Step 1 – Preparing the Pieces

Use an abrasive sanding cloth or wire brush to remove all contaminants from the surface of the pieces to be brazed. Do not touch the cleaned surfaces with your hands as oil from your fingers can prevent the brazing alloy from flowing completely into a joint. Brazing flux is necessary when using high silver content brazing rod, Figure 23-5-1. Apply flux to the male piece before assembly.

Figure 23-5-1 Silver brazing flux and brazing alloy

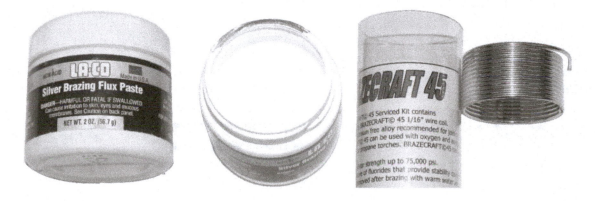

Step 2 – Torch Setup

Prepare the torch according to the manufacturer's instructions. You can refer to Lab 23.1 for air acetylene torches and Lab 23.2 for oxy-acetylene torches.

Step 3 – Light the Torch

Light the torch according to the manufacturer's instructions. You can refer to Lab 23.1 for air-acetylene torches and Lab 23.2 for oxy-acetylene torches. **NEVER USE A BUTANE LIGHTER TO LIGHT A TORCH!**

The only safe devices to use to light any torch are those specifically designed for that purpose, such as spark lighters or flint lighters. When using a flint lighter hold it slightly off to one side of the torch tip. Adjust the torch to achieve a neutral flame. Excessive pressure will cause he flame to lift off the tip. Too little pressure will cause a small flame, close to the tip, which can overheat the torch tip.

Step 4 – **Silver Brazing**

A. Cut a short 3-inch section of 1/2" black iron pipe

B. Cut two short 2-inch sections of 5/8" brass round stock.

C. Make a 5/8" copper swage joint.

D. Clean and flux one end of a brass section and insert it into the 5/8" copper swage joint.

E. Braze the brass and copper joint using high silver content brazing alloy.

F. Clean and flux the other end of the copper swage joint and clean one end of the 1/2" black iron pipe. Insert the 5/8" copper into the 1/2" black iron pipe.

G. Braze the copper and iron joint using high silver content brazing alloy.

H. Clean and flux one end of the remaining brass piece and clean the other end 1/2" black iron pipe. Insert the 5/8" brass piece into the 1/2" black iron pipe.

I. Braze the brass and iron joint using high silver content brazing alloy.

J. The finished project should look like Figure 23-5-2.

Figure 23-5-2 Silver brazing project

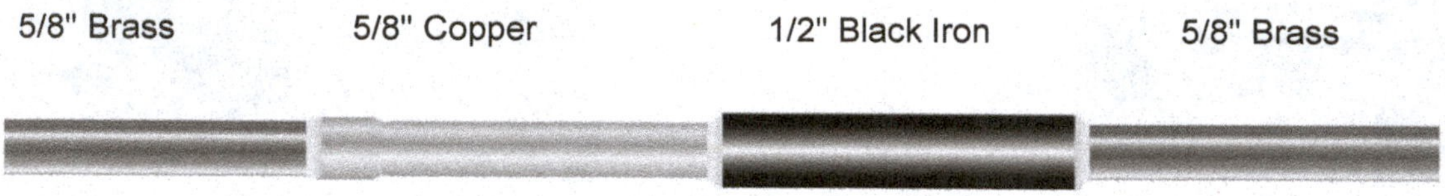

| 5/8" Brass | 5/8" Copper | 1/2" Black Iron | 5/8" Brass |

LAB 24.1 EXAMINE REFRIGERANT SYSTEM PIPING

LABORATORY OBJECTIVE

You will examine refrigeration systems, identify the system refrigeration lines, describe the function of these lines, and describe the condition of the refrigerant in the lines.

***FUNDAMENTALS OF HVACR* TEXT REFERENCE**

Unit 24 Refrigerant System Piping

LABORATORY NOTES

The refrigeration lines are important parts of a refrigeration system. Identifying the line, the refrigerant condition in the line, and understanding the line's function is critical to HVACR technicians.

REQUIRED TOOLS, EQUIPMENT, AND MATERIALS

    A. Safety glasses and gloves

    B. Operating systems, including:

        a. packaged air conditioner

        b. split system air conditioner

        c. mini-split air conditioner

        d. ice maker with a remote air-cooled condenser

    C. Thermometer for reading line temperature

    D. Gauges for reading line pressures

SAFETY REQUIREMENTS

Safety glasses and gloves should always be worn when working on refrigeration systems.

PROCEDURE

Packaged System

    A. With the unit off and secured, examine the packaged unit assigned by the instructor and identify the discharge line, liquid line, and suction line. Be prepared to point them out to the instructor.

    B. Determine where to measure the pressure and temperature on each line.

    C. Operate the unit, measure the pressure and temperature of each line, and complete data Table 24-1-1.

Split System

A. With the unit off and secured, examine the split system assigned by the instructor and identify the liquid line, and suction line. Be prepared to point them out to the instructor.

B. Determine where to measure the pressure and temperature on each line.

C. Operate the unit, measure the pressure and temperature of each line, and complete data Table 24-1-2.

Mini-Split System

A. With the unit off and secured, examine the mini-split system assigned by the instructor and identify the liquid line, and suction line. Be prepared to point them out to the instructor.

B. Determine where to measure the pressure and temperature on each line.

C. Operate the unit, measure the pressure and temperature of each line, and complete data Table 24-1-3.

Ice Maker with Remote Condenser

D. With the unit off and secured, examine the ice-maker assigned by the instructor and identify the discharge line, and liquid line. Be prepared to point them out to the instructor.

E. Determine where to measure the pressure and temperature on each line.

F. Operate the unit, measure the pressure and temperature of each line, and complete data Table 24-1-4.

Table 24-1-1 PACKAGED AIR CONDITIONER				
LINE	MEASURED PRESSURE	SATURATION TEMPERATURE	MEASURED TEMPERATURE	REFRIGERANT CONDITION
DISCHARGE				
LIQUID				
SUCTION				

Table 24-1-2 SPLIT SYSTEM AIR CONDITIONER				
LINE	MEASURED PRESSURE	SATURATION TEMPERATURE	MEASURED TEMPERATURE	REFRIGERANT CONDITION
LIQUID				
SUCTION				

Table 24-1-3 MINI-SPLIT SYSTEM AIR CONDITIONER				
LINE	MEASURED PRESSURE	SATURATION TEMPERATURE	MEASURED TEMPERATURE	REFRIGERANT CONDITION
LIQUID				
SUCTION				

Table 24-1-4 ICE MAKER WITH REMOTE CONDENSER				
LINE	MEASURED PRESSURE	SATURATION TEMPERATURE	MEASURED TEMPERATURE	REFRIGERANT CONDITION
DISCHARGE				
LIQUID				

QUESTIONS

1. Looking at the lines you inspected in all the units, which lines contained superheated gas?

2. What differences did you see between the liquid line on the split system and the liquid line on the mini-split system?

3. Explain why the two lines of the ice maker with a remote condenser are different than the two lines of the split system.

LAB 25.1 SCHRADER CORE REPLACEMENT

LABORATORY OBJECTIVE
The student will demonstrate how to properly replace the core on a refrigeration Schrader valve using a valve core tool.

***FUNDAMENTALS OF HVACR* TEXT REFERENCE**
Unit 25 Accessing Sealed Refrigeration Systems

LABORATORY NOTES
The student will first demonstrate how to remove and replace the Schrader valve core on a system with a nitrogen holding charge. Then they will demonstrate on a system containing refrigerant.

REQUIRED TOOLS, EQUIPMENT, AND MATERIALS
Safety glasses
Gloves
Schrader valve not on a system
Standard valve core removal tool
Core removal tool designed to work on systems under pressure
System with Schrader valve and a nitrogen holding charge
Refrigeration system with Schrader valves

SAFETY REQUIREMENTS
A. Wear safety glasses
B. Wear gloves whenever working on refrigeration systems
C. Standard core tools, such as found on the top of many Schrader valve caps, can only be used while the system is NOT under pressure.
D. A special core removal tool is required to remove and replace the core under pressure (Figure 25-1-2).

PROCEDURE
1. Familiarize yourself with the operation of the Schrader valve. Figure 25-1-1 shows a cutaway view of a typical Schrader valve. Note that the valve opens when the valve stem is depressed. When the stem is depressed, the system is open and at that point refrigerant could leak out of a system or if at a negative pressure, air could leak in.

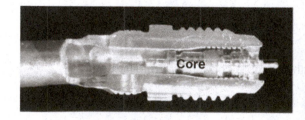

Figure 25-1-1

Figure 25-1-2

2. Use a standard core tool to remove the core from a loose Schrader valve.
3. Examine the core and the valve and reinsert the core into the Schrader valve.
4. Read section 25.4 in *Fundamentals of HVACR* and study Figure 25-1-3 showing the correct sequence.

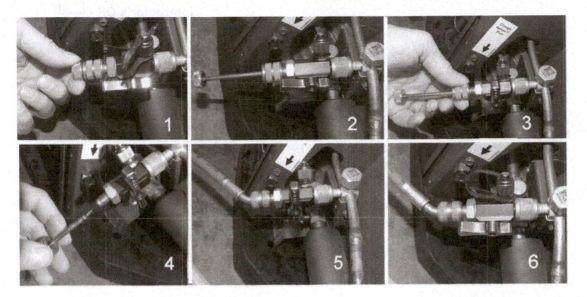

Figure 25-1-3

6. Instructor will demonstrate proper use of core removal tool.
7. You will remove the Schrader core on a system with a nitrogen holding charge
8. Show the core to the instructor.
9. Use the core tool to re-insert the core without losing any of the nitrogen charge.
10. Test valve for leaks using soap bubbles.
11. Repeat steps 7–10 on the system assigned by the instructor.

QUESTIONS

(Circle the letter that indicates the correct answer.)

1. A special tool is required to replace a Schrader valve core while under pressure:
   A. True.
   B. False.

2. Schrader valves opened when:
   A. the valve is turned clockwise.
   B. the valve is turned counterclockwise.
   C. the valve lever is moved to the on position.
   D. the stem on the valve core is depressed.

3. In order to conform to EPA refrigerant management requirements:
   A. always release refrigerant to the atmosphere after servicing the unit.
   B. only vent refrigerant a little at a time.
   C. put service caps back on all access ports.
   D. always run refrigerant through a filter when releasing it to the atmosphere.

LAB 25.2 INSTALLING PIERCING VALVES

LABORATORY OBJECTIVE
The student will demonstrate how to properly install a bolt-on piercing valve and get a reading on the gauges.

***FUNDAMENTALS OF HVACR* TEXT REFERENCE**
Unit 25 Accessing Sealed Refrigeration Systems

LABORATORY NOTES
The student will make a small copper tube assembly, connect it to nitrogen pressure, install a piercing valve on the assembly, and obtain a pressure reading.

REQUIRED TOOLS, EQUIPMENT, AND MATERIALS
Safety glasses and gloves

1/4-inch copper tubing

Two 1/4-inch flare nuts and one 1/4-inch flare plug

Tubing cutter

Flaring tool

Bolt-on piercing valve to fit 1/4-inch tubing

Nitrogen cylinder with pressure regulator

Refrigeration gauge

SAFETY REQUIREMENTS
  A. Wear safety glasses
  B. Nitrogen should ONLY be used with a proper regulator.
  C. See instructor for instructions on pressurizing your assembly.

PROCEDURE
  1. Piercing valves are used to give access to systems that do not have any type of service valves. They are clamped to the tubing, sealed by a bushing gasket, and then they pierce the tube with a tapered needle. Some examples are shown in Figure 25-2-1.

  2. Cut a 6-inch section of 1/4" copper tubing.

  3. Flare both ends of the copper tubing.
     CAUTION: Make sure to put the flare nuts on BEFORE making the last flare.

  4. Plug one end of the assembly using the 1/4-inch flare plug.

  5. Ask the instructor for help pressurizing your assembly.

  6. Check to see that the piercing needle is NOT protruding in the saddle area of the valve. The valve must be installed BEFORE the pin is run down to pierce the copper. Bolt the piercing valve around the tubing.

7. Connect the pressure gauge to the gauge port of the piercing valve.

8. Tighten the piercing needle by turning it in clockwise until it is firmly seated.

9. Back the needle off one full turn to allow pressure to flow to the gauge.

10. The gauge should be reading pressure.

11. Check the piercing valve for leaks using soap bubbles.

12. Turn off the nitrogen and bleed pressure out of the assembly through the gauges.

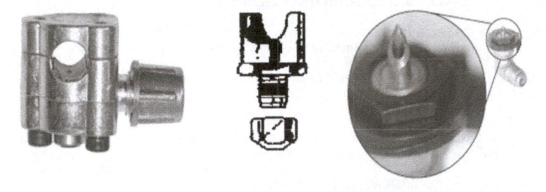

Figure 25-2-1

QUESTIONS

(Circle the letter that indicates the correct answer.)

1. Where are piercing valve used?
   A. Whenever the gauge manifold is attached to a system
   B. On systems without any service valves
   C. Whenever a reading is taken on a service valve body
   D. Primarily on large commercial refrigeration systems

2. A piercing valve can be installed while the system is under pressure:
   A. True.
   B. False.

3. Access-piercing valves should be removed once the source of the sealed system malfunction has been located:
   A. True.
   B. False.

4. When bolting on a piercing valve, the valve stem:
   A. should NOT be protruding, but recessed into the valve.
   B. covered with pipe dope to prevent leakage.
   C. SHOULD be protruding in order to pierce the tubing during installation.
   D. should be heated with a torch to improve its piercing ability.

5. In order to conform to EPA refrigerant management requirements:
   A. always release refrigerant to the atmosphere after servicing the unit.
   B. only vent refrigerant a little at a time.
   C. put service caps back on all access ports.
   D. always run refrigerant through a filter when releasing it to the atmosphere.

LAB 25.3 EXAMINING SERVICE VALVES

LABORATORY OBJECTIVE
The student will demonstrate their knowledge of standard manual service valves.

FUNDAMENTALS OF HVACR TEXT REFERENCE
Unit 25 Accessing Sealed Refrigeration Systems

LABORATORY NOTES
The student will examine a manual stem service valve, identify its parts, demonstrate different valve positions, and explain how manual service valves are used.

REQUIRED TOOLS, EQUIPMENT, AND MATERIALS
Safety glasses
Manual stem service valve
Service valve wrench

SAFETY REQUIREMENTS
Wear safety glasses

PROCEDURE

1. Obtain a manual stem service valve and a service valve wrench from the instructor.

2. Use the service valve wrench to turn he valve stem all the way out counterclockwise. This is the **backseated** position of the valve (Figure 25-3-1).

3. Note that the flow from the center port of the valve to the line port is unrestricted. Also note that the gauge port is now closed.

4. Use the service valve wrench to turn the valve stem one turn in clockwise. This is the **backseat cracked** position (Figure 25-3-2).

5. Note that there is now a crack between the valve disk and the valve seat.

6. Use the valve wrench to turn the stem all the way in clockwise. This is the **frontseated** position.

7. Note that the line port is now closed off but the gauge port is open.

8. Use the valve stem wrench to position the valve stem midway between the front and back seats. This is called **mid-position**. At this position all ports are open (Figure 25-3-3).

9. Look at the valve stem. Is there a packing gland nut?

10. If there is, loosen the packing gland nut and inspect the packing.

11. What type of packing does the valve have?

Backseated	Backseat cracked	Frontseated

Figure 25-3-1	Figure 25-3-2	Figure 25-3-3

Locate the following types of valves on systems in the shop:

Suction service valve _____

Discharge service valve _____

Liquid receiver service valve _____

Liquid line service valve _____

Schrader service valve _____

Piercing valve _____

Split system shutoff valve _____

QUESTIONS
(Circle the letter that indicates the correct answer.)

1. When a service valve is backseated:
   A. the gauge manifold can be attached.
   B. normal flow through the valve will be shut off.
   C. refrigerant can flow through the valve gauge port.
   D. it is in position to act as a backup for the main service valve.

2. What is used to turn the valve stem on a service valve?
   A. An adjustable wrench
   B. Channellocks
   C. A service valve wrench
   D. Any type of wrench or pliers that can grip the valve stem

3. When a service valve is said to be in the mid-position:
   A. there is flow through the service port only.
   B. there is flow through the main valve only.
   C. there is flow through the service port and through the main valve.
   D. there is no flow through either the service port or main valve.

4. In order to conform to EPA refrigerant management requirements:
   A. always release refrigerant to the atmosphere after servicing the unit.
   B. only vent refrigerant a little at a time.
   C. put service caps back on all access ports.
   D. always run refrigerant through a filter when releasing it to the atmosphere.

LAB 25.4 INSTALLING GAUGES ON SCHRADER VALVES

LABORATORY OBJECTIVE
You will demonstrate your ability to safely install gauges on four operating refrigeration systems with Schrader valves, obtain readings, and safely remove the gauges from the systems.

FUNDAMENTALS OF HVACR TEXT REFERENCE
Unit 25 Accessing Sealed Refrigeration Systems

LABORATORY NOTES
You will safely install gauges, get pressure readings, and remove your gauges on four operating systems: 2 packaged air conditioners and 2 split system air conditioners.

REQUIRED TOOLS, EQUIPMENT, AND MATERIALS
Safety glasses
Gloves
6-in-1 tool
Refrigeration gauges
Positive shutoff hose or adapter (optional)
Operating refrigeration systems with Schrader valves

SAFETY REQUIREMENTS
    A. Wear safety glasses and gloves whenever working on refrigeration systems
       Keep hands out of direct stream of refrigerant in the event of a leak or mistake.
    B. When working on R410A systems, use only gauges and hoses designed for R410A.

PROCEDURE
You will check the operating refrigerant pressures on 2 packaged units and 2 split systems assigned by the instructor and complete the data sheet. The procedure is listed below.

Installing Gauges
1. Put on your safety glasses and gloves.
2. Turn the unit off.
3. Check to see that the gauge hand wheels are closed—fully clockwise.
4. Remove the low side valve cap.
5. Take the blue hose end with the Schrader core depressor (usually the angled end) and connect to the low side Schrader valve.
6. Hold the hose with one hand (the left hand for right-handed people) and quickly tighten the knurled end with your most kinesthetically developed hand (the right hand for right-handed people).
7. You should see a reading on your low side gauge (the blue one).

8. Purge the air and residual refrigerant out of your low side hose and manifold by loosening the end of the middle hose on your gauge set and cracking the low side gauge hand wheel for approximately 1 second.
9. You should hear air escaping out of the end of the middle hose.
10. Tighten the middle hose back and close the low side hand wheel.
11. Remove the high side valve cap.
12. Take the red hose end with the Schrader core depressor (usually the angled end) and connect to the high side Schrader valve.
13. You should see a reading on your high side gauge (the red one).
14. Purge the air and residual refrigerant out of your high side hose and manifold by loosening the end of the middle hose on your gauge set and cracking the low side gauge hand wheel for approximately 1 second. You should hear air escaping out of the end of the middle hose.
15. Tighten the middle hose back and close the high side hand wheel.

Reading Pressures

1. The gauge manifold gauge hand wheels should both be closed.
2. Record the pressures on the data sheet.
3. Start the unit.
4. The low side should drop and the high side should rise.
5. Let the unit run long enough for the pressures to stabilize and quit moving, normally 10 to 15 minutes.
6. Record the pressures on the data sheet.

Removing Gauges without Positive Shutoff Adapter

If your gauges have positive shutoff hoses skip down to next procedure

1. Remove the low side (blue) hose with the system running.
2. Turn the unit off and wait for the high side pressure to quit dropping.
3. Quickly disconnect the high side hose from the Schrader valve.
   Refrigerant will spray out. Hold your hand and fingers to the side as you loosen the hose to avoid having them in the direct spray of refrigerant!
4. Loosen the middle hose on your gauges to release the refrigerant trapped in the middle hose.
5. Replace the service valve caps.

Removing Gauges with a Positive Shutoff Adapter

Use this procedure if your gauges have positive shutoff hoses or you have a positive shutoff adapter.

1. The system should remain running.
2. Disconnect the high side hose or close the shutoff valve on the end of the high side hose (depending on the type of positive shutoff hoses you have).
3. Crack both hand wheels on your gauges.
4. The high side pressure should drop, the low side gauge will rise.
5. Both gauges will drop to the operating suction pressure.
6. Remove your low side gauge.
7. Turn the unit off. If this procedure is done correctly, very little refrigerant will escape.
8. Replace the service valve caps.

Data Sheet 25-4-1

Schrader Service Valve Readings				
Unit	Refrigerant	Equalized Pressure	Suction Pressure	Discharge Pressure
Packaged Unit				
Packaged Unit				
Split System				
Split System				

LAB 25.5 INSTALLING MANIFOLD GAUGES ON MANUAL STEM SERVICE VALVES

LABORATORY OBJECTIVE
You will demonstrate your ability to safely install gauges on systems with manual stem service valves, read system pressures, and safely remove gauges from the system.

FUNDAMENTALS OF HVACR TEXT REFERENCE
Unit 25 Accessing Sealed Refrigeration Systems

LABORATORY NOTES
You will first practice installing gauges on an inoperative system with manual service valves. You will then safely install gauges on four operating systems with manual service valves, obtain pressure readings, and safely remove the gauges.

REQUIRED TOOLS, EQUIPMENT, AND MATERIALS
Safety glasses
Gloves
6-in-1 tool
Adjustable wrench
Refrigeration wrench
Refrigeration gauges
Operating refrigeration systems with manual service valves and record your readings in the data sheet.

SAFETY REQUIREMENTS
A. Wear safety glasses and gloves whenever working on refrigeration systems
B. Manual service valves should always be back-seated (fully counterclockwise) before removing gauges. When working on R410A systems, use only gauges and hoses designed for R410A.

PROCEDURE
Installing Gauges

1. Put on your safety glasses and gloves.

2. Turn the unit off.

3. Use the adjustable jaw wrench to remove the valve cover caps.

4. Use the service valve wrench and make certain the valves are back-seated—turned all the way out counterclockwise.

5. Remove the valve caps.

6. Check to see that the gauge hand wheels are closed—fully clockwise.

7. Take the blue hose end with the Schrader core depressor (usually the angled end) and connect to the low side manual service valve.

8. Place the service valve wrench on the valve stem and turn clockwise ½ turn.

9. You should see a reading on your low side gauge (the blue one).

10. Purge the air and residual refrigerant out of your low side hose and manifold by loosening the end of the middle hose on your gauge set and cracking the low side gauge hand wheel for approximately 1 second.

11. You should hear air escaping out of the end of the middle hose.

12. Tighten the middle hose back and close the low side hand wheel.

13. Take the red hose end with the Schrader core depressor (usually the angled end) and connect to the high side manual service valve.

14. Place the service valve wrench on the valve stem and turn clockwise ½ turn.

15. You should see a reading on your high side gauge (the red one).

16. Purge the air and residual refrigerant out of your high side hose and manifold by loosening the end of the middle hose on your gauge set and cracking the low side gauge hand wheel for approximately 1 second. You should hear air escaping out of the end of the middle hose.

17. Tighten the middle hose back and close the high side hand wheel.

Reading Pressures

1. The gauge manifold gauge hand wheels should both be closed.
2. Record the pressures on the data sheet.
3. Start the unit.
4. The low side should drop and the high side should rise.
5. Let the unit run long enough for the pressures to stabilize and quit moving, normally 10 to 15 minutes.
6. Record the pressures on the data sheet.

Removing Gauges (quick and dirty way)

This procedure is legal, quick, and easy, but releases more refrigerant.

1. With the unit running, use the service valve wrench to backseat the suction service valve by turning it counter clockwise. *Note:* This should only take ½ turn.
2. Turn the unit off and wait for the high side pressure to quit dropping.
3. Use the service valve wrench to backset the discharge service valve by turning the valve counterclockwise. *Note:* This should only take ½ turn.
4. Loosen each hose at the service valves to let out trapped refrigerant.
5. Refrigerant will spray out. Hold your hand and fingers to the side as you loosen the hose to avoid having them in the direct spray of refrigerant!
6. Remove the hoses from the service valves.
7. Loosen the middle hose on your gauges to release the refrigerant trapped in the middle hose.
8. Replace the service valve caps and covers.

Removing Gauges by Returning Refrigerant to the System

This procedure takes a little longer, but puts most of the refrigerant back into the system, releasing less refrigerant.

1. The system should remain running.
2. Use the service valve wrench to backseat the discharge service valve.
3. Crack both hand wheels on your gauges.
4. The high side pressure should drop, the low side gauge will rise.
5. Both gauges will drop to the operating suction pressure.
6. Use the service valve wrench to backseat the suction service valve.
7. Turn the unit off.
8. Remove the hoses from the service valves.
9. If this procedure is done correctly, very little refrigerant will escape.
10. Loosen the middle hose on your gauges to release the refrigerant trapped in the middle hose.
11. Replace the service valve caps and covers.

Data Sheet 25-5-1

Manual Service Valve Pressure Readings				
Unit	Refrigerant	Equalized Pressure	Suction Pressure	Discharge Pressure
Unit 1				
Unit 2				
Unit 3				
Unit 4				

LAB 26.1 SYSTEM DEPENDENT LIQUID RECOVERY

LABORATORY OBJECTIVE

The student will demonstrate the correct procedure for the system dependent (passive) method of recovering liquid refrigerant from an operating refrigeration system.

FUNDAMENTALS OF HVACR TEXT REFERENCE
Unit 26 Refrigerant Management and the EPA

LABORATORY NOTES

This lab is intended to help students practice the removal of refrigerant from a system using the system dependent (passive) method of liquid recovery. The liquid is pumped out of the refrigeration unit by using its own compressor. The only recovery equipment required is a recovery cylinder. The entire system charge cannot be recovered this way, but most of it can. The remaining refrigerant would be recovered using a self-contained recovery unit.

REQUIRED TOOLS, EQUIPMENT, AND MATERIALS
Gloves & goggles
Recovery cylinder
Refrigerant scale
Service valve wrench
Gauge manifold
Operating refrigeration system

SAFETY REQUIREMENTS

A. Wear safety goggles and gloves when working with refrigerants. Liquid refrigerant can cause frostbite when in contact with eyes and skin.
B. Use low loss hose fittings to reduce refrigerant spray when connecting and disconnecting refrigeration hoses. Inspect all fittings before attaching hoses.

PROCEDURE

Step 1 – Connect Gauges
Connect the gauge manifold as shown in Figure 26-1-1 and purge the lines of air. The recovery cylinder hose should be connected to the vapor side (blue) valve on the cylinder. The liquid valve on the recovery cylinder adds a significant flow restriction.

Step 2 – Place Cylinder on Scale
After the lines have been connected and purged, place the recovery cylinder on a scale.

Step 3 – Prepare for Recovery

A. The liquid line service valve should be in the mid position to allow flow through the service port.

B. The high side gauge manifold valve should be closed.

C. The vapor valve (blue) on the recovery cylinder should be wide open.

D. The low side gauge manifold valve should be closed.

E. The suction line service valve should be open one turn off its back seat.

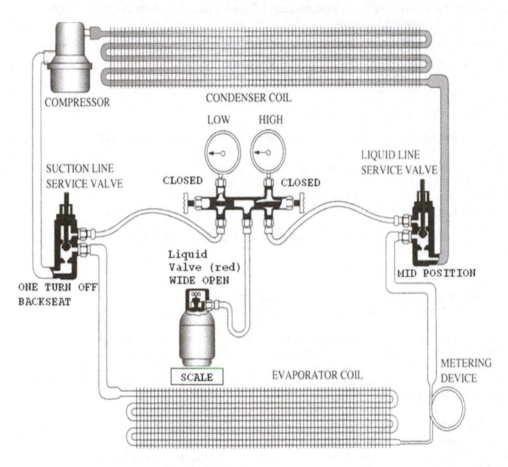

Figure 26-1-1

Step 5 – Recovery

A. Start the system and monitor the high side and low side pressures.

B. Slowly open the gauge manifold high side valve wide open and refrigerant should begin flowing into the recovery cylinder, Figure 26-1-2.

C. When opening the gauge manifold high side valve, the high side gauge pressure should drop to read the tank pressure.

D. Do not allow tank pressure to exceed the normal recovery cylinder maximum pressure (300 psig for a 4BA300 cylinder and 400 psig for a 4BA400 cylinder)

E. Monitor the cylinder weight (Remember no more than 80% full).

F. Cool the recovery cylinder with cold water or ice to help speed recovery if necessary.

G. If possible, monitor the compressor discharge line temperature. Operating on a low charge can make the compressor overheat. If the discharge temperature exceeds 200°F, close the gauge manifold high side valve and shut off the system.

H. It is possible the system compressor may shut down on low pressure. If it does, close the manifold gauge high side valve and shut off the system.

I. If the compressor does not overheat and the system does not shut down on low pressure, continue recovery until the low side pressure reaches 5 psig. You do not want the system to run into a vacuum. Close the gauge manifold high side valve and shut off the system.

J. You cannot get all the system refrigerant out this way. You will need to continue recovery using a self-contained recovery unit.

QUESTIONS

(*Circle the letter that indicates the correct answer.*)

1. The system dependent passive method of recovery with the compressor running is effective at recovering:
   A. most of the liquid charge of the refrigeration unit.
   B. mainly refrigerant vapor.
   C. up to 10% of the total charge of the refrigeration unit.
   D. the entire charge of the refrigeration unit.

2. The system dependent method of recovery with the compressor running moves refrigerant from the system to a recovery cylinder
   A. Using a compressor in the recovery unit
   B. Using gravity
   C. Using pressure developed by the system compressor
   D. By allowing the lighter density refrigerant to flow into the recovery cylinder

3. When using the system dependent method of recovery to recover an entire system charge
   A. The system is kept operating until there is no refrigerant in the system
   B. A self-contained recovery unit will be necessary to recover the last portion of refrigerant
   C. Two recovery cylinders are necessary—one for the low side and one for the high side
   D. It helps to heat the recovery cylinder

4. The system dependent method of recovery will proceed more quickly if
   A. You add nitrogen to the recovery cylinder
   B. You leave the cores in any Schrader valves to help develop escape velocity
   C. You heat the recovery cylinder
   D. You cool the recovery cylinder

5. During system dependent refrigerant recovery, you must be careful to not let the pressure in a 4BA300 recovery cylinder exceed
   A. 100 psig
   B. 250 psig
   C. 300 psig
   D. 400 psig

6. During system dependent recovery you open the high side manifold gauge and the high side gauge pressure reading drops. This is because
   A. The cylinder pressure is lower than the system pressure
   B. There is a leak in the system
   C. You should not be opening the manifold high side valve during system dependent recovery
   D. The recovery process is complete

LAB 26.2 PACKAGED UNIT RECOVERY

LABORATORY OBJECTIVE
You will demonstrate your ability to recover refrigerant from a packaged air conditioning unit.

***FUNDAMENTALS OF HVACR* TEXT REFERENCE**
Unit 26 Refrigerant Management and the EPA

LABORATORY NOTES
You will operate a packaged air conditioning unit and record its operating pressures. Then you will shut off the unit and recover the refrigerant to the EPA specified level.

REQUIRED TOOLS, EQUIPMENT, AND MATERIALS
Safety glasses and gloves
Manifold gauges and extra refrigerant hose
Schrader core removal tools (2)
Operating packaged air conditioning unit
Refrigerant recovery unit, refrigerant recovery cylinder, and extension cord
Scale

SAFETY REQUIREMENTS
    A. Wear safety goggles and gloves when working with refrigerants. Liquid refrigerant can cause frostbite when in contact with eyes and skin.
    B. Inspect all fittings before attaching hoses.

PROCEDURE
(The following assumes the high side gauge port is on the discharge line, not the liquid line.)

Select Proper Equipment

1. Note the type and amount of refrigerant in the system and record on the Data Sheet 26-2-1.
2. Determine the EPA specified recovery level based on the refrigerant type and amount.
3. Record the required recovery level on the Data Sheet 26-2-1.
4. Choose a recovery Unit, recovery cylinder, and extra hose that are compatible with that refrigerant.
5. Calculate the cylinder safe fill weight.
   Tare + (Water Capacity x 0.8 x Refrigerant Specific Gravity)
6. Weigh the cylinder.
7. Determine the amount of space in the cylinder. Safe fill weight – current cylinder weight. Make sure the amount of space in the cylinder exceeds the total unit charge.

Connecting Equipment

1. Connect Schrader valve core removal tools to the high and low side Schrader valves.
   Note: Some packaged units have CoreMax valves. CoreMax valves provide full flow without removing the core. CoreMax valves look similar to Schrader valves, but standard core tools will NOT work with CoreMax valves. Hoses with core depressors MUST be used on CoreMax valves. Figure 26-2-1 shows a comparison of the two valve types.

Figure 26-2-1 CoreMax valve

 Figure 26-2-2 Connections on valve core tools

CoreMax Valve

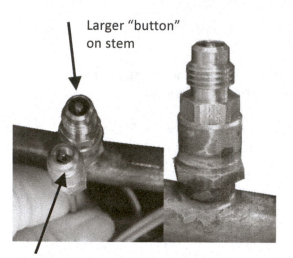

Larger "button" on stem

Schrader Valve   Large Base

Valve Core Tools

2. Connect the high and low side hoses from your gauges to the valve core tools, Figure 26-2-2.
3. Operate the unit long enough for the pressures to stabilize and the compressor to warm up.
4. Read the operating pressures and record on the Data Sheet 26-2-1.

5. Connect the recovery unit and recovery cylinder as shown in Figure 26-2-3.
   A. Connect the middle hose of your gauges to the inlet of the recovery unit.
   B. Connect the outlet of the recovery unit to the recovery cylinder vapor valve.
   C. Invert the cylinder and place it on a scale.
6. Purge the connections to the recovery unit and the recovery unit.
   (This does NOT involve operating the recovery unit or air conditioning unit. The exact procedure to purge the recovery unit and hoses can vary by model. If the recovery unit was last used on the same type of refrigerant, it may not need purging, but the hoses will still need purging.)
7. Zero the scale without the cylinder on it.
8. Place the recovery cylinder on the scale and record the cylinder weight.
9. Zero the scale with the cylinder on it.
10. Set the recovery unit to recover vapor. *(Exact valve positions will vary by model.)*
11. Open the valve on the recovery cylinder.
12. Open both manifold hand wheels.

Figure 26-2-3 Recovery unit connections

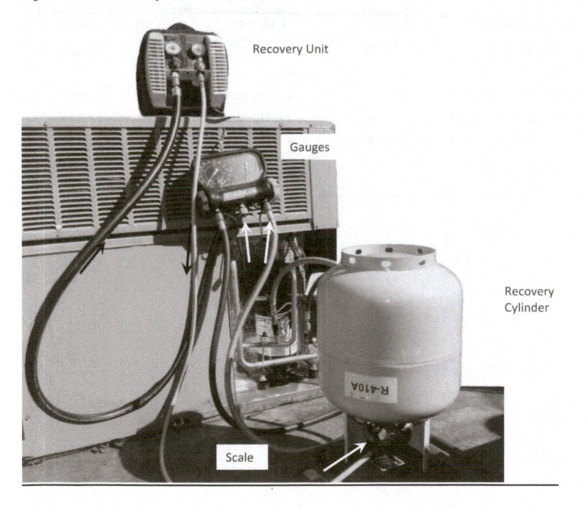

Recovering the Refrigerant

1. Start the recovery unit.
2. Monitor the cylinder weight to ensure that it does not exceed the safe fill level.
3. Operate until the pressure in the air conditioning unit is slightly lower than the EPA specified level. See Table 26-2-1 for EPA specified recovery levels.
4. Close the hand wheels on your gauges and wait to see if the recovery level will hold.
5. Set the recovery unit to purge. *(Exact valve positions will vary by model.)*
6. Operate the recovery unit until it shuts itself off or the inlet gauges indicate a vacuum.
7. Close all valves including the hand wheels on your gauges, the inlet and outlet valves of the recovery unit, and the valve on the recovery cylinder.
8. Turn off the recovery unit.
9. Check and record the amount of refrigerant recovered.
10. Remove Cylinder from the scale and zero the scale.
11. Weigh and record the final cylinder weight.

Table 26-2-1 EPA Specified Recovery Levels

Type of Appliance	Refrigerant Examples	Manufactured before 11/15/1993	Manufactured on or after 11/15/1993
Very high pressure	R-13, R-503	0 psig	0 psig
High pressure < 200 lb	R-22	0 psig	0 psig
High pressure ≥ 200 lb	R-22	4 in Hg vacuum	10 in Hg vacuum
Medium pressure < 200 lb	R-12, R-502	4 in Hg vacuum	10 in Hg vacuum
Medium pressure ≥ 200 lb	R-12, R-502	4 in Hg vacuum	15 in Hg vacuum
Low-pressure appliance	R-11, R-123	25 in Hg vacuum	25 mm Hg absolute pressure

Data Sheet 26-2-1

Package Unit System Refrigerant Recovery	
Refrigerant Type	
Refrigerant Quantity	
Required EPA Recovery Level	
Operating Suction Pressure	
Operating Discharge Pressure	
Cylinder Tare Weight (TW)	
Cylinder Gross Weight	
Water Capacity (WC)	
Safe refrigerant fill weight Tare + (WC x 0.8 x Ref SG)	
Actual Cylinder Weight before Recovery	
Amount of Room in Cylinder (Safe Fill Weight – Actual Weight)	
Amount of Refrigerant Recovered	
Final Cylinder Weight	

LAB 26.3 SPLIT SYSTEM RECOVERY

LABORATORY OBJECTIVE
You will demonstrate your ability to recover refrigerant from a split system air conditioning unit.

FUNDAMENTALS OF HVACR TEXT REFERENCE
Unit 26 Refrigerant Management and the EPA

LABORATORY NOTES
You will operate a split system air conditioning unit and record its operating pressures. Then you will shut off the unit and recover the refrigerant to the EPA specified level.

REQUIRED TOOLS, EQUIPMENT, AND MATERIALS
Safety glasses and gloves
Manifold gauges and extra refrigerant hose
Schrader core removal tools (2)
Operating split system air conditioning unit
Refrigerant recovery unit, refrigerant recovery cylinder, and extension cord
Scale

SAFETY REQUIREMENTS
   A. Wear safety goggles and gloves when working with refrigerants. Liquid refrigerant can cause frostbite when in contact with eyes and skin.
   B. Inspect all fittings before attaching hoses.

PROCEDURE
Select Proper Equipment

   1. Note the type and amount of refrigerant in the system and record on the Data Sheet 26-3-1.
   2. Determine the EPA specified recovery level based on the refrigerant type and amount. Table 26-3-1 shows the EPA specified recovery levels.
   3. Record the required recovery level on Data Sheet 26-3-1.
   4. Choose a recovery Unit, recovery cylinder, and extra hose that are compatible with that refrigerant.
   5. Calculate the cylinder safe fill weight.
      Tare + (Water Capacity x 0.8 x Refrigerant Specific Gravity)
   6. Determine the amount of space in the cylinder.
      Space = Safe fill weight – current cylinder weight.
      Make sure the amount of space in the cylinder exceeds the total unit charge.

230

Connecting Equipment

1. Connect Schrader valve core removal tools to the high and low side Schrader valves.
2. *Note:* Some commercial split systems have CoreMax valves. CoreMax valves provide full flow without removing the core. CoreMax valves look similar to Schrader valves, but standard core tools will NOT work with CoreMax valves. Hoses with core depressors MUST be used on CoreMax valves.
3. Connect the gauges, recovery unit, and recovery cylinder as shown in Figure 26-3-1.
   - A.   Connect core tools to the system high and low side Schrader valves.
   - B.   Connect the high and low side hoses from the gauge manifold to the core tools.
   - C.   Connect the large 3/8-inch hose on the manifold to the recovery unit inlet.
   - D.   Connect the recovery unit outlet to the vapor valve on the recovery cylinder.
4. Operate the unit long enough for the pressures to stabilize and the compressor to warm up.
5. Read the operating pressures and record on the Data Sheet 26-3-1.
6. Purge the connections to the recovery unit and the recovery unit.
   (This does NOT involve operating the recovery unit or air conditioning unit. The exact procedure to purge the recovery unit and hoses can vary by model. If the recovery unit was last used on the same type of refrigerant, it may not need purging, but the hoses will still need purging.)
7. Zero the scale without the cylinder on it.
8. Invert the recovery cylinder and place it on the scale.
9. Record the cylinder weight on Data Sheet 26-3-1.
10. Zero the scale with the cylinder on it.

Figure 26-3-1 Connections for split-system recovery

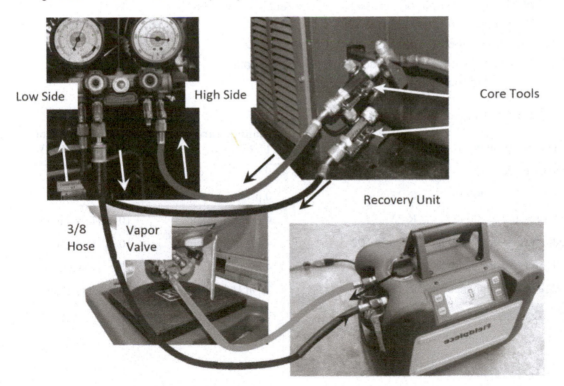

Recovering the Refrigerant

1. Set the recovery unit to recover liquid. *(Exact valve positions will vary by model.)*
2. Fully open the vapor valve on the recovery cylinder with the cylinder inverted on the scale.
3. Crack open only the high side manifold hand wheel, leaving the low side hand wheel closed.
4. Start the recovery unit.
5. You are throttling in liquid through the gauges and/or recovery unit valves.
   Listen to the recovery unit as it recovers the liquid. It will start to complain if you feed too much liquid in at one time.
6. Monitor the cylinder weight to ensure that it does not exceed the safe fill level.
7. Operate in liquid recovery mode until there is no more liquid refrigerant.
   (Connections at the gauge ports will no longer be cold.)
8. Set the recovery unit to recover vapor. *(Exact valve positions will vary by model.)*
9. Open both manifold valve hand wheels.
10. Operate until the pressure in the air conditioning unit is slightly lower than the EPA specified level. Table 26-3-1 shows the specified EPA recovery levels.
11. Close the hand wheels on your gauges, turn off recovery unit, and wait to see if the recovery level will hold.
12. Set the recovery unit to purge. *(Exact valve positions will vary by model.)*
13. Operate the recovery unit until it shuts itself off or the inlet gauges indicate a vacuum.
14. Close all valves including the hand wheels on your gauges, the inlet and outlet valves of the recovery unit, and the valve on the recovery cylinder.
15. Turn off the recovery unit.
16. Check and record the amount of refrigerant recovered.
17. Remove Cylinder from the scale and zero the scale.
18. Weigh and record the final cylinder weight.

Table 26-3-1 EPA Specified Recovery Levels

Type of Appliance	Refrigerant Examples	Manufactured before 11/15/1993	Manufactured on or after 11/15/1993
Very high pressure	R-13, R-503	0 psig	0 psig
High pressure < 200 lb	R-22	0 psig	0 psig
High pressure ≥ 200 lb	R-22	4 in Hg vacuum	10 in Hg vacuum
Medium pressure < 200 lb	R-12, R-502	4 in Hg vacuum	10 in Hg vacuum
Medium pressure ≥ 200 lb	R-12, R-502	4 in Hg vacuum	15 in Hg vacuum
Low-pressure appliance	R-11, R-123	25 in Hg vacuum	25 mm Hg absolute pressure

Data Sheet 26-3-1

Split System Refrigerant Recovery	
Refrigerant Type	
Refrigerant Quantity	
Required EPA Recovery Level	
Operating Suction Pressure	
Operating Discharge Pressure	
Cylinder Tare Weight (TW)	
Cylinder Gross Weight	
Water Capacity (WC)	
Safe Refrigerant Fill Weight Tare + (WC x 0.8 x Ref SG)	
Actual Cylinder Weight before Recovery	
Amount of Room in Cylinder (Safe Fill Weight – Actual Weight)	
Amount of Refrigerant Recovered	
Final Cylinder Weight	

LAB 26.4 PUSH-PULL REFRIGERANT RECOVERY

LABORATORY OBJECTIVE
The student will demonstrate the push-pull procedure for recovering liquid refrigerant from a refrigeration system using a self-contained recovery unit.

***FUNDAMENTALS OF HVACR* TEXT REFERENCE**
Unit 26 Refrigerant Management and the EPA

LABORATORY NOTES
This lab is intended to help students practice the removal of refrigerant from a system using the self-contained active method of liquid recovery. The liquid is forced out of the disabled unit using the recovery machine to lower the pressure in the recovery cylinder and increase the pressure in the disabled unit. This causes rapid movement of the liquid. Push-pull recovery is primary used for large commercial systems. The advantage of the liquid recovery method of transfer is that liquid recovery is much faster than the vapor recovery method. However, the final evacuation still must be done by the vapor recovery method.

REQUIRED TOOLS, EQUIPMENT, AND MATERIALS
Gloves & goggles
Recovery cylinder & hoses
Recovery unit
Refrigerant scale
Disabled refrigeration system

SAFETY REQUIREMENTS
    A. Wear safety goggles and gloves when working with refrigerants. Liquid refrigerant can cause frostbite when in contact with eyes and skin.
    B. Use low loss hose fittings to reduce refrigerant spray when connecting and disconnecting refrigeration hoses. Inspect all fittings before attaching hoses.

PROCEDURE
Step 1 – Recovery Equipment Familiarization

Familiarize yourself with the refrigerant recovery unit that will be used for the lab and carefully read the instructions that came with the unit. It is important to remember that the recovery unit utilizes its own built in compressor (self-contained). Familiarize yourself with the refrigerant recovery cylinder. Notice which valve is the vapor valve and which valve is the liquid valve. Find the cylinder tare weight and water capacity (WC).

***Step 2* – Hose Connections**

A. Connect the recovery unit and recovery cylinder to the system as shown in Figure 26-4-1.
   a. Connect and tighten both ends of a refrigerant hose from the liquid service valve of the disabled unit to the liquid valve (right-hand) on the recovery cylinder.
   b. Connect and tighten both ends of a refrigerant hose from the vapor valve (left-hand) on the recovery cylinder to the inlet of the recovery unit.
   c. Connect and tighten both ends of a refrigerant hose from the outlet of the recovery unit to the discharge service valve of the disabled refrigeration system.
B. Place the cylinder on top of a refrigerant scale.

Figure 26-4-1 Push-pull recovery connections

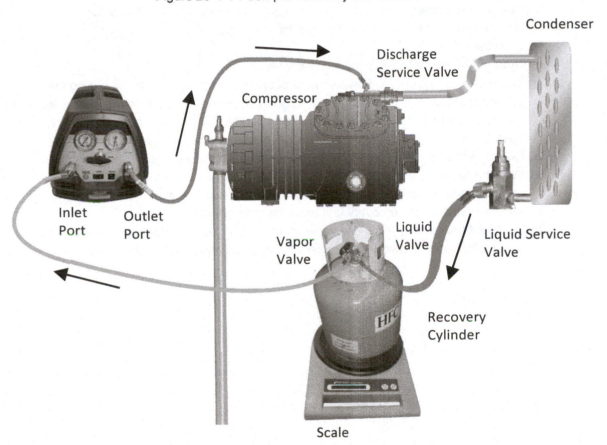

235

Step 6 – Purge the Lines of Air (assuming some refrigerant in recovery cylinder)

A. Place the valves of the recovery unit in the position to recover refrigerant.
B. Loosen the refrigerant hose on the discharge service valve.
C. Crack open the recovery cylinder vapor valve.
D. Air should escape through the hose end on the discharge service valve.
   Purge for 3–5 seconds and then re-tighten the hose.
E. Crack open the cylinder liquid valve.
F. Loosen the hose end on the liquid service valve for 1–2 seconds to purge air out of that
   hose, then re-tighten the hose.
   CAUTION: Re-tighten the hose immediately if liquid escapes.
   You MUST wear gloves and safety glasses!

Step 7 – Recovery

A. Fully open both cylinder valves.
B. Put the discharge service valve and the liquid service valve in mid-position.
C. Put the recovery unit valves in position for vapor recovery.
D. Zero the scale.
E. Start the recovery unit and monitor the scale.
F. The recovery unit will be drawing vapor out of the recovery cylinder, reducing the
   cylinder pressure. Simultaneously, the recovery unit will be pushing high-pressure vapor
   into the system through the discharge service valve. The vapor entering the system will
   push the liquid out of the liquid service valve and into the recovery cylinder.
G. Continue this process until the scale indicates that no more liquid is being removed.
H. With the recovery unit continuing to run, close the cylinder vapor valve.
I. Operate the recovery unit until it shuts off on the low-pressure cut-out, or pulls into a
   vacuum. Shut off the recovery unit and close the cylinder valves.
J. Backseat the discharge service valve and the liquid service valve and connect the
   system for vapor recovery to complete the system recovery, Figure 26-4-2.

Figure 26-4-2 Vapor recovery connections

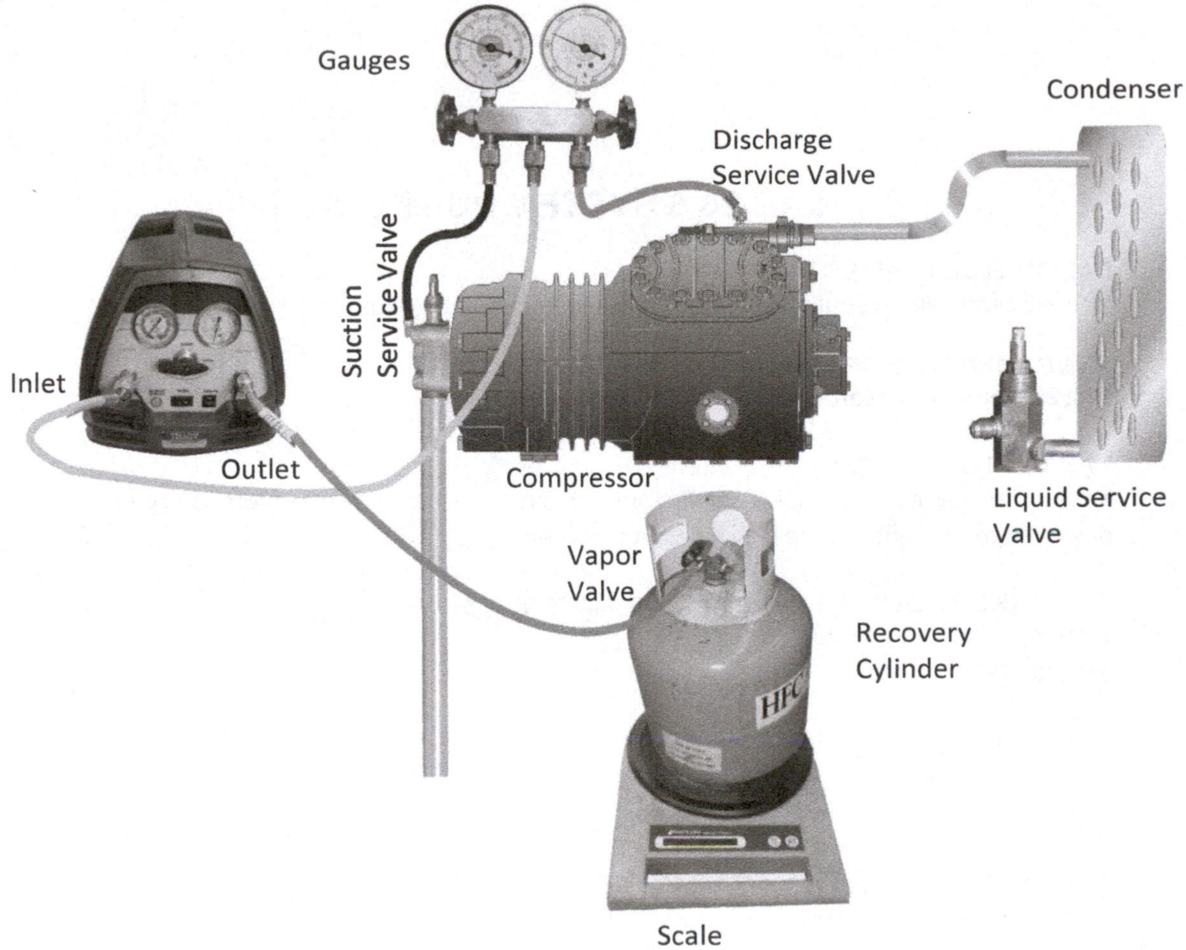

QUESTIONS

(Circle the letter that indicates the correct answer.)

1. The push-pull recovery procedure will recover about:
   A. 45% of the entire charge of the refrigeration unit.
   B. 95% of the entire charge of the refrigeration unit.
   C. 75% of the entire charge of the refrigeration unit.
   D. the entire charge of the refrigeration unit.

2. A push-pull liquid recovery:
   A. is always followed by a vapor recovery.
   B. is followed by a vapor recovery about 50% of the time.
   C. directs liquid into the inlet of the recovery unit.
   D. is slower than a vapor recovery.

3. When performing a push-pull liquid recovery:
   A. a sight glass should be installed in the discharge line of the recovery unit.
   B. a drier should be installed in the discharge line of the recovery unit.
   C. a sight glass should be installed in the liquid line connected to the recovery cylinder.
   D. All of the above are correct.

LAB 26.5 SYSTEM PUMPDOWN

LABORATORY OBJECTIVE
You will demonstrate your ability to recover all the system refrigerant into the high side.

FUNDAMENTALS OF HVACR TEXT REFERENCE
Unit 26 Refrigerant Management and the EPA

LABORATORY NOTES
You will front-seat the king valve or liquid line service valve, operate the system, and pump down the refrigerant into the high side of the system.

REQUIRED TOOLS, EQUIPMENT, AND MATERIALS
Safety glasses and gloves
Adjustable jaw wrench
6- in-1
Refrigeration gauges
Refrigeration valve wrench
Refrigeration system with king valve or liquid line service valve

SAFETY REQUIREMENTS
A. Wear safety glasses and gloves when working on refrigeration systems
B. It is illegal to intentionally vent refrigerants comprised of CFC's and HCFCs.
C. If possible, monitor the high side pressure while pumping the system down.
D. This depends on the system valve configuration.

PROCEDURE
1. Install your gauges with the high-pressure gauge should be installed on the king valve or the liquid line service valve.
2. Purge the air from the hoses and gauge manifold.
3. Front-seat the king valves or liquid line service valve.
4. Operate the unit.
5. The low side should pull into a vacuum.
6. The high side will either pull down to 0 psig, or its pressure will remain, depending upon the orientation of the king valve.
7. Turn off the unit and show the instructor the system after it has pumped down.

LAB 27.1 TESTING WITH NITROGEN AND A TRACE GAS

LABORATORY OBJECTIVE

The student will demonstrate the correct procedure to use nitrogen and a trace gas to check for leaks in a refrigeration system.

FUNDAMENTALS OF HVACR TEXT REFERENCE

Unit 27 Refrigerant Leak Testing

LABORATORY NOTES

After a system has been repaired and before the final charge has been installed, the system needs to be leak tested. This is done by pressurizing the system with an inert gas such as nitrogen or carbon dioxide mixed with a trace amount of 5% or 10% of a non-ozone-depleting refrigerant. You are adding a small amount of refrigerant a leak detector can detect. The only two methods that work with pure nitrogen are soap and bubbles or an ultrasonic leak detector.

Oxygen should not be used, since it can cause an explosion.

Under current EPA regulations, the mixture of a non-ozone-depleting refrigerant and inert gas is not considered to be refrigerant. Therefore, it can be vented to atmosphere following the leak check. However, the EPA does not look favorably on the addition of the inert gas to the refrigerant already present in a system. This is the reason why you must completely evacuate the remaining refrigerant in the system before introducing the trace gas.

This lab is intended to allow student practice on basic leak detection procedures. Since it is not the intention to recharge the system at the completion of this lab, an unrepaired leak could be set up and left for students to find. On such a system it would be impossible to pass a micron evacuation or any vacuum pressure drop test.

REQUIRED TOOLS, EQUIPMENT, AND MATERIALS

Gloves & goggles
Service valve wrench
Gauge manifold
Electronic leak detector
Cylinder of nitrogen gas with regulator
Cylinder of non-refrigerant
Refrigeration system that has a leak

SAFETY REQUIREMENTS

A. Wear safety goggles and gloves when working with refrigerants. Liquid refrigerant can cause frostbite when in contact with eyes and skin.
B. NEVER pressurize a system with oxygen to check for leaks.

PROCEDURE

Step 1
Recover any refrigerant remaining in the system before proceeding further.

Step 2
Connect the non-ozone-depleting refrigerant cylinder and gauge manifold as shown in Figure 27-1-1.

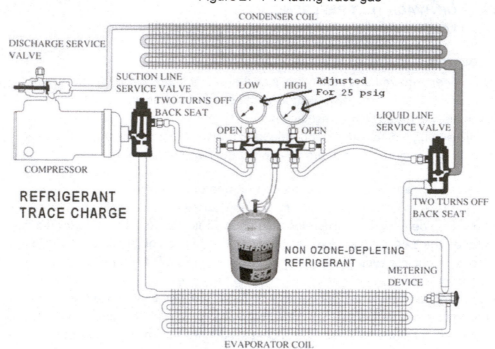

Figure 27-1-1 Adding trace gas

Step 3 – Admit the Trace Gas

A. The center hose is connected to a cylinder of non-ozone-depleting refrigerant.

B. Purge any air from the hose connection to the gauge manifold prior to introducing the refrigerant into the system.

C. After the lines have been purged, slowly admit the refrigerant trace into the system through both the high side and low side valves as a "trace charge" of 25 psig.

D. When the trace charge has been added close the cylinder valve and the gauge manifold valves.

Figure 27-1-2 Adding nitrogen

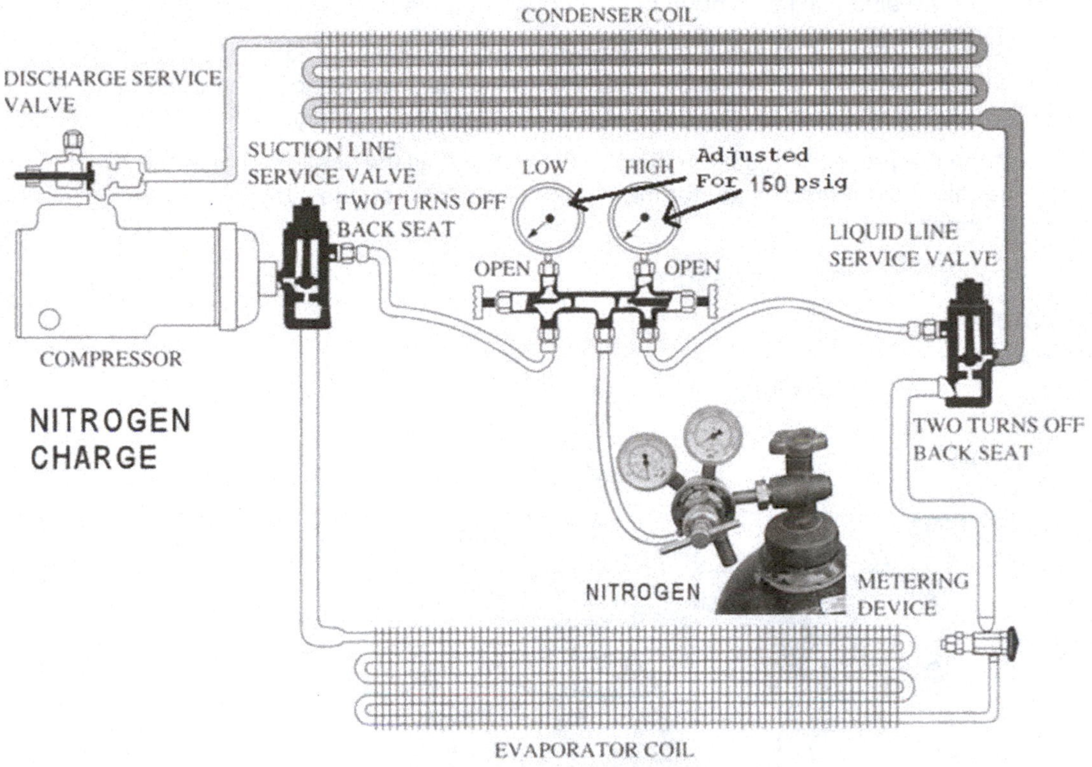

Step 4 – Pressurize with Nitrogen

   A. Connect the nitrogen cylinder and gauge manifold as shown in Figure 27-1-2.

   B. Make sure that the nitrogen cylinder pressure regulator is turned all the way out counterclockwise. Stand to the side and slowly open the cylinder valve counterclockwise until it is fully open, back-seated. **Do not stand in front of the regulator "T" handle.**

   C. The tank pressure should register on the regulator high pressure gauge. The pressure in the tank can be in excess of 2,000 psi.

   D. Slowly turn the regulator "T" handle in (clockwise) until the regulator adjusted pressure reaches approximately 150 psig.

   E. Purge the middle hose, crack the gauge manifold valves, and add nitrogen into the system through both sides of the system until the system is pressurized to 150 psig.

Step 5 – Leak Test

   A. With the system pressure now at 150 psig, you may begin checking for leaks using an electronic leak detector.

   B. Record all leak locations and consult with the Lab Instructor before making any repairs.

Step 6 – Release the Test Gas

    A. After the all of the leaks have been located, release the test gas from the system.

    B. Close the shut off valve on the nitrogen cylinder as well as the gauge manifold valves. Carefully disconnect the hose from the nitrogen cylinder and open the gauge manifold valves to bleed the system pressure to 0 psig.

    C. Once the pressure has been bled, you may back all the way off on the "T" handle (counterclockwise) for the pressure regulator on the nitrogen cylinder.

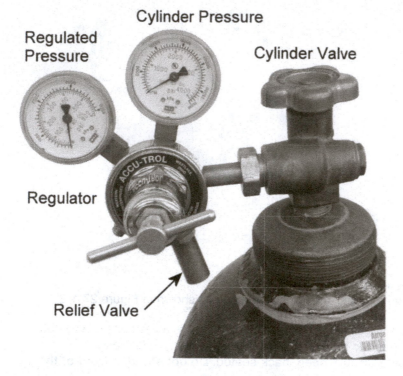

Figure 27-1-3

QUESTIONS

(Circle the letter that indicates the correct answer.)

1. According to EPA regulations:
    A. nitrogen can be added to a charged system for leak testing.
    B. pure CFCs or HCFCs can be released after leak testing.
    C. nitrogen mixed with 10% non-ozone-depleting refrigerant can be used for leak testing.
    D. pure nitrogen is not used because of the explosion hazard.

2. To check for refrigerant leaks using a mixture of HFC refrigerant and nitrogen you may use:
    A. a halide leak detector.
    B. an electronic leak detector.
    C. only soap bubbles.
    D. a sulfur candle.

3. If you use 100% nitrogen for leak testing:
   A. a halide leak detector must be used.
   B. a soap bubble test must be performed.
   C. an electronic leak detector must be used.
   D. you may create an explosion hazard.

4. An ultrasonic type leak detector:
   A. can detect any gas leaking through a small hole.
   B. can only detect larger leaks.
   C. will only work with specially formulated test gas.
   D. requires hearing protection because of the intense high frequency sounds it emits.

5. An electronic leak detector works by:
   A. drawing sample gas over an electronic sensor.
   B. hearing the escape of gas from a small hole.
   C. creating resonant vibrations in the escaping refrigerant.
   D. sensing the magnetic field created by escaping refrigerant.

LAB 27.2 DETECTING LEAKS USING NITROGEN AND SOAP BUBBLES

LABORATORY OBJECTIVE
You will demonstrate your ability to find leaks in refrigeration piping and components using nitrogen and soap bubbles.

FUNDAMENTALS OF HVACR TEXT REFERENCE
Unit 27 Refrigerant Leak Testing

LABORATORY NOTES
You will safely pressurize a system with nitrogen and use soap bubbles to locate leaks in the system.

REQUIRED TOOLS, EQUIPMENT, AND MATERIALS
Safety glasses and gloves
Manifold gauges and refrigeration wrench
Refrigeration system
Nitrogen cylinder and nitrogen pressure regulator
Soap bubbles

SAFETY REQUIREMENTS
   A. Wear safety glasses
   B. Nitrogen should only be used with a regulator
   C. Do not exceed system low side design pressure

PROCEDURE
1. Install your gauges on the high and low sides of the system you are checking.
2. If the pressure is above 0 psig you will need to recover the refrigerant from the system.
3. Find the system's testing pressures on the data plate.
4. Connect the outlet of the nitrogen regulator to the middle hose on your gauges.
5. Adjust the nitrogen regulator T-handle out counterclockwise until there is no spring pressure on it. *Note:* It will come completely out if you adjust it out too far.
6. Stand behind the nitrogen cylinder on the side opposite the regulator and crack the valve open counterclockwise. Once the valve has been cracked open you can continue to open the valve on the nitrogen cylinder the rest of the way.
7. Adjust the T-handle on the regulator in clockwise slowly until the pressure on the regulator gauge is equal to the low side testing pressure of the system.

8. Crack open gauges and allow nitrogen pressure to enter the system until the pressure in the system is at least 50 psig but not over the system's low side testing pressure.
9. Close your gauges and listen. Really large leaks will make an audible hiss.
10. If you do not hear any leaks, apply soap bubbles to all areas likely to have a leak. Be sure to check: Brazed and soldered connections, Mechanical connections, Gaskets and Service valves.
11. Really large leaks will likely just blow the soap solution away because of the velocity of the gas leaving the hole. You should be able to hear these leaks. Moderately large leaks will blow visible bubbles, and very small leaks will form a white froth or foam around the leak after several minutes.
12. Show the instructor the leaks you have located and be prepared to demonstrate how you found them.

LAB 27.3 DETECTING LEAKS USING ULTRASONIC LEAK DETECTOR

LABORATORY OBJECTIVE

The purpose of this lab is to demonstrate your ability to find leaks in refrigeration piping and components using an ultrasonic leak detector.

FUNDAMENTALS OF HVACR TEXT REFERENCE

Unit 27 Refrigerant Leak Testing

LABORATORY NOTES

You will pressurize a system with nitrogen and use the ultrasonic leak detector to locate leaks in the system.

REQUIRED TOOLS, EQUIPMENT, AND MATERIALS

Safety glasses and gloves
Manifold gauges and refrigeration wrench
Nitrogen cylinder and nitrogen pressure regulator
Refrigeration system
Ultrasonic leak detector

SAFETY REQUIREMENTS

Wear safety glasses and gloves

PROCEDURE

1. Install your gauges on the high and low sides of the system you are checking.
2. If the pressure is above 0 psig you will NOT need to add nitrogen. If the system is empty, continue with steps 3–12. If the system already has pressure, proceed to step 10.
3. Find the system's testing pressures on the data plate.
4. Connect the outlet of the nitrogen regulator to the middle hose on your gauges.
5. Adjust the nitrogen regulator T-handle out counterclockwise until there is no spring pressure on it. *Note:* It will come completely out if you adjust it out too far.
6. Stand behind the nitrogen cylinder on the side opposite the regulator and crack the valve open counterclockwise. Once the valve has been cracked open you can continue to open the valve on the nitrogen cylinder the rest of the way.
7. Adjust the T-handle on the regulator in clockwise slowly until the pressure on the regulator gauge is equal to the low side testing pressure of the system.

8. Crack open you gauges an allow nitrogen pressure to enter the system until the pressure in the system is at least 50 psig but not over the system's low side testing pressure.
9. Close your gauges and listen. Really large leaks will make an audible hiss.
10. Use the ultrasonic detector to check all areas likely to have a leak. Be sure to check:
    a. Braze and solder connections
    b. Mechanical connections
    c. Gaskets
    d. Service valves
11. Really large leaks may not be picked up by the ultrasonic detector, you should be able to hear these leaks with your ears.
12. Show the instructor the leaks you have located and be prepared to demonstrate how you found them.

LAB 27.4 DETECTING LEAKS USING AN ELECTRONIC SNIFFER

LABORATORY OBJECTIVE
The purpose of this lab is to demonstrate your ability to find leaks in refrigeration piping and components using an electronic leak detector.

FUNDAMENTALS OF HVACR TEXT REFERENCE
Unit 27 Refrigerant Leak Testing

LABORATORY NOTES
You will demonstrate your ability to use an electronic leak detector for locating leaks in a refrigeration system.

REQUIRED TOOLS, EQUIPMENT, AND MATERIALS
Safety glasses and gloves
Manifold gauges and refrigeration wrench
Refrigeration system
Electronic leak detector(s)

SAFETY REQUIREMENTS
Wear safety glasses and gloves

PROCEDURE
1. Find the system's refrigerant type and testing pressures on the data plate.
2. Verify that the type of refrigerant the system uses can be detected with leak detector(s) you are using.
3. Install your gauges on the high and low sides of the system you are checking.
4. If the pressure is below 50 psig with the system OFF, you will need to add refrigerant.
5. Connect the center hose of your gauges to a refrigerant cylinder and purge your hoses and gauges.
   a. For compound refrigerants and azeotropes, crack open your gauges and allow refrigerant vapor to enter the system.
   b. For zeotropes, invert the cylinder, crack open your high side gauge and allow liquid refrigerant to enter the system.
   c. Add refrigerant until the pressure in the system is at least 50 psig but not over the system's low side testing pressure.
6. Hold the leak detector away from the system and turn it on.

7. It will squeal or tick as it adjusts itself to the surrounding environment, when the continuous squeal or tick stops it is ready to use.
8. Move the tip slowly around all likely leak spots.
9. Be sure to check brazed joints, mechanical connections, gaskets, and service valves
10. When the detector finds a leak, it will squeal loudly. For really small leaks it may just tick faster.
11. After locating a leak, back the detector away from the leak area and wait for the detector calm down. Go back to the same spot again to verify the leak.
12. Show the instructor the leaks you have located and be prepared to demonstrate how you found them using the leak detectors.

LAB 27.5 DETECTING LEAKS USING A FLUORESCENT DYE AND A BLACK LIGHT

LABORATORY OBJECTIVE

The purpose of this lab is to demonstrate your ability to find leaks in refrigeration piping and components using a fluorescent dye and a black light.

FUNDAMENTALS OF HVACR TEXT REFERENCE

Unit 27 Refrigerant Leak Testing

LABORATORY NOTES

You will add fluorescent dye to a system and use a black light to locate system leaks.

REQUIRED TOOLS, EQUIPMENT, AND MATERIALS

Safety glasses, gloves
Manifold gauges, refrigeration wrench
2 small hand valves with 1/4" flare connections, extra refrigeration hose
Dye tube, black light

SAFETY REQUIREMENTS

Wear safety glasses and gloves

PROCEDURE

Dye Injection

1. Find the system's refrigerant type and testing pressures on the data plate.
2. Verify that the type of refrigerant AND the type of lubricant in the system is compatible with the dye you are using.
3. Connect the small hand valves on each end of the dye tube and make sure that the hand valves are closed.
4. Connect a refrigerant hose from the hand valve on the outlet side of the tube to the suction side of the system.
5. Purge the hose by loosening the end connected to the hand valve for 1 second, releasing a small amount of refrigerant and then tightening the hose.
6. Connect a refrigerant hose from the hand valve on the inlet side of the tube to the high side of the system.
7. Purge the hose by loosening the end connected to the hand valve for 1 second, releasing a small amount of refrigerant and then tightening the hose.
8. Operate the system.

9. After the system has operated a few minutes to establish suction and discharge pressures, open the cylinder valve first on the outlet side of the dye tube, and then on the inlet side of the dye tube. This limits the pressure on the dye tube.
10. Watch the dye in the tube. Let the refrigerant gas continue to flow for 15 seconds after all the dye has left the tube to help flush the dye out of the connecting hose.
11. Close the valve on the tube inlet and wait another 15 seconds to allow the tube pressure to drop to the system suction pressure.
12. Close the valve on the tube outlet, shut off the system, and remove the hoses and tube.
13. You will need to clean up the area where the hoses connect. Spray from the dye will cause a false positive leak indication.

Black Light Procedure
1. The system must operate for several hours with the dye in it before the dye will indicate a leak. A full day is recommended.
2. Large mercury vapor black lights must be plugged I and warmed up.
3. Direct black light on suspected leak areas including:
   a. Brazing and solder connections
   b. Mechanical connections
   c. Gaskets
   d. Service valves
4. Leaks will show up as bright yellow.
5. Show the instructor the leaks you have located and be prepared to demonstrate how you found them.

LAB 28.1 CHANGE VACUUM PUMP OIL

LABORATORY OBJECTIVE
The student will demonstrate how to properly use a vacuum pump.

FUNDAMENTALS OF HVACR TEXT REFERENCE
Unit 28 Refrigerant System Evacuation

LABORATORY NOTES
For this lab exercise there should be a typical vacuum pump for the student to inspect.

REQUIRED TOOLS, EQUIPMENT, AND MATERIALS
Vacuum pump
Vacuum pump oil
Container to catch dirty oil

SAFETY REQUIREMENTS
Wear safety glasses

PROCEDURE
Step 1 – **Vacuum Pump Figures**

Study Figures 28-1-1, 28-1-2, and 28-1-3 to learn the components of different styles of vacuum pumps.

Figure 28-1-1   Figure 28-1-2

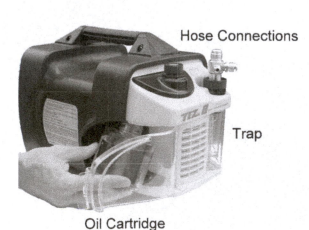

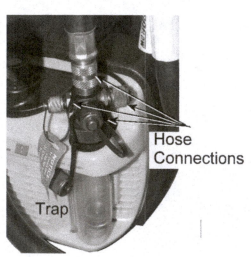

Figure 28-1-3

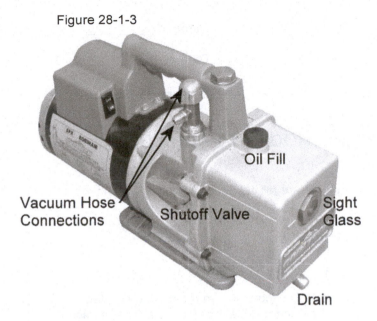

Oil Fill

Vacuum Hose
Connections

Shutoff Valve

Sight
Glass

Drain

Step 2 – **Vacuum Pump Components**

Compare your vacuum pump to Figures 28-1-1, 28-1-2, and Figure 28-1-3 to locate the
following components:

A. Drain

B. Oil Fill

C. Oil Sight Glass

D. Oil Cartridge

E. Shutoff Valve

F. Trap

Step 3 – **Change Vacuum Pump Oil**

Change the vacuum pump oil using the type and quantity of oil recommended by the vacuum
pump manufacturer.

QUESTIONS

1. What is the rating of the vacuum pump in CFM?

2. What is the lowest vacuum that it can pull down to under ideal conditions?

3. How many and what size vacuum hose connections does it have?

4. What type of oil does it use?

5. What quantity of oil does it use?

6. Describe the process for changing the oil.

LAB 28.2 DEEP METHOD OF EVACUATION

LABORATORY OBJECTIVE
The student will demonstrate the correct procedure for the deep method of evacuation of a refrigeration system.

FUNDAMENTALS OF HVACR TEXT REFERENCE
Unit 28 Refrigerant System Evacuation

LABORATORY NOTES
This lab is intended to help students practice the evacuation of a refrigeration system. Proper evacuation of a system will remove non-condensable gases and water while assuring a tight, dry system before charging.

REQUIRED TOOLS, EQUIPMENT, AND MATERIALS
Gloves & goggles
Deep vacuum pump and micron vacuum gauge
Service valve wrench
Gauge manifold, vacuum manifold, or vacuum "Y" fitting
Vacuum hoses
Refrigeration system

SAFETY REQUIREMENTS
Wear safety goggles and gloves when working on refrigeration systems. Oil in the system can become acidic over time and can cause acid burns to the skin and eyes.

PROCEDURE
Step 1
Familiarize yourself with the deep vacuum pump. It is somewhat like an air compressor in reverse. Vacuum pumps are rated according to free air displacement in cubic feet per minute (CFM), or liters per minute (l/pm) in the SI system. The degree of vacuum the pump can achieve is expressed in microns of mercury absolute pressure. One micron is equivalent to 1/1000 of a millimeter of mercury, or approximately 1/25,400 of an inch of mercury. The micron vacuum reading reveals how much pressure remains. A reading of 0 microns would be a perfect vacuum. Deep vacuum pumps achieve levels of 50 microns. A deep vacuum is considered to be a vacuum of 500 microns or lower.

These pressures are too small to read on a standard gauge manifold. Some vacuum pumps include a rough vacuum gauge as shown in Figure 28-2-1, but this is still not adequate for measuring a deep vacuum. A deep vacuum gauge is required, as shown in Figure 28-2-2. Table 28-2-1 compares different vacuum levels in various units of measure.

This is the main purpose of evacuation is to remove non-condensable gasses and to reduce the pressure enough to vaporize any water in the system and then pump it out. Table 28-2-1 shows the dramatic change in the boiling point of water as the pressure decreases.

Figure 28-2-1   Figure 28-2-2

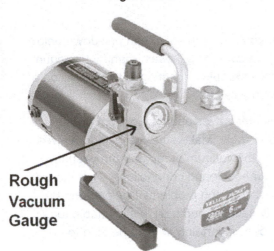

Rough
Vacuum
Gauge

Table 28-2-1 Water Boiling Point at Different Vacuum Levels

Water Boiling Point	Absolute Pressure		Inches of Vacuum
	PSIA	Microns of Mercury	
212°F	14.7	760,000	0
79°F	0.5	25,400	28.9
72°F	0.4*	20,080	29.1*
32°F	0.09*	4,579	29.7*
−12°F	0.009*	500	29.9*
−25°F	0.005*	250	29.91*
−60°F	0.0005*	25	29.92*

* Too small a change to see on a mechanical gauge

Step 2 – System Must Be Empty

There should be no refrigerant in the system or portion of the system to be evacuated. For new split-system installations, the lines and indoor coil will need to be evacuated prior to charging. On repairs, such as compressor replacements, the system must be evacuated prior to charging.

Step 3 – System Leak Test

You should leak test the system with nitrogen before starting the deep evacuation. It is not possible to evacuate a leaky system. Any time spent trying to evacuate a leaky system is wasted. The ability to pull and hold a deep vacuum on a system does prove it is clean and tight.

Step 4 – Verify Vacuum Pump and Connections

To avoid wasting time, it is a good idea to verify that your vacuum setup can pull down below 500 microns. To do this you want to connect your valve core tools, hoses, manifold, vacuum gauge, and vacuum pump as you would on the system except connecting to the valves on a recovery cylinder instead of the system, as shown in Figure 28-2-3.

The core tools should be connected to the cylinder valves and the vacuum gauge should be connected to the side port of one of the core tools. Make sure the cylinder valves are closed and the core tool valves are open. Turn on the vacuum pump and the vacuum gauge. You are basically pulling a vacuum on just your hoses and connections. The vacuum should quickly drop to 200 microns or lower. The most common problems that prevent the setup from pulling down are leaks in the setup or dirty vacuum pump oil. If it is unable to pull down below 500 microns when just evacuating the hoses it certainly will not be able to evacuate a system under 500 microns.

Figure 28-2-3 Core tools and vacuum gauge

Step 5 – Vacuum Pump Connections

There are many ways to make the connections between the vacuum pump, the vacuum gauge, and the system. The method used will depend somewhat upon the equipment available. Keep in mind that the most effective way to speed up evacuation is to reduce restriction in the connections. Fewer connections using, shorter, larger diameter hoses are preferable. Figure 28-2-4 shows how to connect a 4-valve manifold to a vacuum pump using a short, large diameter vacuum hose connected to the 3/8" vacuum port on the manifold. Figure 28-2-5 shows how to connect a large-bore vacuum manifold and hoses to a vacuum pump. Figure 28-2-6 shows how to use a vacuum "Y" fitting to connect the vacuum hoses to the vacuum pump. Core removal tools should be used when connecting to Schrader valves because the Schrader valve cores reduce flow to a trickle, regardless of all other considerations. The vacuum gauge should be connected as close to the system as possible. Core tools help with this, allowing a vacuum gauge to be connected to the side port of the core tool. Figure 28-2-7 shows how to connect the hoses and vacuum gauge to the Schrader valves on a split-system.

Figure 28-2-5 Vacuum pump connected to a large-bore vacuum manifold

Figure 28-2-4 Vacuum pump connected to a four-valve manifold using a short, large diameter vacuum hose.

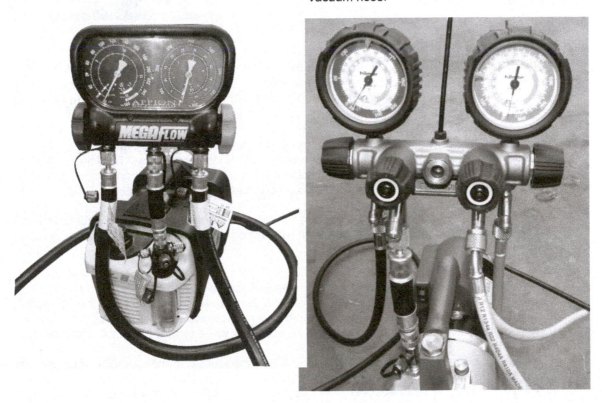

Figure 28-2-6 Vacuum pump connected to vacuum hoses using vacuum rated "Y" fitting

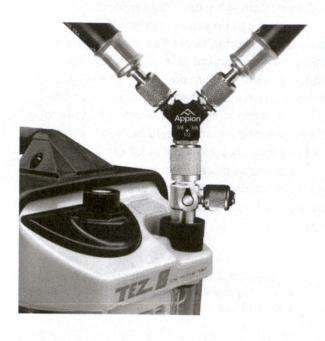

Figure 28-2-7 Core tools and vacuum gauge connected to Schrader valves

Step 6 – Evacuation

A. Open any valves in your vacuum setup so that the vacuum pump is evacuating both sides of the system. Depending upon the equipment being used this would include the vacuum pump isolation valve, the gauge manifold high side and low side valves, the gauge manifold vacuum valve, and the core tool valves.

B. Open the ballast port on vacuum pumps equipped with a ballast valve. Removing one of the unused connections on vacuum pumps with multiple connections will help the vacuum pump start. After the vacuum pump gets started you can cap the extra connection that you opened. For ballasted pups you should close the ballast port after the initial pull-down.

C. Operate the vacuum pump until the vacuum gauge indicates less than 500 microns. In practice, it is a good idea to leave the pump running as long as the vacuum reading is continuing to drop.

D. Valve off the system after the vacuum pulls down. If you are using core tools you can simply close the valves on the core tools. Turn off the vacuum pump after valving off. You should NOT leave a vacuum pump connected to an evacuated system. On vacuum pumps with isolation valves, make sure and close the isolation valve before turning the pump off.

E. The vacuum should hold under 500 microns. Either moisture or trace amounts of refrigerant are indicted by the vacuum increasing above 500 microns but leveling off. Free water in the system will cause a rise to about 20,000 microns. A leak will cause the micron gauge to rise steadily, eventually reaching atmospheric pressure if you wait long enough.

QUESTIONS

(Circle the letter that indicates the correct answer.)

1. Vacuum pumps are rated according to:

   A. amperage.
   B. horsepower.
   C. free air displacement(CFM).
   D. refrigerant type.

2. One micron is equivalent to:

   A. 1/25,400 of an inch of mercury.
   B. 1/100 of an inch of water.
   C. 0.05 psia.
   D. 0.1 psig.

3. A short evacuation hose with a larger interior diameter:

   A. is better for charging than evacuation.
   B. speeds up the evacuation time.
   C. obstructs the flow from the system.
   D. cannot be used for evacuation purposes.

4. Proper evacuation will remove:

   A. air.
   B. water vapor.
   C. non condensable gases.
   D. All of the above are correct.

5. A deep vacuum is considered to be:

   A. 500 microns or less.
   B. 500 microns or more.
   C. 28 inches Hg vacuum.
   D. 29 inches Hg vacuum.

6. The main reason for a deep vacuum is to:

   A. suck out any refrigerant oil in the system.
   B. suck out any water in the system.
   C. suck out any copper oxides.
   D. suck out any refrigerant.

7. A system is considered to be dry and leak free if it can maintain a vacuum of less than:

   A. 1200 microns.
   B. 500 microns.
   C. 20,000 microns.
   D. 29 inches vacuum.

8. A reading of 0 microns would be:

    A. atmospheric pressure.
    B. 0 psig.
    C. a perfect vacuum.
    D. read on the micron gauge at the beginning of the evacuation.

9. At a pressure of 250 microns, water will boil at a temperature of:

    A. 212°F.
    B. 100°C.
    C. 0°F.
    D. −25°F.

10. If the micron gauge begins to rise steadily once the vacuum pump is shut off this usually indicates:

    A. a leak.
    B. that everything is normal.
    C. that the micron gauge is faulty.
    D. the presence of moisture.

11. Free water in the system will cause a rise to about 20,000 microns:

    A. True.
    B. False.

LAB 28.3 TRIPLE EVACUATION

LABORATORY OBJECTIVE
The student will demonstrate the correct procedure for the triple evacuation method of dehydrating a refrigeration system.

FUNDAMENTALS OF HVACR TEXT REFERENCE
Unit 28 Refrigerant System Evacuation

LABORATORY NOTES
This lab is intended to help students practice the evacuation of a refrigeration system. Proper evacuation of a system will remove non-condensable gases and water while assuring a tight, dry system before charging.

The triple evacuation procedure consists of three consecutive evacuations spaced by two dilutions of a dry gas. Nitrogen is preferred but helium and CO_2 can also be used. The clean dry gas will act as a carrier, mixing with system contamination (air & water) and carry it out with the subsequent evacuation. It is a time-consuming procedure, but effective in obtaining a clean dry system.

REQUIRED TOOLS, EQUIPMENT, AND MATERIALS
Gloves & goggles
Deep vacuum pump and micron vacuum gauge
Service valve wrench
Gauge manifold, vacuum manifold, or vacuum "Y" fitting
Vacuum hoses
Cylinder of nitrogen gas with regulator
Refrigeration system

SAFETY REQUIREMENTS
Wear safety goggles and gloves when working on refrigeration systems. Oil in the system can become acidic over time and can cause acid burns to the skin and eyes.

PROCEDURE
Step 1 – The System Should Be Empty
There should be no refrigerant in the system or portion of the system to be evacuated. For new split-system installations, the lines and indoor coil will need to be evacuated prior to charging. On repairs, such as compressor replacements, the system must be evacuated prior to charging.

Step 2 – System Leak Test

You should leak test the system with nitrogen before starting the evacuation. It is not possible to evacuate a leaky system. Any time spent trying to evacuate a leaky system is wasted. The ability to pull and hold a deep vacuum on a system does prove it is clean and tight.

Step 3 – Vacuum Pump Connections

There are many ways to make the connections between the vacuum pump, the vacuum gauge, and the system. The use of a four-valve manifold makes swapping between evacuating the system and adding nitrogen much easier. The vacuum pump is connected to the large 3/8" vacuum hose and the nitrogen regulator is connected to the 1/4" middle hose, Figure 28-3-1.

Step 4 – System Connections

Core removal tools should be used when connecting to Schrader valves because the Schrader valve cores reduce flow to a trickle, regardless of all other considerations. The vacuum gauge should be connected as close to the system as possible. Core tools help with this, allowing a vacuum gauge to be connected to the side port of the core tool. Figure 28-3-1 shows how to connect the hoses and vacuum gauge to the Schrader valves on a split-system.

Figure 28-3-1 Triple evacuation connections

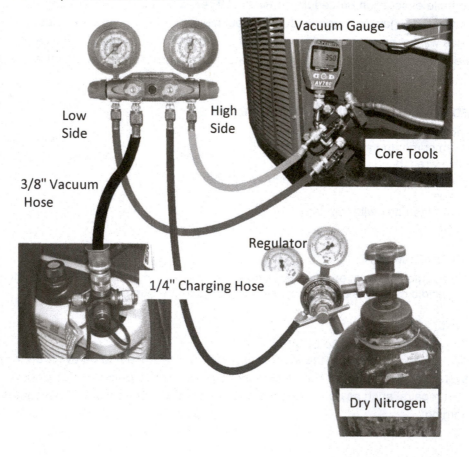

Step 5 – **First Evacuation**

    A. Open any valves in your vacuum setup so that the vacuum pump is evacuating both sides of the system. Depending upon the equipment being used this would include the vacuum pump isolation valve, the gauge manifold high side and low side valves, the gauge manifold vacuum valve, and the core tool valves.

    B. Open the ballast port on vacuum pumps equipped with a ballast valve. Removing one of the unused connections on vacuum pumps with multiple connections will help the vacuum pump start. After the vacuum pump gets started you can cap the extra connection that you opened. For ballasted pups you should close the ballast port after the initial pull-down.

    C. Operate the vacuum pump until the vacuum gauge indicates less than 2000 microns. Record this level in Data Sheet 28-3-1.

       Note that this level varies depending upon manufacturer. Daikin recommends 4000 microns, LG recommends 2000 microns, and Trane recommends 750 microns.

    D. Valve off the system. Close the pump isolation valve as well as the manifold valves. Vacuum pump isolation valves are typically not trustworthy. Turn off the vacuum pump. Wait for 5 minutes and watch to see if the vacuum level holds below 2000 microns. If it does not, start the vacuum pump, open the valves, and pull down again until the vacuum will hold below 2000 microns.

    E. Make sure the manifold vacuum valve is closed, open the manifold charging valve and pressurize the system to 50 psig with nitrogen.

    F. Let the system sit at this pressure for 5 minutes and then release the nitrogen.

Step 6 – **Second Evacuation**
Repeat step 5 but pull down and hold at or below 1000 microns. Again, this level varies by manufacturer. Daikin recommends 1500 microns, LG recommends 1000 microns, and Trane recommends 500 microns.

Step 7 – **Third Evacuation**
Repeat step 5 but pull down and hold at or below 500 microns. This final level is more commonly agreed upon. Most manufacturers consider a vacuum that will hold below 500 microns acceptable.

Data Sheet 28-3-1

Evacuation	Initial Micron Level	Micron Hold Level	Nitrogen Charge	Blotting Period
First Evacuation				
Second Evacuation				
Third Evacuation				

QUESTIONS

(Circle the letter that indicates the correct answer.)

1. Triple evacuation refers to:
   A. using three separate vacuum pumps.
   B. evacuating a refrigeration system three consecutive times.
   C. an evacuation level that increases by thirds.
   D. an evacuation level that decreases by thirds.

2. Dry gases that can be used in a triple evacuation process include:
   A. air.
   B. oxygen.
   C. system refrigerant.
   D. nitrogen.

3. Nitrogen cylinder pressures:
   A. are never above 150 psig.
   B. are generally between 350 to 400 psig.
   C. can be as high as 2,000 psig.
   D. can sometimes be negative (vacuum).

4. Proper evacuation will remove
   A. All system contaminants.
   B. Only on-condensable gases.
   C. Only water.
   D. Only contaminants that can be removed as a gas.

5. The reason for introducing a clean dry gas between evacuations is to:
   A. mix with the system contaminants.
   B. destroy the system contaminants.
   C. pre-charge the system before adding refrigerant.
   D. check for leaks.

LAB 28.4 VACUUM DECAY TEST

LABORATORY OBJECTIVE

The student will demonstrate the correct procedure for the vacuum decay test to check for leaks in a refrigeration system.

FUNDAMENTALS OF HVACR TEXT REFERENCE

Unit 28 Refrigerant System Evacuation

LABORATORY NOTES

To prepare for this lab, students must complete an evacuation procedure first. The timed vacuum pressure drop test is typically done at the conclusion of a deep evacuation or a triple evacuation to verify the system is free of contaminants and does not leak.

REQUIRED TOOLS, EQUIPMENT, AND MATERIALS

Gloves and goggles

Vacuum pump

Gauge manifold and service valve wrench

Refrigeration system

SAFETY REQUIREMENTS

Wear safety goggles and gloves when working on refrigeration systems. Oil in the system can become acidic over time and can cause acid burns to the skin and eyes.

PROCEDURE

Step 1 – Evacuation

Perform a deep evacuation on the assigned refrigeration system. Refer to Lab 28.2 for details on performing a deep evacuation.

Step 2 – Valving off

After performing the evacuation, close both gauge manifold valves and the vacuum pump isolation valve. Turn off the vacuum pump. Record the ultimate vacuum level accomplished in Data Sheet 28-4-1.

Step 3 – Waiting

Wait 10 minutes and see if the micron level quits rising. The vacuum normally rises after valving off, but the rise should level off. Ideally you want to see the vacuum hold under 500 microns. A small vacuum level rise simply indicates that the vacuum in the system is equalizing. The pressure near where the hoses are connected is lower than further back in the system. This very slight pressure difference is simply equalizing. However, a rise to over 1000 microns might indicate moisture. It could also simply indicate a small amount of refrigerant trapped in the compressor oil, a filter drier, an accumulator, or a receiver. A continued rise that does not level off indicates a leak.

Data Sheet 28-4-1

Initial pulldown Micron level	Micron level after 10 minutes

QUESTIONS

(Circle the letter that indicates the correct answer.)

1. A vacuum decay test:
   A. will be able to pinpoint the location of a leak.
   B. will be able to provide an approximate location of a leak.
   C. indicates that there is a leak somewhere.
   D. will verify the accuracy of the vacuum pump.

2. A vacuum leak can be detected by:
   A. an electronic leak detector.
   B. a halide torch.
   C. bubble test.
   D. an ultrasonic type leak detector.

3. Ideally, the vacuum should hold under:
   A. 500 microns.
   B. 50 microns.
   C. 10 microns.
   D. 2 microns.

4. When a system is evacuated:
   A. water in the system will vaporize.
   B. water vapor in the system will condense.
   C. refrigeration oil will vaporize.
   D. refrigeration oil will condense.

5. Compound gauges:
   A. can only indicate vacuum in inches of mercury.
   B. can read vacuum in microns.
   C. are the preferred method to check the system for vacuum.
   D. are more accurate because of their compound accuracy.

6. If the Schrader valve core is pulled from a service port:
   A. a vacuum can be pulled much quicker.
   B. the vacuum takes longer.
   C. an evacuation is no longer necessary.
   D. the level of system evacuation can be determined by direct inspection.

7. After valving off, a continual rise in the micron reading on the vacuum gauge indicates:
   A. The presence of water in the system.
   B. Small amounts of refrigerant dissolved in the compressor oil.
   C. The vacuum in the system is equalizing.
   D. The system has a leak.

8. A system is evacuated down to 250 microns and valves off. The reading on the micron vacuum gauge rises to 400 microns and then holds. This most likely indicates:
   A. That the system is dry and does not leak.
   B. That the system leaks.
   C. That the system is still contaminated with moisture.
   D. That the system has non-condensable gas in it.

LAB 29.1 WEIGHING IN LIQUID CHARGE WITH COMPRESSOR OFF

LABORATORY OBJECTIVE

The student will demonstrate the correct procedure for performing a liquid charge on a small refrigeration unit using a digital scale.

FUNDAMENTALS OF HVACR TEXT REFERENCE

Unit 29 Refrigerant System Charging

LABORATORY NOTES

Liquid charging is always much faster than charging with vapor. In this lab you will be charging liquid refrigerant into the high side of an evacuated system with the compressor off. Prior to charging, the system must be leak tested and evacuated. The refrigerant will be drawn into the evacuated system due to the difference in pressure. The liquid charge sometimes stops before the full charge is completed. In this case, the charge must be completed using the vapor method for compounds and azeotropes, or by throttling in small amounts of liquid refrigerant in the case of zeotropic refrigerant. Refer to Lab 29.2 for more information on vapor charging or Lab 29.3. For more information on charging with zeotropic refrigerant.

REQUIRED TOOLS, EQUIPMENT, AND MATERIALS

Safety glasses and gloves
Gauge Manifold, extra refrigeration hose, and service valve wrench
Digital scale
Cylinder of refrigerant
Operating refrigeration system

SAFETY REQUIREMENTS

A. Wear safety goggles and gloves when working with refrigerants. Liquid refrigerant can cause frostbite when in contact with eyes and skin.

B. Use low loss hose fittings to reduce refrigerant spray when connecting and disconnecting refrigeration hoses. Inspect all fittings before attaching hoses.

PROCEDURE

Step 1 – Identify Refrigerant

The refrigerant type and required amount of charge can be found on the refrigeration system nameplate as shown below in Figure 29-1-1.

Figure 29-1-1 System data plate

M/N XP19 – 024 – 230 – 01	
S/N 5805F38025	
CONTAINS HFC – 410A	DESIGN PRESSURE
FACTORY CHARGE	HI 446 PSIG
12 LBS 0 OZS	LO 236 PSIG
ELECTRICAL RATING	NOMINAL VOLTS: 208/230

Step 2 – Connections

   A. Connect the manifold gauges to the high and low side of the system and connect the middle hose to the refrigerant cylinder, as shown in Figure 29-1-2.

   B. Purge the air from the hoses and manifold.

   C. Normally the refrigerant cylinder must be turned upside down (inverted) for liquid. It is a good practice to purge the hoses and fully open the cylinder valve prior to turning it over.

Figure 29-1-1

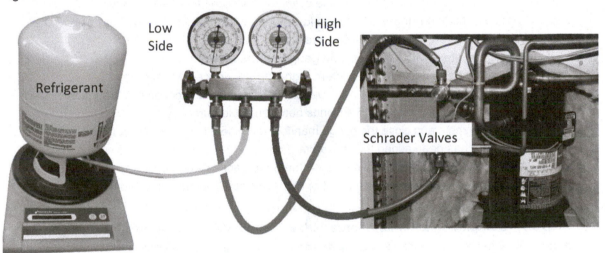

Low Side   High Side

Refrigerant

Schrader Valves

Digital Scale

Step 3 – Charging

   A. Zero the scale after inverting the cylinder on the scale. You may need to wait a minute for the scale reading to stabilize.

   B. Open the manifold high side valve and leave the low side manifold valve closed. You do NOT want to dump liquid into the low side! The system should already have been evacuated and in a vacuum. When the high side service valve is opened, the refrigerant should flow freely into the high side of the system as a liquid.

C. Watch the scale display as you charge. The scale should be displaying the amount of refrigerant that has been removed from the cylinder. Typically, it will display a negative sign in front of the number to let you know that much weight has been removed.

D. Start close the manifold high side valve when the charge gets within a few ounces, but leave it cracked open.

E. Close the manifold high side valve completely once the scale indicates that the charge is complete.

F. Record the amount of liquid refrigerant added on Data Sheet 29-1-1.

Procedure to Complete a Partial Charge

The full charge can frequently be achieved in liquid with the system off for with small to medium sized residential systems. Larger systems often require additional charge because the pressure in the system rises so quickly that it equals the cylinder pressure before the charge is complete. If the full charge is not complete after adding the liquid, then the charge must be completed with the compressor operating to reduce the low side pressure. The specific procedure used varies depending upon the type of refrigerant. Zeotropic refrigerants must leave the cylinder as a liquid, so the cylinder must stay inverted. For zeotropic refrigerants, follow procedure 4a. Compounds and azeotropic refrigerants may be removed from the cylinder as a gas. To finish a charge in vapor form with either a compound or an azeotropic refrigerant, follow procedure 4b.

Step 4a – Completing a Partial Charge with Zeotropic Refrigerant
For any 400 series zeotropic refrigerant the cylinder must stay inverted. You will meter in small quantities of liquid refrigerant into the low side in short bursts. The idea is to allow the liquid to vaporize before it gets to the compressor.

A. Start the system with both manifold gauge valves closed.
   The gauges should show the system high side and low side pressures.

B. Crack open the manifold low side valve for not more than 5 seconds or less and then close it. Repeat this, waiting at least 15 seconds between liquid bursts.

C. Watch the scale and close the gauge manifold low side valve when the correct amount of refrigerant has been added to the system.

D. With the system continuing to run, close the refrigerant cylinder valve and crack open the manifold low side valve to allow most of the refrigerant trapped in the middle hose to be pulled into the system.

Step 4b – Completing a Partial Charge with a Compound or Azeotropic Refrigerant
If the refrigerant is a compound or an azeotrope, the charge may be completed in vapor with the compressor operating. Calculate the remaining refrigerant required to complete the charge.

Charge Remaining = Total Charge Required – Liquid Added

A. For compounds or azeotropes turn the cylinder back over so that it is upright for vapor.

B. Start the system with both manifold gauge valves closed.
   The gauges should show the system high side and low side pressures.

C. Slowly open the manifold low side valve until it is fully open.

D. Watch the scale and close the gauge manifold low side valve when the correct amount of refrigerant has been added to the system.

E. With the system continuing to run, close the refrigerant cylinder valve and open the manifold low side valve to allow most of the refrigerant trapped in the middle hose to be pulled into the system.

Step 5 – Finishing up

A. Keep the gauge manifold connected and operate the system until the pressures to stabilize.

B. Verify that the system pressures and temperatures agree with the manufacturer's recommendations and record them on Data Sheet 29-1-1.

Data Sheet 29-1-1

Refrigerant Type on Data Plate	
Refrigerant Quantity on Data Plate	
Total Charge Added	
Operating Suction Pressure	
Operating Discharge Pressure	
Manufacturer's Performance Specification	
Measured Performance Specification	

QUESTIONS

(Circle the letter that indicates the correct answer.)

1. If the refrigeration system is overcharged:
   A. the discharge pressure will be higher than expected.
   B. the discharge pressure will be lower than expected.
   C. the evaporator temperature will be too cold.
   D. the condenser temperature will be too cold.

2. If the refrigeration system is undercharged:
   A. the system subcooling will be higher than expected.
   B. the system subcooling will be lower than expected.
   C. the evaporator superheat will be too low.
   D. the compressor amp draw will increase.

3. When liquid charging:
   A. never allow liquid refrigerant to enter the compressor.
   B. the compressor shell should be filled with liquid before the unit is started.
   C. charge into the suction side of the system.
   D. charge into liquid into both sides of an evacuated system.

4. Throttling liquid refrigerant through the gauge manifold:
   A. will increase its temperature.
   B. will allow some of it to flash to vapor.
   C. will damage the gauge manifold.
   D. will damage the gauge manifold valve seats.

5. To speed up the charging process:
   A. heat the cylinder with a propane torch.
   B. Cool the cylinder with ice.
   C. weigh in liquid.
   D. weigh in the entire charge in vapor.

6. When charging, if the gauge manifold valves are closed (front seated):
   A. no flow will occur through the manifold.
   B. system pressures can still be read.
   C. the system is isolated from the charging cylinder.
   D. All of the above are correct.

LAB 29.2 VAPOR CHARGING WITH A PARTIAL CHARGE

LABORATORY OBJECTIVE
The student will demonstrate the correct procedure for adding a partial charge using vapor.

FUNDAMENTALS OF HVACR TEXT REFERENCE
Unit 29 Refrigerant System Charging

LABORATORY NOTES
The student will add a partial charge to a system which is low on refrigerant. The refrigerant vapor will be added to the low side with the unit operating. When the system runs, the low side pressure is lower than the cylinder pressure, drawing vapor refrigerant into the unit due to the difference in pressure.

When charging zeotropic refrigerants, such as R-410A, they must leave the cylinder as a liquid to prevent fractionation of the refrigerant. 400 series zeotropic refrigerants can NOT be vapor charged. See Lab 29.3 for details on charging zeotropic refrigerants.

REQUIRED TOOLS, EQUIPMENT, AND MATERIALS
Gloves and safety glasses

Gauge manifold and service valve wrench

Digital scale

Cylinder of refrigerant

Operating refrigeration system

SAFETY REQUIREMENTS
A. Wear safety goggles and gloves when working with refrigerants. Liquid refrigerant can cause frostbite when in contact with eyes and skin.
B. Use low loss hose fittings to reduce refrigerant spray when connecting and disconnecting refrigeration hoses. Inspect all fittings before attaching hoses.

PROCEDURE
1. Locate the refrigerant type and charging chart on the equipment, Figure 29-2-1.
2. Connect the gauges to the high and low sides of the system, Figure 29-2-1.
3. Connect the middle hose to the refrigerant cylinder, Figure 29-2-1.
4. Purge the air from the gauge hoses and the manifold.
5. Turn on system.
6. Check the system charge against the manufacturer's charging chart.
7. If the system is low on charge, proceed with steps 8–10.

8. Leave the cylinder in the upright position and open the valve on the refrigerant cylinder.
9. Open the low side of the manifold, allowing refrigerant to flow into the system.
10. Periodically stop adding refrigerant and check pressures and temperatures against manufacturer's charging chart.
11. Repeat steps 9 and 10 until the system is operating according to the manufacturer's charging chart.
12. The cylinder will get cold as vapor is removed, causing the cylinder pressure to drop. The Low side gauge should rise when the low side manifold valve is opened because it will be showing the cylinder pressure. If it does not rise when opened, there is no pressure difference between the cylinder and the system. The cylinder pressure must be higher than the system low side operating pressure in order to vapor charge. Heating the cylinder with warm water can help keep the cylinder pressure up high enough to continue charging.

    Safety Note: NEVER use excessive heat sources such as live steam or a torch to heat the refrigerant cylinder. They could cause a rapid pressure increase leading to a catastrophic cylinder rupture.

Figure 29-2-1 Unit data plate

CONTAINS HCFC – 22	DESIGN PRESSURE	
FACTORY CHARGE	278	HI PSIG
12 LBS 8 OZS	144	LO PSIG

Figure 29-2-2 Connections for vapor charge

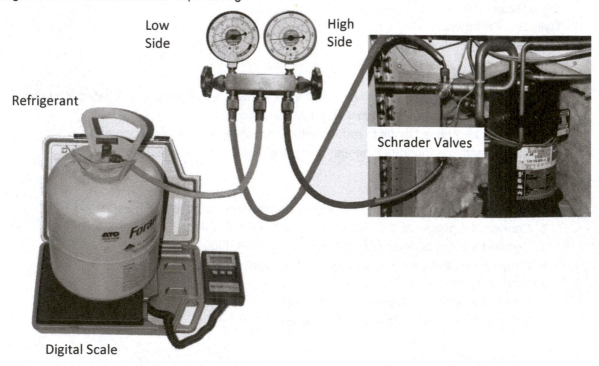

Low Side

High Side

Refrigerant

Schrader Valves

Digital Scale

LAB 29.3 CHARGING WITH BLENDS (ZEOTROPES)

LABORATORY OBJECTIVE

The student will demonstrate the correct procedure for charging a blended (zeotropic) refrigerant into an operating system.

FUNDAMENTALS OF HVACR TEXT REFERENCE

Unit 29 Refrigerant System Charging

LABORATORY NOTES

Zeotropic refrigerant, such as R-410A, must leave the refrigerant cylinder as a liquid to prevent fractionation (separation) of the refrigerant components. Frequently it is necessary to complete a charge with the system operating in order to draw refrigerant out of the cylinder and into the system. However, the low side of the system normally is only charged with vapor because we want to avoid returning liquid to the compressor. The procedure for charging zeotropic refrigerant into an operating system involves throttling small quantities of liquid into the suction line of an operating system and allowing it to flash off to vapor.

REQUIRED TOOLS, EQUIPMENT, AND MATERIALS

Safety glasses and gloves
Gauge manifold and service valve wrench
Digital scale
Cylinder of zeotropic refrigerant
Operating refrigeration system that uses zeotropic refrigerant

SAFETY REQUIREMENTS

A. Wear safety goggles and gloves when working with refrigerants. Liquid refrigerant can cause frostbite when in contact with eyes and skin.

B. Use low loss hose fittings to reduce refrigerant spray when connecting and disconnecting refrigeration hoses. Inspect all fittings before attaching hoses.

PROCEDURE

Step 1 – Identify Refrigerant

The refrigerant type and required amount of charge can be found on the refrigeration system nameplate as shown below in Figure 29-3-1.

Figure 29-3-1 Data plate

M/N XP19 – 024 – 230 – 01	
S/N 5805F38025	
CONTAINS HFC – 410A	DESIGN PRESSURE
FACTORY CHARGE	HI 446 PSIG
12 LBS 0 OZS	LO 236 PSIG

Step 2 – Connections

A. Connect the manifold gauges to the high and low side of the system and connect the middle hose to the refrigerant cylinder, as shown in Figure 29-3-2.

B. Purge the air from the hoses and manifold.

C. Normally the refrigerant cylinder must be turned upside down (inverted) for liquid. It is a good practice to purge the hoses and fully open the cylinder valve prior to turning it over.

Figure 29-3-2

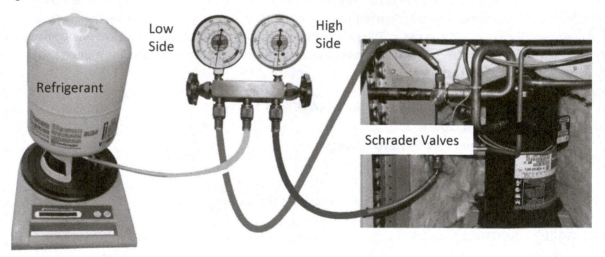

Digital Scale

Step 3 – Charging

A. You must never allow liquid refrigerant to enter the compressor, however the zeotropic refrigerant blend must leave the cylinder as a liquid and the refrigerant must enter the suction line while the system is running.

B. To accomplish this, you will introduce liquid to the gauge manifold, but you will throttle the manifold low side valve to allow for only a very slight flow as in Figure 29-3-3.

C. The liquid refrigerant will flash to vapor as it passes through the gauge manifold valve.

D. Using a flow restrictor on the outlet of the cylinder can also help flash the refrigerant. Figure 29-3-4 shows a cylinder with a flow restrictor.

E. Start the system in the normal operating mode. The system pressures should register on the gauge manifold. You can start throttling in refrigerant into the low side after the low side pressure drops below the cylinder pressure. If the cylinder pressure is higher than the operating suction pressure the low side gauge reading will increase every time you open the low side manifold valve.

F. Be patient, it is easier to overcharge a system using this method. Introduce refrigerant in short bursts of a few seconds at a time and leave time between bursts for the refrigerant to flash.

G. Crack open the manifold low side valve for not more than 5 seconds and then close it. Repeat this, waiting at least 15 seconds between liquid bursts.

H. Watch the scale and close the gauge manifold low side valve when the correct amount of refrigerant has been added to the system.

I. With the system continuing to run, close the refrigerant cylinder valve and crack open the manifold low side valve to allow most of the refrigerant trapped in the middle hose to be pulled into the system.

J. Allow the pressures and temperatures to stabilize before comparing the operating characteristics against the manufacturer's specifications.

Figure 29-3-3 Flashing liquid through manifold

Figure 29-3-4 Restriction device used to help flash liquid refrigerant

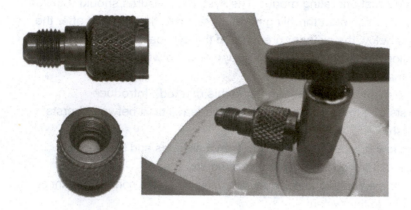

Step 4 – Finishing up

A. Keep the gauge manifold connected and operate the system until the pressures to stabilize.
B. Verify that the system pressures and temperatures agree with the manufacturer's recommendations and record them on Data Sheet 29-3-1.

Data Sheet 29-3-1

Refrigerant Type on Data Plate	
Refrigerant Quantity on Data Plate	
Operating Suction Pressure	
Operating Discharge Pressure	
Total Charge Added	
Manufacturer's Performance Specification	
Actual Operating Performance	

QUESTIONS
(Circle the letter that indicates the correct answer.)

1. Zeotropic blends:
   A. can experience "temperature glide."
   B. are numbered in the 500 series.
   C. have one single boiling point.
   D. must leave the cylinder as a vapor.

2. Fractionation occurs when:
   A. refrigerants are mixed with oil.
   B. charging blends as a liquid.
   C. charging blends as a vapor.
   D. charging blends as a liquid.

3. When liquid charging with the system running:
   A. never allow liquid refrigerant to enter the compressor.
   B. liquid refrigerant blends can be allowed to enter the compressor.
   C. always fractionate the refrigerant first.
   D. liquid refrigerant is charged into the high side with the compressor running.

4. Throttling liquid refrigerant through the gauge manifold when charging a liquid blend:
   A. will increase its temperature.
   B. will allow some of it to flash to vapor.
   C. will damage the gauge manifold.
   D. will damage the gauge manifold valve seats.

5. What types of refrigerants must be charged as a liquid?
   A. Azeotropic blends.
   B. Zeotropic blends.
   C. Near-azeotropic blends.
   D. Compounds.

6. Zeotropic blends are identified as the:
   A. Z series.
   B. 500 series.
   C. 400 series.
   D. 700 series.

LAB 29.4 LIQUID CHARGING WITH COMPRESSOR RUNNING

LABORATORY OBJECTIVE

The student will demonstrate the correct procedure for performing a liquid charge on an operating refrigeration system with a king valve.

FUNDAMENTALS OF HVACR TEXT REFERENCE

Unit 29 Refrigerant System Charging

LABORATORY NOTES

In this lab you will be charging liquid refrigerant into an operating system. This is typically done when adding refrigerant to a large system that already has a partial refrigerant charge. On large systems, a king valve located between the condenser and the metering valve offers a convenient means of charging the system on the high side, with the compressor running. The refrigeration unit must have a manual king valve located after the condenser. The king valve must be positioned so that when it is front-seated the gauge port is open to the refrigerant line, not the liquid receiver.

REQUIRED TOOLS, EQUIPMENT, AND MATERIALS

Safety glasses and gloves
Gauge manifold and service valve wrench
Digital scale
Cylinder of refrigerant
Operating refrigeration system

SAFETY REQUIREMENTS

A. Wear safety goggles and gloves when working with refrigerants. Liquid refrigerant can cause frostbite when in contact with eyes and skin.

B. Use low loss hose fittings to reduce refrigerant spray when connecting and disconnecting refrigeration hoses. Inspect all fittings before attaching hoses.

PROCEDURE

Step 1 – Identify Refrigerant

The refrigerant type and quantity can be found on the refrigeration system nameplate as shown in Figure 29-4-1.

Figure 29-4-1 Unit data plate

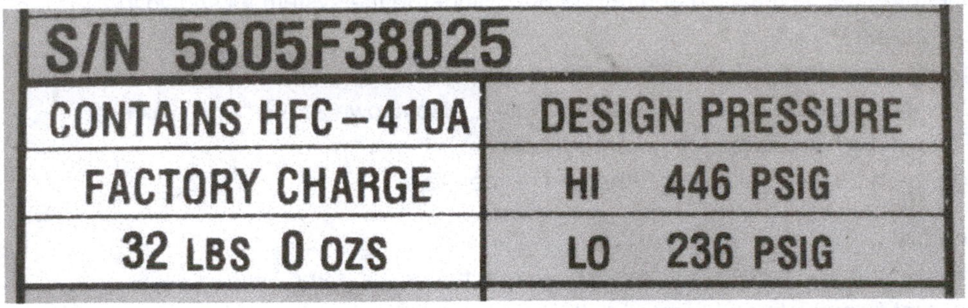

S/N 5805F38025	
CONTAINS HFC—410A	DESIGN PRESSURE
FACTORY CHARGE	HI 446 PSIG
32 LBS 0 OZS	LO 236 PSIG

Figure 29-4-2 Connections

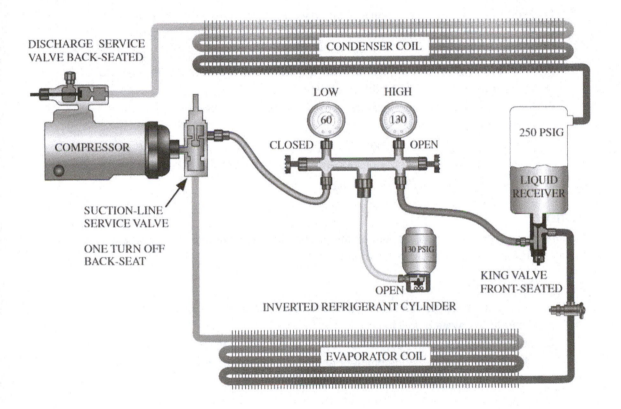

Step 2 – Connections

A. Connect the gauge manifold and refrigerant cylinder to the system as shown in Figure 29-4-2.
   a. the low side hose to the suction service valve
   b. the high side hose to the liquid service valve, also known as the "king" valve.
   c. The middle hose to the refrigerant cylinder.
B. Purge the air from the lines and the gauge manifold.

Step 3 – Charging

A. Front-seat the king valve. This blocks flow from the receiver and instead, pulls refrigerant from the line connected to the gauge port (Figure 29-4-3).
B. Make sure that the high and low side gauge manifold valves are closed to start.
C. Open the refrigerant cylinder fully and then turn it over and balance it on a digital scale.
D. Zero the scale once the bottle is inverted and balanced on the scale.
E. Back-seat crack the suction service valve and front-seat the KING VALVE, Figure 29-4-3.
F. Start the system and run it in the normal operating mode and with the king valve front-seated. The liquid line pressure will drop.
G. Open the manifold high side valve to allow refrigerant from the cylinder to be drawn into the system.
H. Use the high side gauge manifold valve to control the rate of charge and continue charging until the proper amount of refrigerant has been added as indicated by the reading on the digital scale.

CAUTION

You cannot use the pressures indicated on the gauges to judge when to stop charging. It is easy to overcharge a system using this method.

Step 4 – Finishing up

A. Keep the gauge manifold connected and backseat crack the king valve.
B. Operate the system until the pressures to stabilize.
C. Verify that the suction and discharge pressures of the refrigeration system agree with the manufacturer's recommendations and record the pressures on Data Sheet 29-4-1.

Data Sheet 29-4-1

Refrigerant Type on Data Plate	
Refrigerant Quantity on Data Plate	
Liquid Quantity Added	
Manufacturer's Specified Suction Pressure	
Manufacturer's Specified Discharge Pressure	
Operating Suction Pressure	
Operating Discharge Pressure	

Figure 29-4-3

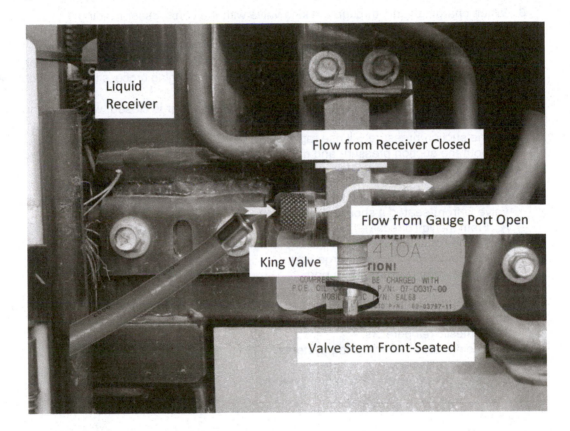

QUESTIONS

(Circle the letter that indicates the correct answer.)

1. When charging through the liquid side with the compressor running:
   A. the refrigerant cylinder should be in the upright position.
   B. the liquid should be added directly into the suction.
   C. the king valve should be front-seated.
   D. the king valve should be back-seat cracked.

2. The king valve is located:
   A. in the suction line.
   B. in the liquid line right after the condenser or receiver.
   C. right opposite the queen valve.
   D. at the outlet of the evaporator.

3. When charging liquid through the king valve with the compressor running:
   A. The liquid is drawn through the king valve into the compressor suction.
   B. The liquid line pressure to drops because it is isolated from the liquid receiver.
   C. The pressures showing on the gauges tell the technician when to stop charging.
   D. You must be patient because this is a much slower charging process than vapor charging.

4. If the king valve is front-seated and the refrigerant cylinder is closed, when the unit operates:
   A. the discharge pressure will rise rapidly.
   B. the suction pressure will rise rapidly.
   C. both the suction pressure and discharge pressure will rise rapidly.
   D. the unit will stop on the low pressure cut out.

5. For a king valve to be useful for liquid charging with the system operating, when the king valve is front-seated it should close off flow:
   A. From the liquid receiver.
   B. To the liquid line leading to the expansion valve.
   C. To the compressor discharge port.
   D. To the compressor suction port.

6. Liquid charging through the king valve with the system operating is mainly used when charging:
   A. Domestic refrigerators.
   B. Residential air conditioning systems.
   C. New installations.
   D. Large commercial refrigeration systems.

LAB 29.5 CHARGING PACKAGED UNIT BY WEIGHT

LABORATORY OBJECTIVE
You will demonstrate your ability to weigh a charge into an evacuated air conditioning unit.

FUNDAMENTALS OF HVACR TEXT REFERENCE
Unit 29 Refrigerant System Charging

LABORATORY NOTES
You will weigh in the manufacturer's specified refrigerant charge into an evacuated packaged unit.

REQUIRED TOOLS, EQUIPMENT, AND MATERIALS
Safety glasses and gloves
Manifold gauges, refrigeration wrench, and extra refrigeration hose
Packaged air conditioning unit
Refrigerant
Scale

SAFETY REQUIREMENTS
A. Wear safety goggles and gloves when working with refrigerants. Liquid refrigerant can cause frostbite when in contact with eyes and skin.
B. Use low loss hose fittings to reduce refrigerant spray when connecting and disconnecting refrigeration hoses. Inspect all fittings before attaching hoses.

PROCEDURE
Step 1 – Identify Refrigerant
The refrigerant type and quantity can be found on the refrigeration system nameplate.
Step 2 – Connections
A. The system should be evacuated with your gauges connected to the low side and high side gauge ports with the gauges closed.
B. Turn on the scale and zero it. Place the refrigerant cylinder on the scale, weigh the cylinder, and record its weight in Data Sheet 29-5-1.
C. Connect the center hose to the refrigerant cylinder and purge just the center hose. The others should still be in a vacuum.
D. Open the refrigerant cylinder valve and place the cylinder on the scale in a position to deliver liquid. That is upside down for most non-refillable cylinders. For refrigerant recovery cylinders, that is upright with the hose connected to the liquid valve.
E. Zero the scale with the cylinder on the scale.

Step 3 – Charging

A. The system should NOT be operating.

B. Fully open the high side manifold gauge.

C. Monitor the scale and close the high side gauge when the correct weight is read on the scale.

D. Close the valve on the charging cylinder.

Step 4 – Verification

A. Operate the system.

B. Compare the operating system characteristics to the manufacturer's specifications.

C. Crack the low side gauge and meter in the refrigerant trapped in the gauge hoses into the low side.

D. Remove you gauges and replace the valve caps.

E. Complete Data Sheet 29-5-1.

Data Sheet 29-5-1

Charging Refrigerant System by Weight	
Type of Refrigerant	
Amount of Refrigerant	
Cylinder Weight before Charging	
Cylinder Weight after Charging	
Amount of Refrigerant Charged into Unit	
Operating Suction Pressure	
Operating Discharge Pressure	

LAB 29.6 CHARGING SPLIT SYSTEM BY WEIGHT

LABORATORY OBJECTIVE

You will demonstrate your ability to determine the correct charge for a split system and weigh it into an evacuated split system air conditioning unit.

FUNDAMENTALS OF HVACR TEXT REFERENCE

Unit 29 Refrigerant System Charging

LABORATORY NOTES

Although split system condensing units have a factory charge stated on the data plate, the correct charge for any particular installation varies depending upon the refrigerant line length. You will determine the correct charge for a split system and then weigh it into an evacuated split system.

REQUIRED TOOLS, EQUIPMENT, AND MATERIALS

Safety glasses and gloves
Manifold gauges, refrigeration wrench, and an extra refrigeration hose
Split system air conditioning unit
Scale
Refrigerant

SAFETY REQUIREMENTS

A. Wear safety goggles and gloves when working with refrigerants. Liquid refrigerant can cause frostbite when in contact with eyes and skin.
B. Use low loss hose fittings to reduce refrigerant spray when connecting and disconnecting refrigeration hoses. Inspect all fittings before attaching hoses.

PROCEDURE

Step 1 – Identify Refrigerant

The refrigerant type and quantity can be found on the refrigeration system nameplate as shown in Figure 29-6-1.

Figure 29-6-1 Unit data plate

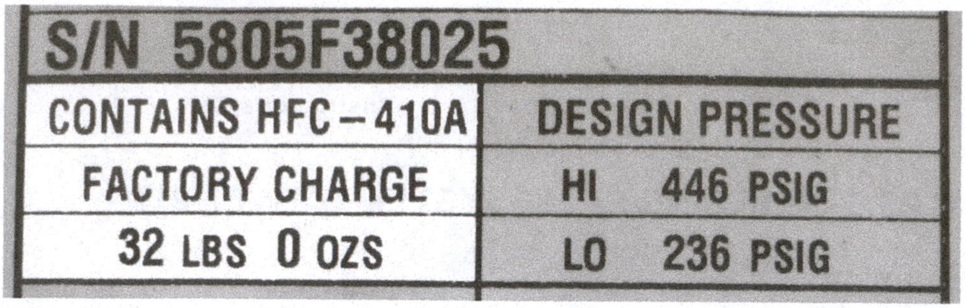

S/N 5805F38025

CONTAINS HFC – 410A	DESIGN PRESSURE	
FACTORY CHARGE	HI	446 PSIG
32 LBS 0 OZS	LO	236 PSIG

Step 2 – Calculate Total Charge

Calculate the correct refrigerant charge using the manufacturer's instructions and record the information on Data Sheet 29-6-1. Table 29-6-1 shows charge adjustment factors for both R22 and R410A for different diameter liquid lines.

A typical formula is: Condensing Unit Charge + (0.62 oz x (actual line length – 15 ft))

(Note: This is for units with a 3/8" liquid line and a factory charge for 15 feet of lines.)

Table 29-6-1 Line Length Charge Adjustment Factors

Liquid Line Diameter	R22	R410A
1/4 inch	0.23	0.19
5/16 inch	0.40	0.33
3/8 inch	0.62	0.51
1/2 inch	1.12	1.01
5/8 inch	1.81	1.64

Step 3 – Connections

A. The system should be evacuated with your gauges connected to the low side and high side gauge ports with the gauges closed.

B. Turn on the scale and zero it. Place the refrigerant cylinder on the scale, weigh the cylinder, and record its weight in Data Sheet 29-6-1.

C. Connect the center hose to the refrigerant cylinder and purge just the center hose. The others should still be in a vacuum.

D. Open the refrigerant cylinder valve and place the cylinder on the scale in a position to deliver liquid. That is upside down for most non-refillable cylinders. For refrigerant recovery cylinders, that is upright with the hose connected to the liquid valve.

E. Zero the scale with the cylinder on the scale.

Step 4 – Charging
A. The system should NOT be operating.
B. Fully open the high side manifold gauge.
C. Monitor the scale and close the high side gauge when the correct weight is read on the scale.
D. Close the valve on the charging cylinder.

Step 5 – Verification
A. Operate the system.
B. Compare the operating system characteristics to the manufacturer's specifications.
C. Crack the low side gauge and meter in the refrigerant trapped in the gauge hoses into the low side.
D. Remove you gauges and replace the valve caps.
E. Complete Data Sheet 29-6-1.

Data Sheet 29-6-1

Charging Split System by Weight	
Type of Refrigerant from Data Plate	
Line Length	
Charge Adjustment (Line Length – Allowance (15)) x per ft allowance	
Factory Charge from Data Plate	
Total Charge Factory Charge + Charge Adjustment	
Cylinder Weight before Charging	
Cylinder Weight after Charging	
Amount of Refrigerant Charged into Unit	
Operating Suction Pressure	
Operating Discharge Pressure	

LAB 29.7 PRESSURE-TEMPERATURE CHARGE CHECK

LABORATORY OBJECTIVE

You will demonstrate your ability to check refrigerant charge in three refrigeration systems using a pressure-temperature chart.

FUNDAMENTALS OF HVACR TEXT REFERENCE

Unit 29 Refrigerant System Charging

LABORATORY NOTES

You will use common temperature relationships and a pressure-temperature chart to determine the correct charge for three different refrigeration systems. You will then check the charge and compare the actual temperatures and pressures to the calculated target temperatures and pressures.

REQUIRED TOOLS, EQUIPMENT, AND MATERIALS

Safety glasses and gloves
Manifold gauges and refrigeration wrench
Thermometer
Three operating refrigeration systems

SAFETY REQUIREMENTS

A. Wear safety goggles and gloves when working with refrigerants. Liquid refrigerant can cause frostbite when in contact with eyes and skin.
B. Use low loss hose fittings to reduce refrigerant spray when connecting and disconnecting refrigeration hoses. Inspect all fittings before attaching hoses.

PROCEDURE

Step 1 – Equipment Identification

Examine the unit and record important unit characteristics in the data sheet, including:

A. Type of Unit (air conditioner, freezer, etc.)

B. Type of Refrigerant

C. Metering Device (TEV or fixed restriction)

D. Condenser (air or water)

E. Evaporator (air or water)

Step 2 – Operating Conditions

Measure and record the relevant operating conditions, including:

    A. Ambient Temperature for air cooled condensers

    B. Inlet water temperature for water cooled condensers

    C. Return Air Temperature for air conditioning units

    D. Return Air We Bulb Temperature for air conditioning units

    E. Box Temperature for commercial refrigeration

Step 3 – Determine the Target Condenser Temperature and Pressure

Estimate the target condenser temperature by adding the estimated temperature difference between the air or water flowing over the condenser and the refrigerant. Typical differences are listed below.

    A. 18 SEER Air Cooled Condenser Temp = Ambient Temp + 10°F

    B. 16 SEER Air Cooled Condenser Temp = Ambient Temp + 15°F

    C. 14 SEER Air Cooled Condenser Temp = Ambient Temp + 18°F

    D. 13 SEER Air Cooled Condenser Temp = Ambient Temp + 20°F

    E. Water Cooled Condenser Temperature = inlet water temperature + 20°F

Use a pressure-temperature chart or a PT app to determine the target condenser pressure from the target condenser temperature. Record the target condenser temperature and pressure on the Data Sheet 29-7-1.

Step 4 – Determine the Target Evaporator Temperature and Pressure

Estimate the target evaporator temperature by subtracting the estimated temperature difference between the air or water flowing over the evaporator and the refrigerant. Typical differences are listed below.

    A. 18 SEER Evaporator Temperature = Return Air Temp – 25°F

    B. 13 to 16 SEER Evaporator Temperature = Return Air Temp – 30°F

    C. 12 SEER and older Evaporator Temperature = Return Air Temp – 35°F

    D. Commercial Refrigeration Evaporator Temperature = Box Temperature – 10°F

Use a pressure-temperature chart or a PT app to determine the target evaporator pressure from the target evaporator temperature. Record the target evaporator temperature and pressure on the Data Sheet 29-7-1.

Step 5 – Reading the Actual Temperatures and Pressures

Connect your manifold gauges to the system and operate the system until the pressures stabilize and the compressor warms up. Record the actual operating condenser and evaporator pressures on the data sheet. Use a PT Chart or App to determine the actual operating condenser and evaporator temperatures and record them on the Data Sheet 29-7-1.

Step 6 – Assessment

Compare the estimated target temperatures and pressures to the actual temperatures and pressures. Determine if the system is charged correctly, undercharged, overcharged, or has another refrigeration problem. Complete Data Sheet 29-7-1.

Step 7 – Unit 2

Repeat the entire process on Unit 2 and complete Data Sheet 29-7-2.

Step 8 – Unit 3

Repeat the entire process on Unit 3 and complete Data Sheet 29-7-3.

LAB 29.7 PRESSURE–TEMPERATURE CHARGING DATA SHEET 29-7-1	
TYPE OF UNIT (*Air conditioner, freezer, etc.*)	
TYPE OF CONDENSER (*Air or Water cooled*)	
TYPE OF EVAPORATOR (*Cools Air or Water*)	
TYPE OF METERING DEVICE (*Expansion valve, cap tube, orifice*)	
TYPE OF REFRIGERANT (*R22, R410A, R134a, etc.*)	
OUTDOOR AMBIENT/ WATER TEMPERATURE (*Temp of condenser air or water*)	
DESIRED CONDENSER TEMPERATURE (*Condenser Saturation from formula*)	
DESIRED CONDENSER PRESSURE (*Condenser Pressure from PT chart*)	
RETURN AIR TEMP/ BOX TEMP (*Temp of air in house or refrigerated box*)	
RETURN AIR WET BULB (*Measured with digital psychrometer*)	
DESIRED EVAPORATOR TEMPERATURE (*Evaporator saturation using formula*)	
DESIRED EVAPORATOR PRESSURE (*Evaporator Pressure from PT chart*)	
ACTUAL CONDENSER PRESSURE (*High side pressure actually read on gauges*)	
ACTUAL CONDENSER TEMPERATURE (*High side pressure converted to temp by PT*)	
ACTUAL EVAPORATOR PRESSURE (*Low side pressure on gauges*)	
ACTUAL EVAPORATOR TEMPERATURE (*Low side pressure converted to temp by PT*)	
SYSTEM CONDITION (*overcharged, undercharged, or correct*)	

LAB 29.7 PRESSURE–TEMPERATURE CHARGE CHECK DATA SHEET 29-7-2	
TYPE OF UNIT (*Air conditioner, freezer, etc.*)	
TYPE OF CONDENSER (*Air or Water cooled*)	
TYPE OF EVAPORATOR (*Cools Air or Water*)	
TYPE OF METERING DEVICE (*Expansion valve, cap tube, orifice*)	
TYPE OF REFRIGERANT (*R22, R410A, R134a, etc.*)	
OUTDOOR AMBIENT/ WATER TEMPERATURE (*Temp of condenser air or water*)	
DESIRED CONDENSER TEMPERATURE (*Condenser Saturation from formula*)	
DESIRED CONDENSER PRESSURE (*Condenser Pressure from PT chart*)	
RETURN AIR TEMP/ BOX TEMP (*Temp of air in house or refrigerated box*)	
RETURN AIR WET BULB (*Measured with digital psychrometer*)	
DESIRED EVAPORATOR TEMPERATURE (*Evaporator saturation using formula*)	
DESIRED EVAPORATOR PRESSURE (*Evaporator Pressure from PT chart*)	
ACTUAL CONDENSER PRESSURE (*High side pressure actually read on gauges*)	
ACTUAL CONDENSER TEMPERATURE (*High side pressure converted to temp by PT*)	
ACTUAL EVAPORATOR PRESSURE (*Low side pressure on gauges*)	
ACTUAL EVAPORATOR TEMPERATURE (*Low side pressure converted to temp by PT*)	
SYSTEM CONDITION (*overcharged, undercharged, or correct*)	

LAB 29.7 PRESSURE–TEMPERATURE CHARGING DATA SHEET 29-7-3

TYPE OF UNIT (*Air conditioner, freezer, etc.*)	
TYPE OF CONDENSER (*Air or Water cooled*)	
TYPE OF EVAPORATOR (*Cools Air or Water*)	
TYPE OF METERING DEVICE (*Expansion valve, cap tube, orifice*)	
TYPE OF REFRIGERANT (*R22, R410A, R134a, etc.*)	
OUTDOOR AMBIENT/ WATER TEMPERATURE (*Temp of condenser air or water*)	
DESIRED CONDENSER TEMPERATURE (*Condenser Saturation from formula*)	
DESIRED CONDENSER PRESSURE (*Condenser Pressure from PT chart*)	
RETURN AIR TEMP/ BOX TEMP (*Temp of air in house or refrigerated box*)	
RETURN AIR WET BULB (*Measured with digital psychrometer*)	
DESIRED EVAPORATOR TEMPERATURE (*Evaporator saturation using formula*)	
DESIRED EVAPORATOR PRESSURE (*Evaporator Pressure from PT chart*)	
ACTUAL CONDENSER PRESSURE (*High side pressure actually read on gauges*)	
ACTUAL CONDENSER TEMPERATURE (*High side pressure converted to temp by PT*)	
ACTUAL EVAPORATOR PRESSURE (*Low side pressure on gauges*)	
ACTUAL EVAPORATOR TEMPERATURE (*Low side pressure converted to temp by PT*)	
SYSTEM CONDITION (*overcharged, undercharged, or correct*)	

LAB 29.8 PARTIAL CHARGE – PRESSURE/TEMPERATURE

LABORATORY OBJECTIVE

You will demonstrate your ability to safely add refrigerant vapor to an operating unit until the system operating characteristics meet the manufacturer's specifications.

FUNDAMENTALS OF HVACR TEXT REFERENCE

Unit 29 Refrigerant System Charging

LABORATORY NOTES

You will add refrigerant vapor to the low side with the system operating until the system operating characteristics match the desired pressures determined by common temperature relationships and a pressure-temperature chart.

REQUIRED TOOLS, EQUIPMENT, AND MATERIALS

Safety glasses and gloves
Manifold gauges and refrigeration wrench
Thermometer
Operating refrigeration system and refrigerant
Scale

SAFETY REQUIREMENTS

A. Wear safety goggles and gloves when working with refrigerants. Liquid refrigerant can cause frostbite when in contact with eyes and skin.

B. Use low loss hose fittings to reduce refrigerant spray when connecting and disconnecting refrigeration hoses. Inspect all fittings before attaching hoses.

PROCEDURE

Step 1 – Equipment Identification

Examine the unit and record important unit characteristics in the data sheet, including:

A. Type of Unit (air conditioner, freezer, etc.)

B. Type of Refrigerant

C. Metering Device (TEV or fixed restriction)

D. Condenser (air or water)

E. Evaporator (air or water)

Step 2 – **Operating Conditions**

Measure and record the relevant operating conditions, including:

    A. Ambient Temperature for air cooled condensers

    B. Inlet water temperature for water cooled condensers

    C. Return Air Temperature for air conditioning units

    D. Return Air Wet Bulb Temperature for air conditioning units

    E. Box Temperature for commercial refrigeration

Step 3 – **Determine the Target Condenser Temperature and Pressure**

Estimate the target condenser temperature by adding the estimated temperature difference between the air or water flowing over the condenser and the refrigerant. Typical differences are listed below.

    A. 18 SEER Air Cooled Condenser Temp = Ambient Temp + 10°F

    B. 16 SEER Air Cooled Condenser Temp = Ambient Temp + 15°F

    C. 14 SEER Air Cooled Condenser Temp = Ambient Temp + 18°F

    D. 13 SEER Air Cooled Condenser Temp = Ambient Temp + 20°F

    E. Water Cooled Condenser Temperature = inlet water temperature + 20°F

Use a pressure-temperature chart or a PT app to determine the target condenser pressure from the target condenser temperature. Record the target condenser temperature and pressure on the Data Sheet 29-8-1.

Step 4 – **Determine the Target Evaporator Temperature and Pressure**

Estimate the target evaporator temperature by subtracting the estimated temperature difference between the air or water flowing over the evaporator and the refrigerant. Typical differences are listed below.

    E. 18 SEER Evaporator Temperature = Return Air Temp – 25°F

    F. 13 to 16 SEER Evaporator Temperature = Return Air Temp – 30°F

    G. 12 SEER and older Evaporator Temperature = Return Air Temp – 35°F

    H. Commercial Refrigeration Evaporator Temperature = Box Temperature – 10°F

Use a pressure-temperature chart or a PT app to determine the target evaporator pressure from the target evaporator temperature. Record the target evaporator temperature and pressure on the Data Sheet 29-8-1.

Step 5 – **Reading the Actual Temperatures and Pressures**

Connect your manifold gauges to the system and operate the system until the pressures stabilize and the compressor warms up. Record the actual operating condenser and evaporator pressures on the data sheet. Use a PT Chart or App to determine the actual operating condenser and evaporator temperatures and record them on the Data Sheet 29-8-1.

Step 6 – **Assessment and Adjustment**

Compare the estimated target temperatures and pressures to the actual temperatures and pressures. Determine if the system is charged correctly, undercharged, overcharged, or has another refrigeration problem.

If the system is undercharged and the refrigerant is a compound or azeotropic then

1. Connect the middle gauge hose to the vapor valve on a refrigerant cylinder.
2. *(Make sure that the refrigerant listed on the data plate is the same as in the cylinder.)*
3. Open the cylinder valve and purge the middle hose at the manifold.
4. Turn the scale on.
5. Place the cylinder in the upright position on the scale and zero the scale.
6. With the unit operating, open the low side manifold hand wheel.
7. The compound gauge should show a higher pressure – this is the cylinder pressure.
8. Periodically close the low side manifold hand wheel to see what the system operating pressure is.
9. Monitor the low side and high side pressures.
10. When the operating pressures are similar to the desired pressures, you are finished.
11. Record the system operating pressures and amount of refrigerant used on Data Sheet 29-8-1.

Else if the system is undercharged and the refrigerant is zeotropic

1. Connect the middle gauge hose to the vapor valve on a refrigerant cylinder.
2. *(Make sure that the refrigerant listed on the data plate is the same as in the cylinder.)*
3. Open the cylinder valve and purge the middle hose at the manifold.
4. Turn the scale on.
5. Place the cylinder in the inverted (upside down) position on the scale and zero the scale.
6. With the unit operating, **crack** the low side manifold hand wheel for no more than 5 seconds.
7. Wait at least 15 seconds before opening the valve again.
8. The compound gauge should show a higher pressure when the hand wheel is opened - this is the cylinder pressure.
9. Monitor the low side and high side pressures.
10. When the operating pressures are similar to the target pressures, you are finished.
11. Record the system operating pressures and amount of refrigerant used on Data Sheet 29-8-1.

Else If the system is overcharged then

1. Connect the middle gauge hose to the vapor valve on a refrigerant recovery cylinder.
2. *(Make sure that the refrigerant listed on the data plate is the same as in the cylinder.)*
3. Open the cylinder valve and purge the middle hose at the manifold.
4. Turn the scale on.
5. Place the cylinder in the upright position on the scale and zero the scale.

6. With the unit operating, open the high side manifold hand wheel for a few seconds and then close it.

7. The high side gauge should show a lower pressure when the hand wheel is opened – this is the cylinder pressure.

8. Monitor the low side and high side pressures

9. Periodically open the high side manifold hand wheel for a few seconds to let refrigerant out of the system into the recovery cylinder.

10. When the operating pressures are similar to the desired pressures, you are finished.

11. Record the system operating pressures and amount of refrigerant used on Data Sheet 29-8-1.

CHARGING DATA SHEET 29-8-1	
TYPE OF UNIT (Air conditioner, freezer, etc.)	
TYPE OF CONDENSER (Air or Water cooled)	
TYPE OF EVAPORATOR (Cools Air or Water)	
TYPE OF METERING DEVICE (Expansion valve, cap tube, orifice)	
TYPE OF REFRIGERANT (R22, R410A, R134a, etc.)	
OUTDOOR AMBIENT/ WATER TEMPERATURE (Temp of condenser air or water)	
DESIRED CONDENSER TEMPERATURE (Condenser Saturation from formula)	
DESIRED CONDENSER PRESSURE (Condenser Pressure from PT chart)	
RETURN AIR TEMP/ BOX TEMP (Temp of air in house or refrigerated box)	
RETURN AIR WET BULB TEMP (Measured with a psychrometer)	
DESIRED EVAPORATOR TEMPERATURE (Evaporator saturation using formula)	
DESIRED EVAPORATOR PRESSURE (Evaporator Pressure from PT chart)	
ACTUAL CONDENSER PRESSURE (High side pressure actually read on gauges)	
ACTUAL CONDENSER TEMPERATURE (High side pressure converted to temp by PT)	
ACTUAL EVAPORATOR PRESSURE (Low side pressure on gauges)	
ACTUAL EVAPORATOR TEMPERATURE (Low side pressure converted to temp by PT)	

LAB 29.9 SUPERHEAT CHARGE CHECK

LABORATORY OBJECTIVE

You will demonstrate your ability to check refrigerant charge in three refrigeration systems using system superheat.

***FUNDAMENTALS OF HVACR* TEXT REFERENCE**

Unit 29 Refrigerant System Charging

LABORATORY NOTES

The capacity of air conditioning systems which have fixed restriction metering devices varies depending upon the outdoor ambient temperature and the indoor wet bulb temperature. This capacity difference is reflected in the system operating superheat. You will determine the manufacturer's recommended superheat, check system superheat using the system suction pressure and the suction line temperature, and compare the actual operating superheat to the manufacturer's recommended superheat. Finally, you will determine if the system is overcharged, undercharged, or correctly charged by comparing the recommended superheat with the actual superheat.

REQUIRED TOOLS, EQUIPMENT, AND MATERIALS

Safety glasses and gloves
Manifold gauges and refrigeration wrench
Thermometer
Three operating refrigeration systems with fixed restriction metering devices

SAFETY REQUIREMENTS

A. Wear safety goggles and gloves when working with refrigerants. Liquid refrigerant can cause frostbite when in contact with eyes and skin.

B. Use low loss hose fittings to reduce refrigerant spray when connecting and disconnecting refrigeration hoses. Inspect all fittings before attaching hoses.

PROCEDURE

Step 1 – Equipment Identification

Examine the unit and record important unit characteristics in the data sheet, including:

A. Type of Refrigerant

B. Metering Device (TEV or fixed restriction)

Note: Superheat charge checking is only used for systems with fixed restriction metering devices.

Step 2 – Operating Conditions

Measure and record the relevant operating conditions, including outdoor ambient temperature, return air temperature, and return air wet bulb temperature.

Step 3 – Determine the Manufacturer's Specified System Superheat

Use the manufacturer's information to determine the desired superheat and record it on the data sheet. The outdoor ambient temperature and the return air wet bulb temperature are the most commonly specified operating conditions used to determine system's specified operating superheat.

Step 4 – Check Operating Superheat

Install gauge manifolds on system. Place a clamp or strap type electronic thermometer on the suction line entering the condensing unit. Operate system until pressures stabilize and compressor is warm. Read and record the suction line temperature and the suction line pressure. Determine the evaporator saturation temperature using a PT Chart. Calculate the superheat using the formula

Operating Superheat = Suction Line Temperature – Evaporator Saturation Temperature

Step 5 – Assessment

Compare the operating superheat to the desired superheat to determine if the unit is overcharged, undercharged, or correct. If the operating superheat is lower than the manufacturer's specified superheat the system is overcharged. If the operating superheat is higher than the manufacturer's specified superheat the system is undercharged. However, there are operating conditions that can cause system superheat to be either high or low. Make certain airflow is correct before adjusting the system charge.

LAB 29.9 SUPERHEAT CHARGING DATA SHEET 29-9-1	
TYPE OF METERING DEVICE (*Expansion valve, cap tube, orifice*)	
TYPE OF REFRIGERANT (*R22, R410A, R134a, etc.*)	
OUTDOOR AMBIENT TEMPERATURE (*Temp of air entering condenser*)	
RETURN AIR DRY BULB TEMP (*Regular return air temperature*)	
RETURN AIR WET BULB TEMP (*Temp measurement from psychrometer*)	
DESIRED SYSTEM SUPERHEAT (*Superheat from manufacturer's chart*)	
SUCTION LINE TEMPERATURE (*Actual measured suction line temp*)	
EVAPORATOR PRESSURE (*Low side pressure on gauges*)	
EVAPORATOR SATURATION TEMP (*Low side pressure converted to temp by PT*)	
SYSTEM SUPERHEAT (*Suction line temp minus evaporator saturation temp*)	
SYSTEM CONDITION (*overcharged, undercharged, or correct*)	

LAB 29.9 SUPERHEAT CHARGING DATA SHEET 20-9-2

TYPE OF METERING DEVICE (*Expansion valve, cap tube, orifice*)	
TYPE OF REFRIGERANT (*R22, R410A, R134a, etc.*)	
OUTDOOR AMBIENT/ WATER TEMPERATURE (*Temp of condenser air or water*)	
RETURN AIR DRY BULB TEMP (*Regular return air temperature*)	
RETURN AIR WET BULB TEMP (*Temperature measurement from psychrometer*)	
DESIRED SYSTEM SUPERHEAT (*Superheat from manufacturer's chart*)	
SUCTION LINE TEMPERATURE (*Actual measured suction line temperature*)	
EVAPORATOR PRESSURE (*Low side pressure on gauges*)	
EVAPORATOR SATURATION TEMP (*Low side pressure converted to temp by PT*)	
SYSTEM SUPERHEAT (*Suction line temp minus evaporator saturation temp*)	
SYSTEM CONDITION (*overcharged, undercharged, or correct*)	

LAB 29.9 SUPERHEAT CHARGING DATA SHEET 29-9-3

TYPE OF METERING DEVICE (*Expansion valve, cap tube, orifice*)	
TYPE OF REFRIGERANT (*R22, R410A, R134a, etc.*)	
OUTDOOR AMBIENT/ WATER TEMPERATURE (*Temp of condenser air or water*)	
RETURN AIR DRY BULB TEMP (*Regular return air temperature*)	
RETURN AIR WET BULB TEMP (*Temperature measurement from psychrometer*)	
DESIRED SYSTEM SUPERHEAT (*Superheat from manufacturer's chart*)	
SUCTION LINE TEMPERATURE (*Actual measured suction line temperature*)	
EVAPORATOR PRESSURE (*Low side pressure on gauges*)	
EVAPORATOR SATURATION TEMP (*Low side pressure converted to temp by PT*)	
SYSTEM SUPERHEAT (*Suction line temp minus evaporator saturation temp*)	
SYSTEM CONDITION (*overcharged, undercharged, or correct*)	

LAB 29.10 PARTIAL CHARGE – SUPERHEAT

LABORATORY OBJECTIVE

You will demonstrate your ability to safely add refrigerant to an operating air conditioning system until the operating superheat meets the manufacturer's specifications.

FUNDAMENTALS OF HVACR TEXT REFERENCE

Unit 29 Refrigerant System Charging

LABORATORY NOTES

The capacity of air conditioning systems which have fixed restriction metering devices varies depending upon the outdoor ambient temperature and the indoor wet bulb temperature. This capacity difference is reflected in the system operating superheat. You will connect your gauges, operate an air conditioning system with a fixed restriction metering device, and check its charge according to the manufacturer's superheat specifications. You will add refrigerant to the low side with the operating system until the operating superheat meets the manufacturer's specification.

REQUIRED TOOLS, EQUIPMENT, AND MATERIALS

Safety glasses and gloves
Manifold gauges and extra hose
Refrigeration wrench
Thermometer
Operating refrigeration system
Refrigerant
Scale

SAFETY REQUIREMENTS

  A. Wear safety goggles and gloves when working with refrigerants. Liquid refrigerant can cause frostbite when in contact with eyes and skin.
  B. Use low loss hose fittings to reduce refrigerant spray when connecting and disconnecting refrigeration hoses. Inspect all fittings before attaching hoses.

PROCEDURE

Step 1 – Equipment Identification

Examine the unit and record important unit characteristics in the data sheet, including:
  A. Type of Refrigerant
  B. Metering Device (TEV or fixed restriction)

Note: Superheat charge checking is only used for systems with fixed restriction metering devices.

Step 2 – Operating Conditions

Measure and record the relevant operating conditions, including outdoor ambient temperature, return air temperature, and return air wet bulb temperature.

Step 3 – Determine the Manufacturer's Specified System Superheat

Use the manufacturer's information to determine the desired superheat and record it on the data sheet. The outdoor ambient temperature and the return air wet bulb temperature are the most commonly specified operating conditions used to determine system's specified operating superheat.

Step 4 – Check Operating Superheat

Install gauge manifolds on system. Place a clamp or strap type electronic thermometer on the suction line entering the condensing unit. Operate system until pressures stabilize and compressor is warm. Read and record the suction line temperature and the suction line pressure. Determine the evaporator saturation temperature using a PT Chart. Calculate the superheat using the formula

Operating Superheat = Suction Line Temperature – Evaporator Saturation Temperature

Step 5 – Assessment

Compare the operating superheat to the desired superheat to determine if the unit is overcharged, undercharged, or correct. If the operating superheat is lower than the manufacturer's specified superheat the system is overcharged. If the operating superheat is higher than the manufacturer's specified superheat the system is undercharged. However, there are operating conditions that can cause system superheat to be either high or low. Make certain airflow is correct before adjusting the system charge.

Step 6 – Adjusting Charge

If the system is undercharged and the refrigerant is a compound or azeotropic then

    A. Connect the middle gauge hose to the vapor valve on a refrigerant cylinder.
          (Make sure that the refrigerant listed on the data plate is the same as in the cylinder.)
    B. Open the cylinder valve and purge the middle hose at the manifold.
    C. Turn the scale on, place the cylinder upright on the scale and zero the scale.
    D. With the unit operating, open the low side manifold hand wheel.
    E. The compound gauge should show a higher pressure – this is the cylinder pressure.
    F. Periodically close the low side manifold hand wheel to see what the system operating pressure is.
    G. Monitor the suction pressure and suction line temperature, recalculate the superheat each time you add refrigerant.
    H. Repeat until the operating superheat equals the manufacturer's specified superheat.
    I. Record the system operating data and amount of refrigerant used on Data Sheet 29-10-1.

Else if the system is undercharged and the refrigerant is zeotropic

    A. Connect the middle gauge hose to the vapor valve on a refrigerant cylinder.

    B. *(Make sure that the refrigerant listed on the data plate is the same as in the cylinder.)*

    C. Open the cylinder valve and purge the middle hose at the manifold.

    D. Turn the scale on, place the cylinder in the inverted (upside down) position on the scale and zero the scale.

    E. With the unit operating, **crack** the low side manifold hand wheel for no more than 5 seconds.

    F. Wait at least 15 seconds before opening the valve again.

    G. The compound gauge should show a higher pressure when the hand wheel is opened this is the cylinder pressure.

    H. Monitor the suction line temperature and suction pressure, recalculating superheat each time you add refrigerant.

    I. Repeat until the operating superheat equals the manufacturer's specified superheat.

    J. Record the system operating data and amount of refrigerant used on Data Sheet 29-10-1.

Else If the system is overcharged then

    A. Connect the middle gauge hose to the vapor valve on a refrigerant recovery cylinder. *(Make sure that the refrigerant listed on the data plate is the same as in the cylinder.)*

    B. Open the cylinder valve and purge the middle hose at the manifold.

    C. Turn the scale on, place the cylinder in the upright position on the scale and zero the scale.

    D. With the unit operating, open the high side manifold hand wheel for a few seconds and then close it.

    E. The high side gauge should show a lower pressure when the hand wheel is opened – this is the cylinder pressure.

    F. Monitor the suction line temperature and suction pressure, recalculating superheat each time you remove refrigerant.

    G. Repeat until the operating superheat equals the manufacturer's specified superheat.

    H. Record the system operating data and amount of refrigerant recovered on Data Sheet 29-10-1.

DATA SHEET 29-10-1 CHARGING BY SUPERHEAT	
TYPE OF METERING DEVICE (*Expansion valve, cap tube, orifice*)	
TYPE OF REFRIGERANT (*R22, R410A, R134a, etc.*)	
OUTDOOR AMBIENT TEMPERATURE (*Temp of air over condenser*)	
RETURN AIR DRY BULB TEMP (*Regular return air temperature*)	
RETURN AIR WET BULB TEMP (*Temperature measurement from psychrometer*)	
TARGET SYSTEM SUPERHEAT (*Superheat from manufacturer's chart*)	
SUCTION LINE TEMPERATURE (*Actual measured suction line temperature*)	
EVAPORATOR PRESSURE (*Low side pressure on gauges*)	
EVAPORATOR SATURATION TEMP (*Low side pressure converted to temp by PT*)	
SYSTEM SUPERHEAT (*Suction line temp minus evaporator saturation temp*)	

LAB 29.11 SUBCOOLING CHARGE CHECK

LABORATORY OBJECTIVE
You will demonstrate your ability to check refrigerant charge in three refrigeration system using condenser subcooling.

FUNDAMENTALS OF HVACR TEXT REFERENCE
Unit 29 Refrigerant System Charging

LABORATORY NOTES
You will first determine the manufacturer's recommended subcooling. You will check system subcooling using the system liquid pressure and the liquid line temperature and compare the actual operating subcooling to the manufacturer's recommended subcooling. Finally, you will determine if the system is overcharged, undercharged, or correctly charged by comparing the recommended subcooling with the actual subcooling.

REQUIRED TOOLS, EQUIPMENT, AND MATERIALS
Safety glasses and gloves
Manifold gauges and refrigeration wrench
Thermometer
Three operating refrigeration systems with TEV metering devices

SAFETY REQUIREMENTS
A. Wear safety goggles and gloves when working with refrigerants. Liquid refrigerant can cause frostbite when in contact with eyes and skin.
B. Use low loss hose fittings to reduce refrigerant spray when connecting and disconnecting refrigeration hoses. Inspect all fittings before attaching hoses.

PROCEDURE

Step 1 – Equipment Identification
Examine the unit and record important unit characteristics in the data sheet, including:
A. Type of Refrigerant
B. Metering Device (TEV or fixed restriction)
   Note: Subcooling charge checking is only used for systems with thermostatic expansion valve metering devices.

Step 2 – Operating Conditions
Measure and record the relevant operating conditions, including outdoor ambient temperature, return air temperature, and return air wet bulb temperature.

Step 3 – Determine the Manufacturer's Specified System Subcooling

Use the manufacturer's information to determine the desired subcooling and record it on the data sheet. The outdoor ambient temperature is the most commonly specified operating condition used to determine a system's specified operating subcooling. Many manufacturers specify the subcooling on the unit data plate.

Step 4 – Check Operating Subcooling

Install gauge manifolds on system. Place a clamp or strap type electronic thermometer on the liquid line leaving the condensing unit. Operate system until pressures stabilize and compressor is warm. Read and record the liquid line temperature and the liquid line pressure. Determine the condenser saturation temperature using a PT Chart. Calculate the subcooling using the formula

Operating Subcooling = Condenser Saturation Temperature – Liquid Line Temperature

Step 5 – Assessment

Compare the operating subcooling to the target subcooling to determine if the unit is overcharged, undercharged, or correct. If the operating subcooling is higher than the manufacturer's specified subcooling the system is overcharged. If the operating subcooling is lower than the manufacturer's specified subcooling the system is undercharged. However, there are operating conditions that can cause system subcooling to be either high or low. Make certain airflow is correct before adjusting the system charge.
Record the system data on data sheet.

Step 6 – System 2

Repeat and record data on Data Sheet 29-11-2.

Step 7 – System 3

Repeat and record data on Data Sheet 29-11-3.

DATA SHEET 29-11-1 SUBCOOLING CHARGE CHECK	
TYPE OF METERING DEVICE (*Expansion valve, cap tube, orifice*)	
TYPE OF REFRIGERANT (*R22, R410A, R134a, etc.*)	
OUTDOOR AMBIENT TEMPERATURE (*Temp of air over condenser*)	
RETURN AIR DRY BULB TEMP (*Regular return air temperature*)	
DESIRED SYSTEM SUBCOOLING (*Subcooling from manufacturer's chart*)	
CONDENSER PRESSURE (*High side pressure on gauges*)	
CONDENSER SATURATION TEMP (*High side pressure converted to temp by PT*)	
LIQUID LINE TEMPERATURE (*Actual measured liquid line temperature*)	
SYSTEM SUBCOOLING (*Condenser saturation temp minus liquid line temp*)	
SYSTEM CONDITION (*overcharged, undercharged, or correct*)	

DATA SHEET 29-11-2 SUBCOOLING CHARGE CHECK

TYPE OF METERING DEVICE (*Expansion valve, cap tube, orifice*)	
TYPE OF REFRIGERANT (*R22, R410A, R134a, etc.*)	
OUTDOOR AMBIENT TEMPERATURE (*Temp of air over condenser*)	
RETURN AIR DRY BULB TEMP (*Regular return air temperature*)	
DESIRED SYSTEM SUBCOOLING (*Subcooling from manufacturer's chart*)	
CONDENSER PRESSURE (*High side pressure on gauges*)	
CONDENSER SATURATION TEMP (*High side pressure converted to temp by PT*)	
LIQUID LINE TEMPERATURE (*Actual measured liquid line temperature*)	
SYSTEM SUBCOOLING (*Condenser saturation temp minus liquid line temp*)	
SYSTEM CONDITION (*overcharged, undercharged, or correct*)	

DATA SHEET 29-11-3 SUBCOOLING CHARGE CHECK

TYPE OF METERING DEVICE (*Expansion valve, cap tube, orifice*)	
TYPE OF REFRIGERANT (*R22, R410A, R134a, etc.*)	
OUTDOOR AMBIENT TEMPERATURE (*Temp of air over condenser*)	
RETURN AIR DRY BULB TEMP (*Regular return air temperature*)	
DESIRED SYSTEM SUBCOOLING (*Subcooling from manufacturer's chart*)	
CONDENSER PRESSURE (*High side pressure on gauges*)	
CONDENSER SATURATION TEMP (*High side pressure converted to temp by PT*)	
LIQUID LINE TEMPERATURE (*Actual measured liquid line temperature*)	
SYSTEM SUBCOOLING (*Condenser saturation temp minus liquid line temp*)	
SYSTEM CONDITION (*overcharged, undercharged, or correct*)	

LAB 29.12 CHARGE CHECK USING MANUFACTURER'S CHARGING CHART

LABORATORY OBJECTIVE

You will demonstrate your ability to check refrigerant charge using a manufacturer's charging chart in three refrigeration systems.

***FUNDAMENTALS OF HVACR* TEXT REFERENCE**

Unit 29 Refrigerant System Charging

LABORATORY NOTES

You will inspect the unit information to determine the correct method for checking the refrigerant charge for three different refrigeration systems. You will then check the charge and compare the actual operating performance to the manufacturer's specified system performance and determine if the unit is overcharged, undercharged, or correctly charged.

REQUIRED TOOLS, EQUIPMENT, AND MATERIALS

Safety glasses and gloves
Manifold gauges and refrigeration wrench
Thermometer
Three operating refrigeration systems

SAFETY REQUIREMENTS

A. Wear safety goggles and gloves when working with refrigerants. Liquid refrigerant can cause frostbite when in contact with eyes and skin.
B. Use low loss hose fittings to reduce refrigerant spray when connecting and disconnecting refrigeration hoses. Inspect all fittings before attaching hoses.

PROCEDURE

Step 1 – Equipment Identification

Examine the unit and record important unit characteristics in the data sheet, including:

A. Type of Unit (air conditioner, freezer, etc.)
B. Type of Refrigerant
C. Metering Device (TEV or fixed restriction)
D. Condenser (air or water)
E. Evaporator (air or water)

Step 2 – Operating Conditions

Measure and record the relevant operating conditions, including:
   A. Ambient Temperature for air cooled condensers
   B. Inlet water temperature for water cooled condensers
   C. Return Air Temperature for air conditioning units
   D. Return Air We Bulb Temperature for air conditioning units
   E. Box Temperature for commercial refrigeration

Step 3 – Determine the Manufacturer's Specified Target Data

Examine the system charging information to determine the manufacturer's recommendation. Measure and record the relevant operating conditions specified by the manufacturer *(Note: Different manufacturers may specify different data. The data sheet contains places for more data than most systems will require. You ONLY need to measure and record data required by the manufacturer.)*

Step 4 – Measure and Record the System Operating Data

Measure the appropriate data and record in the data sheet.

Step 5 – Assess the System Operating Data

Compare the measured operating characteristics to the manufacturer's desired operating characteristics to determine if the system is overcharged, undercharged, or charged correctly.

Note: Only the data required by the manufacturer is necessary to collect.

MANUFACTURER'S CHARGING DATA SHEET 29-12-1	
TYPE OF UNIT (*Air conditioner, freezer, etc.*)	
TYPE OF CONDENSER (*Air or Water cooled*)	
TYPE OF EVAPORATOR (*Cools Air or Water*)	
TYPE OF METERING DEVICE (*Expansion valve, cap tube, orifice*)	
TYPE OF REFRIGERANT (*R22, R410A, R134a, etc.*)	
OUTDOOR AMBIENT/ WATER TEMPERATURE (*Temp of condenser air or water*)	
RETURN AIR DRY BULB TEMP (*Regular return air temperature*)	
RETURN AIR WET BULB TEMP (*Temperature measurement from psychrometer*)	
DESIRED SUPERHEAT, SUBCOOLING, OR APPROACH *(from manufacturers chart)*	
CONDENSER PRESSURE (*High side pressure on gauges*)	
CONDENSER SATURATION TEMP (*High side pressure converted to temp by PT*)	
LIQUID LINE TEMPERATURE *(Actual measured liquid line temperature)*	
CONDENSER APPROACH TEMP *(liquid line temp minus outdoor ambient)*	
SYSTEM SUBCOOLING *(Condenser saturation temp minus liquid line temp)*	
SYSTEM SUPERHEAT *(Suction line temp minus evaporator saturation temp)*	
DESIRED EVAPORATOR PRESSURE *(Evaporator Pressure from Manufacturers chart)*	
DESIRED CONDENSER PRESSURE *(Condenser Pressure from Manufacturer's chart)*	
SYSTEM CONDITION *(overcharged, undercharged, or correct)*	

LAB 29.13 WATER COOLED SYSTEM CHARGE CHECK

LABORATORY OBJECTIVE
You will demonstrate your ability to check refrigerant charge in a system with a water cooled condenser using the manufacturer's charging chart.

FUNDAMENTALS OF HVACR TEXT REFERENCE
Unit 29 Refrigerant System Charging

LABORATORY NOTES
You will inspect the unit information to determine the correct method for checking the refrigerant charge. You will then check the charge and compare the actual conditions to the desired conditions and determine if the unit is overcharged, undercharged, or correctly charged.

REQUIRED TOOLS, EQUIPMENT, AND MATERIALS
Safety glasses and gloves
Manifold gauges and refrigeration wrench
Thermometer
Operating water cooled refrigeration system

SAFETY REQUIREMENTS
A. Wear safety goggles and gloves when working with refrigerants. Liquid refrigerant can cause frostbite when in contact with eyes and skin.
B. Use low loss hose fittings to reduce refrigerant spray when connecting and disconnecting refrigeration hoses. Inspect all fittings before attaching hoses.

PROCEDURE

Step 1 – Equipment Identification
Examine the unit and record important unit characteristics in the data sheet, including:
A. Type of Unit (air conditioner, freezer, etc.)
B. Type of Refrigerant
C. Metering Device (TEV or fixed restriction)

Step 2 – Operating Conditions
Measure and record the relevant operating conditions, including:
A. Inlet condenser water temperature
B. Return Air Temperature for air conditioning units
C. Return Air We Bulb Temperature for air conditioning units
D. Box Temperature for commercial refrigeration

312

Step 3 – **Determine the Manufacturer's Specified Target Data**

Examine the system charging information to determine the manufacturer's recommendation.

      A. Water cooled systems typically specify

      B. Waterflow trough the condenser

      C. Inlet water temperature

      D. Water temperature rise

      E. Condenser pressure

      F. Evaporator pressure

Step 4 – **Measure and Record the System Operating Data**

Measure the appropriate data and record in the data sheet.

Step 5 – **Assess the System Operating Data**

Compare the measured operating characteristics to the manufacturer's desired operating characteristics to determine if the system is overcharged, undercharged, or charged correctly. Condenser waterflow is crucial! Low condenser waterflow can make the pressure look like the system is overcharged, high condenser waterflow can make the pressures look like the system is undercharged.

Note: **Only the data required by the manufacturer is necessary to collect.**

LAB 29.13 WATER COOLED SYSTEM CHARGING DATA SHEET	
TYPE OF REFRIGERANT (*R22, R410A, R134a, etc.*)	
SPECIFIED CONDENSER WATER FLOW (*water flow from manufacturer's chart*)	
ACTUAL CONDENSER WATER FLOW (*water flow from flow meter*)	
RETURN AIR DRY BULB TEMP (*Regular return air temperature*)	
RETURN AIR WET BULB TEMP (*Temperature measurement from psychrometer*)	
SPECIFIED WATER TEMP DIFFERENCE (*water temp difference from manufacturer's chart*)	
ENTERING WATER TEMPERATURE (*Temperature of entering water*)	
LEAVING WATER TEMPERATURE (*Temperature of leaving water*)	
ACTUAL WATER TEMP DIFFERENCE (*leaving water minus entering water*)	
DESIRED CONDENSER PRESSURE (*Condenser Pressure from manufacturer chart*)	
ACTUAL CONDENSER PRESSURE (*Condenser Pressure from Gauges*)	
DESIRED EVAPORATOR PRESSURE (*Evaporator Pressure from manufacturer chart*)	
ACTUAL EVAPORATOR PRESSURE (*Evaporator Pressure from Gauges*)	
SYSTEM CONDITION (*overcharged, undercharged, or correct*)	

LAB 30.1 USING NON-CONTACT VOLTAGE DETECTORS

LABORATORY OBJECTIVE

You will use a non-contact voltage detector to safely distinguish between energized electrical components and wires and non-energized components and wires.

FUNDAMENTALS OF HVACR TEXT REFERENCE

Unit 30 Electrical Safety

LABORATORY NOTES

You will hold your non-contact voltage detector near components and wires to determine which are energized and which are not energized.

REQUIRED TOOLS, EQUIPMENT, AND MATERIALS

Safety glasses
Non-contact voltage detector
Energized electrical components and wires
De-energized electrical components and wires

SAFETY REQUIREMENTS

A. Wear safety glasses
B. In general, avoid touching things. Do not touch ANYTHING without first testing with your non-contact voltage detector.
C. A circuit does not have to be operating to be energized with a potentially fatal voltage.
D. Always assume that ALL circuits can kill you.

PROCEDURE

1. Hold your non-contact voltage detector near the components and wires assigned by the instructor.
2. Do NOT open electrical panels or stick your hand near electrical equipment that the instructor has not assigned you to inspect.
3. Most testers have some type of confidence check, such as a light flash or beep when the button is first pressed. Become familiar with yours and do not use it when it fails to beep or flash when first activated.
4. Non-contact voltage detectors normally do not indicate through grounded metal enclosures or shielded cables.
5. Be prepared to show the instructor the difference between an energized circuit and a de-energized circuit.

LAB 30.2 ELECTRICAL SAFETY PROCEDURES

LABORATORY OBJECTIVE

You will identify the location of the electric disconnect switches in the shop and demonstrate your ability to turn off a disconnect switch.

FUNDAMENTALS OF HVACR TEXT REFERENCE

Unit 30 Electrical Safety

LABORATORY NOTES

You will identify the disconnect switch for an air conditioning unit and demonstrate how to turn it off. You will then locate the electric panel that feeds that disconnect switch and identify the circuit breaker that feeds the disconnect switch. Finally, you will locate the main switchgear that feeds the electric panel.

REQUIRED TOOLS, EQUIPMENT, AND MATERIALS

Safety glasses

Electrical disconnects

SAFETY REQUIREMENTS

A. Wear safety glasses
B. In general, avoid touching things. Do not touch ANYTHING without first testing with your non-contact voltage detector.

PROCEDURE

1. Locate the power conduit leaving the unit.
2. Follow it to the disconnect switch. It should be within sight of the unit, often right next to it.
3. Check with your non-contact voltage detector to make sure the box is not energized.
4. Demonstrate how to turn the disconnect switch off.
5. The disconnect switch should be labeled, including the name of the electric panel feeding it.
6. Locate that electric panel.
7. Find the circuit breaker that controls the disconnect switch. It should be labeled on the panel.
8. Find the main switch gear for the shop.
9. Find the disconnect switch for the panel.

LAB 30.3 CHANGING FUSES

LABORATORY OBJECTIVE

You will demonstrate your ability to safely replace a cartridge fuse in a disconnect switch using non-conductive fuse pullers.

FUNDAMENTALS OF HVACR TEXT REFERENCE

Unit 30 Electrical Safety

LABORATORY NOTES

You will check the disconnect switch to make sure it is de-energized, turn off the disconnect switch, open the disconnect switch, remove a cartridge fuse using a non-conductive fuse puller, replace the cartridge fuse using a non-conductive fuse puller, and close the disconnect switch.

REQUIRED TOOLS, EQUIPMENT, AND MATERIALS

Safety glasses

Non-contact voltage detector

Non-conductive fuse puller

Electrical disconnect with cartridge fuses

SAFETY REQUIREMENTS

A. Wear safety glasses
B. In general, avoid touching things. Do not touch ANYTHING without first testing with your non-contact voltage detector.
C. A circuit does not have to be operating to be energized with a potentially fatal voltage. Always assume that ALL circuits can kill you!
D. Do NOT stand directly in front of the disconnect when opening it – stand to the side.
E. Disconnect should be turned OFF before opening.
F. Use ONLY non-conductive fuse pullers rated for the disconnect voltage.
G. Do NOT use pliers or finger to remove the fuses.

PROCEDURE

1. Check the disconnect box with your non-contact voltage detector to make sure it is not energized.
2. Turn the switch off.
3. Open the disconnect switch. Do NOT stand directly in front when opening it.
4. Use non-conductive fuse pullers to remove a cartridge fuse from the disconnect switch.
5. Safety Note: Do NOT use pliers or fingers!
6. Replace the cartridge fuse.
7. Close the disconnect switch.

LAB 31.1 SERIES AND PARALLEL CIRCUITS

LABORATORY OBJECTIVE

You will demonstrate your understanding of series and parallel circuits by safely wiring series and parallel circuits to specification.

FUNDAMENTALS OF HVACR TEXT REFERENCE

Unit 31 Basic Electricity

LABORATORY NOTES

In this lab exercise you will learn the characteristics of series, parallel, and series-parallel circuits by designing, wiring, and operating circuits using a 110-volt source.

REQUIRED TOOLS, EQUIPMENT, AND MATERIALS

Safety glasses
Multimeter
Wire stripper
Crimp tool
Wire cutter
6-in-1 screwdriver

120-volt power cord
Wire
Electrical terminals
Wire nuts
Three 120-volt light sockets and lights
Three SPST toggle switches

SAFETY REQUIREMENTS

A. Wear safety glasses
B. Power cord should be unplugged during all wiring.
C. Do NOT energize ANY circuit without instructor's explicit instruction to do so.

PROCEDURE

You should first draw each circuit using the following symbols:

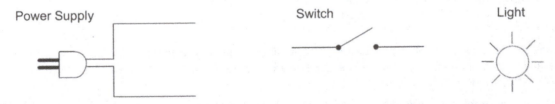

Power Supply   Switch   Light

Have the instructor check your drawing before wiring the circuit.

Wire each circuit according to your drawing.

All circuits should be checked by the instructor BEFORE they are energized.

CIRCUIT #1

1. Design and draw a circuit with 1 switch controlling 1 load. The switch should be wired in series with the load.

2. Wire and operate the circuit. (Have instructor check the circuit before operation.)

3. What happens when the switch is opened?

4. What happens when the switch is closed?

CIRCUIT #2

1. Design and draw a circuit with 2 switches controlling 1 load. The switches should be wired in series with each other and in series with the load.

2. Wire and operate the circuit. (Have the instructor check the circuit before operation.)

3. Close both switches. What happens?

4. Open 1 switch. What happens?

5. Close the first switch and open the other switch. What happens?

6. Summarize the operation of this circuit.

CIRCUIT #3

1. Design and draw a circuit with 2 switches controlling 1 load. The switches should be wired in parallel with each other but in series with the load.

2. Wire and operate the circuit. (Have instructor check the circuit before operation.)

3. Open both switches. What happens?

4. Close 1 switch. What happens?

5. Open the first switch and close the other switch. What happens?

6. Summarize the operation of this circuit.

CIRCUIT #4

1. Design and draw a circuit with 1 switch controlling 2 loads. The loads should be wired in parallel with each other but in series with the switch.

2. Wire and operate the circuit. (Have instructor check the circuit before operation.)

3. Summarize the operation of this circuit.

CIRCUIT #5

1. Design and draw a circuit with 2 switches controlling 2 loads. The loads should be wired in parallel with each other. The switches should be wired in series with each other. The switches should be wired in series with the loads.

2. Wire and operate the circuit. (Have instructor check the circuit before operation.)

3. Close both switches. What happens?

4. Open 1 switch. What happens?

5. Close the first switch and open the other switch. What happens?

6. Summarize the operation of this circuit.

CIRCUIT #6

1. Design and draw a circuit with 2 switches controlling 2 loads. The loads should be wired in parallel with each other. The switches should be wired in parallel with each other. The switches should be wired in series with the loads.

2. Wire and operate the circuit. (Have instructor check circuit before operation.)

3. Close both switches. What happens?

4. Open 1 switch. What happens?

5. Close the first switch and open the other switch. What happens?

6. Summarize the operation of the circuit.

CIRCUIT #7

1. Design and draw a circuit with 2 switches controlling 2 loads. Each load should be wired in series with one switch. The switches should be wired so that each switch controls one load and the loads operate independently.

2. Wire and operate the circuit. (Have instructor check circuit before operation.)

3. Close both switches. What happens?

4. Open 1 switch. What happens?

5. Close the first switch and open the other switch. What happens?

6. Summarize the operation of this circuit.

CIRCUIT #8

1. Design and draw a circuit with 1 switch controlling 2 loads. The loads should be wired in series with each other and in series with the switch.

2. Wire and operate the circuit. (Have the instructor check the circuit before operation.)

3. What do you notice about the amount of light the bulbs are producing?

4. Why is this?

5. With the circuit operating, CAREFULLY unscrew one of the light bulbs. What happens? Why?

6. Summarize the operation of this circuit.

CIRCUIT #9

1. Design and draw a circuit with 1 switch controlling 2 loads. The loads should be wired in series with each other and in series with the switch. Wire another switch in parallel with one light.

2. Wire and operate the circuit? (Have instructor check the circuit before operating.)

3. Close both switches. What happens?

4. Open the switch that is wired in parallel with one light. What happens?

5. Explain the operation of the parallel-wired switch.

6. How does this contrast to the operation of the series wired switch?

CIRCUIT #10

1. Design and draw a circuit with 3 loads in series. One of the loads should have a switch wired in parallel to it. A second load should also have a switch wired in parallel to it. The remaining load will operate continuously while the other 2 loads will each be controlled by their own switch.

2. Wire and operate the circuit. (Have instructor check circuit before operation.)

3. Close both switches. What happens?

4. Open 1 switch. What happens?

5. Close the first switch and open the other switch. What happens?

6. Summarize the operation of this circuit.

SUMMARY

1. Explain how switches wired in series with loads control the loads.

2. Explain how switches wired in parallel with loads control the loads.

3. Explain how loads wired in series with each other behave.

4. Explain how loads wired in parallel with each other behave.

5. Explain the operation of a circuit with one load wired in series with several switches wired in series with each other.

6. Explain the operation of a circuit with one load wired in series with two switches that are wired in parallel with each other.

7. How are MOST loads in air conditioning systems wired with respect to each other?

8. How are MOST loads in air conditioning systems wired with respect to switches?

9. How are MOST switches in air conditioning systems wired with respect to each other?

LAB 31.2 APPLY OHMS LAW TO SERIES CIRCUIT VOLTAGE CHANGES

LABORATORY OBJECTIVE

You will demonstrate your understanding of ohms law in series circuits by safely wiring series circuits to specification and measuring the circuit characteristics.

***FUNDAMENTALS OF HVACR* TEXT REFERENCE**

Unit 31 Basic Electricity

LABORATORY NOTES

In this lab exercise you will build a simple SERIES circuit using a variable power supply and two electric strip heaters. You will measure the resistance, voltage, and current at different voltage levels and compare your measurements using Ohm's law.

REQUIRED TOOLS, EQUIPMENT, AND MATERIALS

Safety glasses	120-volt power cord
Multimeter	Adjustable power supply
Ammeter	Wire
Wire stripper/crimper	Electrical terminals
Wire cutter	Wire nuts
6-in-1 screwdriver	Two electric heaters

SAFETY REQUIREMENTS

A. Wear safety glasses
B. Power cord should be unplugged during all wiring.
C. Do NOT energize ANY circuit without instructor's explicit instruction to do so.

PROCEDURE

Step 1

WITH THE CIRCUIT DISCONNECTED FROM THE POWER SUPPLY, measure the resistance of each heater and the total circuit resistance. Record your measurements.

Note: The power cord must be unplugged from the adjustable power supply to get a correct reading because the coil in the power supply will affect the reading.

Step 2

Use ohms law to calculate the total circuit resistance, the circuit current and voltage drop across each heater for the voltages listed in the chart. Record your calculations.

Step 3

Adjust the voltage source to produce the first voltage shown on the chart below.

Step 4

Measure the circuit current with a clamp-on ammeter. Use a volt meter to measure the voltage drop across each heater and across the entire circuit. Record your measurements.
Note: You may need to wrap the wire around the jaw of the ammeter several times to get a reading.

Step 5

Repeat steps 3 and 4 for each of the other voltages listed in the chart.

LAB 31.2 SERIES HEATERS DATA TABLE				
	100 Volt Setting Measured	100 Volt Setting Calculated	50 Volt Setting Measured	50 Volt Setting Calculated
Heater 1 Resistance				
Heater 2 Resistance				
Total Circuit Resistance				
Heater 1 Volts				
Heater 2 Volts				
Total Circuit Volts		100 volts		50 volts
Heater 1 Amps				
Heater 2 Amps				
Total Circuit Amps				

LAB 31.3 OHMS LAW & PARALLEL CIRCUIT VOLTAGE CHANGES

LABORATORY OBJECTIVE

You will demonstrate your understanding of ohms law in parallel circuits by safely wiring parallel circuits to specification and measuring the circuit characteristics.

FUNDAMENTALS OF HVACR TEXT REFERENCE

Unit 31 Basic Electricity

LABORATORY NOTES

In this lab exercise you will build a simple PARALLEL circuit using a variable power supply and two electric strip heaters. You will measure the resistance, voltage, and current at different voltage levels and compare your measurements using Ohm's law.

REQUIRED TOOLS, EQUIPMENT, AND MATERIALS

Safety glasses	120-volt power cord
Multimeter	Adjustable power supply
Ammeter	Wire
Wire stripper/crimper	Electrical terminals
Wire cutter	Wire nuts
6-in-1 screwdriver	Two electric heaters

SAFETY REQUIREMENTS

A. Wear safety glasses
B. Power cord should be unplugged during all wiring.
C. Do NOT energize ANY circuit without instructor's explicit instruction to do so.

PROCEDURE
Step 1

BEFORE CONSTRUCTING THE CIRCUIT, measure the resistance OF EACH HEATER.

Note: You cannot read the resistance of an individual component in a parallel circuit without removing it from the circuit. If the device is still connected in parallel, you will read the total resistance of the entire parallel circuit, not just the individual device.

Step 2

Wire the two heaters in parallel and read the total circuit resistance BEFORE connecting the circuit to the adjustable power supply.

Note: The power cord must be unplugged from the adjustable power supply to get a correct reading because the coil in the power supply will affect the reading.

Step 3

Use ohms law to calculate the total circuit resistance, circuit current and current draw of each heater for the voltages listed in the chart. Record your calculations.

Step 4

Adjust the voltage source to produce the first voltage shown on the chart below.

Step 5

Measure the total circuit current and the individual heater currents with a clamp-on ammeter. Measure the voltage across the total circuit and across each heater with a volt meter. Record your measurements.

Step 6

Repeat steps 4 and 5 for each of the other voltages listed in the chart.

LAB 31.3 PARALLEL HEATERS DATA TABLE				
	50 Volt Setting Measured	50 Volt Setting Calculated	25 Volt Setting Measured	25 Volt Setting Calculated
Heater 1 Resistance				
Heater 2 Resistance				
Total Circuit Resistance				
Heater 1 Volts				
Heater 2 Volts				
Total Circuit Volts		50 volts		25 volts
Heater 1 Amps				
Heater 2 Amps				
Total Circuit Amps				

324

LAB 32.1 ALTERNATING CURRENT PRINCIPLES

LABORATORY OBJECTIVE

You will demonstrate the effects of capacitive and inductive reactance in alternating current circuits.

FUNDAMENTALS OF HVACR TEXT REFERENCE

Unit 32 Alternating Current Fundamentals

LABORATORY NOTES

In this portion of the lab exercise you will wire a 20-microfarad capacitor in series with a 120-volt light, measure the circuit current and the voltage drop across the light and the capacitor. You will then repeat this using a 10-microfarad capacitor and compare the results.

REQUIRED TOOLS, EQUIPMENT, AND MATERIALS

Safety glasses	120-volt power cord
Multimeter	60-watt incandescent light
Ammeter	20 microfarad 370 volt run capacitor
Wire cutter	10 microfarad 370 volt run capacitor
	Wire crimper/stripper
120-volt coil	Iron center to insert in coil
6-in-1 screwdriver	120-volt shaded pole motor

SAFETY REQUIREMENTS

A. Wear safety glasses

B. Power cord should be unplugged during all wiring.

C. Do NOT energize ANY circuit without instructor's explicit instruction to do so.

PROCEDURE

Capacitive Reactance

1. Wire a 20-microfarad run capacitor in series with a 60-watt light. Connect the circuit to a 110-volt AC source.
2. Measure the circuit amp draw._____
3. Measure the voltage across the light._____
4. Measure the voltage across the capacitor._____
5. Measure the source voltage._____

6. Discharge the capacitor by unplugging the circuit and shorting the plug ends across a piece of metal or wire. Test the capacitor charge by measuring the voltage across the capacitor terminals with the DC scale on your volt meter. If there is no charge, proceed to the next step. If there is a voltage, wait until the meter discharges the capacitor and the voltage is close to 0 volts DC. What we are doing here is removing the charge on the capacitor so that it is safe to handle.

7. Replace the 20 MFD capacitor with a 10 MFD capacitor and operate the circuit.
8. Measure the circuit amp draw. _____
9. Measure the voltage across the light._____
10. Measure the voltage across the capacitor. _____
11. Measure the source voltage._____
12. Which capacitor produces a higher amp draw?
13. Which capacitor appears to have a higher capacitive reactance, the 20 MFD or the 10 MFD? Why?

OVERVIEW

Inductive Reactance of Coil

In this portion of the lab exercise you will wire a coil in series with a 120-volt light, measure the circuit current and the voltage drop across the light and the coil. You will then insert an iron center, measure the voltages and current again, and compare the results with the iron center in and out of the coil.

1. Wire a 120-volt solenoid coil in series with a 60-watt light. The solenoid should have its core removed.
2. Measure the amp draw of the circuit. _____
3. Measure the voltage across the light. _____
4. Measure the voltage across the solenoid coil. _____
5. Slowly slide an iron or steel core into the solenoid while the circuit is operating.
6. What happens to the light?
7. Measure the amp draw of the circuit. _____
8. Measure the voltage across the light. _____
9. Measure the voltage across the solenoid coil. _____
10. Explain you results.
11. When does the solenoid appear to have a higher inductive reactance? Why?

OVERVIEW

Inductive Reactance of Motor

In this portion of the lab exercise you will wire a shaded pole motor in series with a 120-volt light, measure the circuit current and the voltage drop across the light and the motor. You will then stall the motor, measure the voltages and current again, and compare the results with the motor spinning and the motor stalled.

1. Replace the solenoid with a small 110-volt shaded pole motor and operate the circuit.
2. Measure the amp draw of the circuit. _____
3. Measure the voltage across the light. _____
4. Measure the voltage across the motor. _____
5. Explain your results.
6. Hold the rotor of the motor so that it can't turn and operate the circuit.
7. *Note:* You should be using a LOW TORQUE MOTOR. This would NOT BE SAFE with a high torque motor. If you do not know the difference, ASK THE INSTRUCTOR!
8. Now, release the motor.
9. What happens?
10. Does the motor appear to have more inductive reactance when it is turning or when it is stationary? Why?

LAB 32.2 ALTERNATING CURRENT DEMONSTRATION

LABORATORY OBJECTIVE
(This lab is demonstrated by the instructor.)
The instructor will demonstrate the effects of capacitive and inductive reactance in a resonant alternating current circuit.

FUNDAMENTALS OF HVACR TEXT REFERENCE
Unit 32 Alternating Current
Fundamentals

LABORATORY NOTES
The instructor will wire a circuit with a light, capacitor, and motor in series. One SPST switch is wired in parallel with the capacitor and another SPST switch is wired in parallel with the motor. With both switches off, all three devices are in the circuit. Turning either switch on shorts the component it is wired in parallel with out of the circuit. This allows the instructor to show the effects of pure resistance, inductive reactance, capacitive reactance, and resonance where the inductive and capacitive reactance offset each other. A dual-trace oscilloscope will be connected to the incoming voltage and to the voltage across the light. The voltage across the light will be in phase with the circuit current, so the current phase can be seen with it. Comparing the two waveforms you will see the waveforms shift as the capacitive and inductive reactance is added or subtracted from the circuit.

REQUIRED TOOLS, EQUIPMENT, AND MATERIALS
Multimeter and ammeter
Wire cutter and wire crimper/stripper
6-in-1 screwdriver
120-volt power cord and some wire
60-watt incandescent light
10 microfarad 370 volt run capacitor
Small 120-volt shaded pole motor (8-watt unit bearing motor)
Two SPST switches

SAFETY REQUIREMENTS
Wear safety glasses

PROCEDURE

Resonance

1. Refer to Figure 32-2-1 for wiring diagram. Description follows.
2. Wire a small 230-volt shaded pole motor in series with a 20-MFD capacitor and a 60-watt light bulb.
3. Wire a SPST toggle switch in parallel to the capacitor.
4. Wire another SPST toggle switch in parallel to the motor.
5. Connect the circuit to a 110-volt AC. power source. The neutral, or ground, should be connected to the light and the 120 volt "hot" wire should be connected to the motor.
6. While operate the circuit with both toggle switches in the OFF position.
7. Measure the voltage across the motor. _____
8. Measure the voltage across the light. _____
9. Measure voltage across the capacitor. _____
10. Turn the toggle switch wired in parallel to the motor ON but leave the other switch OFF.
11. Measure the voltage across the motor. _____
12. Measure the voltage across the light. _____
13. Measure the voltage across the capacitor. _____
14. Turn the toggle switch wired in parallel to the capacitor ON but turn the other witch OFF.
15. Measure the voltage across the motor. _____
16. Measure the voltage across the light. _____
17. Measure the voltage across the capacitor. _____
18. With a dual-trace oscilloscope, place lead A on the incoming voltage. This lead will show the phase angle of the voltage in the circuit.
19. Place lead B between the light bulb socket and the capacitor. This lead will measure the phase angle of the current in the circuit.
20. Operate the circuit. Switch each toggle switch on and off and observe the effect on the oscilloscope.

EXPLANATION

An oscilloscope is a meter that reads voltage. However, rather than giving a number which corresponds to the effective voltage of an AC circuit, it shows the entire waveform. We see the entire sine wave. Lead A, on the incoming voltage, will show us the sine wave of the voltage being supplied to the circuit. Lead B will read the sine wave of the voltage drop across the light, our resistive load in this circuit. Remember, in resistive circuits the current and voltage are in phase with each other. Therefore, by reading the voltage across the resistive load, we are also showing the phase of the current in the circuit. By comparing the two sine waves produced, we can visually see the phase shift between voltage and current in an alternating current circuit.

Figure 32-2-1

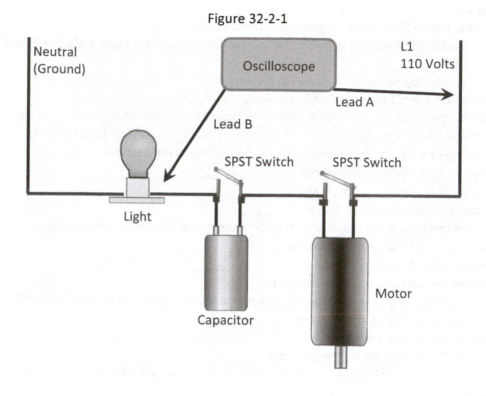

Neutral
(Ground)

Oscilloscope

L1
110 Volts

Lead A

Lead B

SPST Switch

SPST Switch

Light

Capacitor

Motor

Name_____

Date_____

Instructor's OK ☐

LAB 33.1 ELECTRICAL METER FAMILIARIZATION

LABORATORY OBJECTIVE
The student will demonstrate knowledge of electrical meter types, functions, and settings.

FUNDAMENTALS OF HVACR TEXT REFERENCE
Unit 33 Electrical Measuring and Test Instruments

LABORATORY NOTES
You will recognize different styles of electrical meters, identify their functions, and explain how these functions are used.

REQUIRED TOOLS, EQUIPMENT, AND MATERIALS
Analog VOM
Digital multimeter
Analog clamp-on ammeter
Digital clamp-on ammeter
Variable transformer
Resistors
Small light, motor, or heater that can be plugged in to measure current

SAFETY REQUIREMENTS
A. Wear safety glasses and gloves
B. Never tough energized circuit components

BACKGROUND INFORMATION
ANALOG VOMS
VOM stands for volt, ohm, milliammeter. Analog VOMs can measure volts, ohms, and small amounts of inline current. Analog meters use a needle that moves across a set of scales to indicate the reading. Technicians must understand which scale to read and how to interpret the reading to use an analog meter, Figure 33-1-1 shows scales on a typical analog meter.

Figure 33-1-1 Analog VOM scales

Figure 33-1-2 Ohm scales

This meter has a mirrored scale. The purpose of the mirror is to avoid parallax. Parallax is a condition in which objects appear in different places in relation to each other than they actually are because of the viewing angle. When looking at the meter straight on you will only see one needle. When looking from either side you will see both the needle and the needle's reflection in the mirror, telling you that you are not viewing the meter correctly.

Ohms

The ohms scale is on the top with infinite ohms being on the left with the needle at rest and 0 ohms being full scale on the right. The resistance reading decreases the further to the right the needle moves. The ohms scale is logarithmic, meaning the value of the marks is not even. For example, there are 5 marks between 100 ohms and 50 ohms, making each one worth 10 ohms. There are also 5 marks between 5 and 0, making each one worth 1 ohm.

Analog meters typically have multiple ohm scales. They are listed as multipliers such as R x1, R x 10, R x 1000 (1K), and R x 10,000 (10K). Figure 33-1-2 shows the scales for a typical analog meter. For these scales you read the meter and multiply that reading times the multiplier to get the actual reading. A reading of 5000 ohms would have the needle pointing to 5 while on the R x 1000 scale.

Volts

On this meter, volts are read underneath the mirror. There are multiple voltage scales for reading voltages in different ranges. The voltage scales run from 0 volts on the left with the needle at rest, to the maximum voltage reading for each scale on the right at full scale. The marked voltage scales are 1200, 60, 15, and 3. Which scale you read depends upon the selection you make with the range selection switch. The selection switch offers options of 1200, 600, 300, 60, 15, 3, and 0.6 volts. These are the maximum voltage that can be read on each setting. Should you try to read a voltage higher than the maximum for that scale, you will either blow a fuse in the meter, or damage the meter. That is why you generally start at the highest voltage scales and work down. Notice that there are more possible selections than there are scales. You read one of the marked scales and apply a simple math transformation to the reading to make the highest number on the scale equal the highest possible reading. For example, to read voltage on the 600 volt setting you read the 60 scale and add a 0.

Milliamps

Analog meters similar to this one can only read small amounts of current, and only with the meter connected in series with the load. HVACR technicians seldom do either of these. There is one instance where HVAC technicians read a small current flow with the meter connected in series: microamp measurements on furnace flame sensors. Milliamps are read on this meter using the same scales as voltage, but setting the selector switch to the desired milliamp scale.

DIGITAL MULTIMETERS

Digital multimeters can perform many more measurements than just volts, ohms, and milliamps. Figure 33-1-3 shows the functions of one popular digital multimeter. Reading clockwise from the OFF position, the functions are: volts, non-contact volt detection, ohms, frequency, temperature, microamps, milliamps, and in-line current up to 10 amps. A few settings can perform multiple functions. The function button allows technicians to change between different functions.

For example, the meter can read resistance, continuity, perform a diode test, or check a capacitor all at the ohms setting. The function button is used to toggle through the different functions. Many of these functions have ranges. Normally, the technician does not have to set the range because the meter is auto-ranging. That means it can determine on its own what range is needed based on the input it is receiving. It is possible to select a range manually by pushing the range button. Digital meters have a safety rating known as a category rating. The category rating describes the type of service for which the meter is designed. The higher the category rating, the safer the meter. Category I is only appropriate for working on small appliances that plug into a receptacle, while Category IV meters can be used on high energy measurements, such as the entrance cable to a house. Meters used in HVACR should be at least category III rated.

Figure 33-1-3 Digital multimeter functions   Figure 33-1-4 Analog clamp-on ammeter

ANALOG CLAMP METERS

Analog clamp-on ammeters measure amps by sensing the magnetic field around a wire, turning the magnetism into induced voltage, and moving a needle to indicate the current. Most clamp-on ammeters like the one shown in Figure 33-1-4 can read AC current, AC volts, and continuity. The different scales are printed on a cylinder inside the meter and are selected by rotating the cylinder. The clamp should only be placed around one wire. Analog clamp-on ammeters have largely been replaced by digital clamp-on multimeters. However, analog meters work better for times when you don't need a specific reading, just the indication of the presence or absence of current and the speed of the meter indication is important. For example, when checking the current through a compressor starting capacitor. You want to see an indication of current when the compressor starts and then no current after it gets started. This shows as a needle deflection on an analog ammeter. Many digital clamp-on ammeters are not fast enough indicating to see this.

DIGITAL CLAMP MULTIMETERS

Some digital clamp-on ammeters are simply a digital version of an analog meter, measuring amps, volts, and limited ohms. For technicians who carry a separate multimeter and ammeter, these digital clamp-on ammeters work well. Other digital clamp-on meters are more like a digital multimeter with an amp-clamp added. For technicians that only carry one meter, these digital clamp-on ammeters are ideal. The clamp-on meter shown in Figure 33-1-5 has most scales that an HVACR technician would need. The one limitation is that the maximum reading on the ohms scale is limited to 9999 ohms. For most HVACR needs this is not an issue, but there are some resistance measurements you would not be able to take, such as the resistance of some potential relay coils. The functions are selected by rotating the dial. The meter is auto-ranging, meaning that the range for each function is determined by the meter.

Figure 33-1-5 Digital clamp-on ammeter

PROCEDURE

Step 1 – Analog VOM

A. List the voltage ranges on the meter.

B. List the factors needed for each range. For example, in Figure 33-1-1 the 600 volt range is read on the 60 scale with a factor of 10.

C. List the value of each mark on each scale. For example, in Figure 33-1-1, the value of each mark on the 60 scale is 2 because there are 10 spaces between 40 and 60.

D. Select the meter's highest voltage range and read the voltage of the variable transformer output. Select a lower voltage range if necessary to read the voltage.

E. List the resistance ranges on the meter.

F. Select R x 1, place the leads together, and adjust the meter until the needle reads 0 ohms. This step is necessary before taking any resistance reading with an analog meter.

G. Read and record the resistance of the electrical component provided by the instructor.

Step 2 – Digital Multimeter

A. List the functions that this meter performs.

B. Is this an auto-ranging meter?

C. Does this meter have a non-contact voltage detection function?
    If so, put the meter to NCV and check the power cord to the variable transformer.

D. Select volts and read the voltage of the variable transformer output.

E. While reading voltage use the range button to manually change ranges. Watch the position of the decimal and any leading or trailing zeros as you change ranges.

F. Put the meter back to auto ranging. This is normally done by pushing and holding the range button.

G. Select Ohms and read the resistance of the electrical component provided by the instructor.

H. While reading resistance use the range button to manually change ranges. Watch the position of the decimal and any leading or trailing zeros as you change ranges.

I. Put the meter back to auto ranging. This is normally done by pushing and holding the range button.

J. Can this meter read temperature?
    If so, how? Insert the temperature probe into the meter and read the shop temperature. Use the F/C selector to change the reading to Celsius and then back to Fahrenheit.

K. Can this meter read capacitor capacitance?
    If so, select the capacitance function and read the capacitance of the capacitor provided by the instructor.

Step 3 – Analog Clamp-on Ammeter

A. List the amperage ranges on the meter.

B. List the voltage range on the meter.

C. List the value of each mark on each scale. For example, in Figure 33-1-4, the value of each mark on the 6 scale is 0.2 because there are 5 spaces between 5 and 6.

D. Select the meter's highest voltage range and read the voltage of the variable transformer output. Select a lower voltage range if necessary to read the voltage.

E. Select the meter's highest amperage range.

F. Plug in the electrical component provided by the instructor and read the amperage by snapping the jaw around one wire going to the component. Select a lower amperage range if necessary to read the amperage.

Step 4 – Digital Clamp-on Ammeter

A. List the functions that this meter performs.

B. Is this an auto-ranging meter?

C. Does this meter have a non-contact voltage detection function?
If so, put the meter to NCV and check the power cord to the variable transformer.

D. Select Amps on the meter, plug in the electrical component provided by the instructor and read the amperage by snapping the jaw around one wire going to the component.

E. Select volts and read the voltage of the variable transformer output.

F. Select Ohms and read the resistance of the electrical component provided by the instructor.

G. Can this meter read temperature?
If so, how? Insert the temperature probe into the meter and read the shop temperature. Use the F/C selector to change the reading to Celsius and then back to Fahrenheit.

H. Can this meter read capacitor capacitance?
If so, select the capacitance function and read the capacitance of the capacitor provided by the instructor.

LAB 33.2 MEASURING VOLTS

LABORATORY OBJECTIVE

You will demonstrate your ability to safely read voltages at different locations using a multimeter.

FUNDAMENTALS OF HVACR TEXT REFERENCE

Unit 33 Electrical Measuring and Test Instruments

LABORATORY NOTES

Voltmeters are connected in parallel with the load to read the voltage difference between two points, often referred to as the voltage drop. Always start to measure voltage using the highest range on the meter when using manual ranging meters

REQUIRED TOOLS, EQUIPMENT, AND MATERIALS

Safety glasses

Multimeter

6-in-1 screwdriver

Equipment with voltage to read

SAFETY REQUIREMENTS

In general, avoid touching things. Do not touch ANYTHING without first testing with your non-contact voltage detector. A circuit does not have to be operating to be energized with a potentially fatal voltage. Always assume that ALL circuits can kill you!

PROCEDURE

Set the meter to read AC voltage. Set manual ranging meters to the highest voltage range. Place the meter where you can see it, you should not hold it in your hand. Watch the leads as you place them on the location to be read. Look at the meter display once the leads are securely on that location. Read and record the voltage of the locations listed in the table.

Read Volts On	Volt Reading
Voltage of Wall Receptacle	
A/C Condensing Unit Disconnect	
A/C Condensing Unit Contactor	
Voltage across Run Capacitor while operating	
Control Transformer Primary	
Control Transformer Secondary	
Leg to Leg on Single-Phase Box	
Leg to Ground on Single-Phase Box	
Leg to Leg on Three-Phase Box	
Leg to Ground on Three-Phase Box	

LAB 33.3 MEASURING AMPS

LABORATORY OBJECTIVE
You will demonstrate your ability to safely read current (amps) using a clamp-on ammeter.

FUNDAMENTALS OF HVACR TEXT REFERENCE
Unit 33 Electrical Measuring and Test Instruments

LABORATORY NOTES
You will use an ammeter to check current at the different locations indicated by the chart. Loads must be energized and operating in order to read amps. Always start with the highest possible scale and then work down to the most appropriate scale. Never put the clamp around two different wires at the same time.

REQUIRED TOOLS, EQUIPMENT, AND MATERIALS
Safety glasses
Non-contact voltage detector
Ammeter
Electrical tools
Operating equipment for reading current

SAFETY REQUIREMENTS
In general, avoid touching things. Do not touch ANYTHING without first testing with your non-contact voltage detector. A circuit does not have to be operating to be energized with a potentially fatal voltage. Always assume that ALL circuits can kill you!

PROCEDURE
A circuit does have to be operating to read current. Clamp the meter jaws around a single wire. For three-phase loads, multiply the meter reading times the square root of 3 (1.73) to get the full current being used. Read and record the current (amps) of 10 operating electric loads listed in the table.

Load	Amps
Power wire of operating air conditioner	
Black wire on operating compressor	
Red wire on operating compressor	
Yellow wire on operating compressor	
Red secondary transformer wire on operating unit	
Wire to operating furnace fan motor	
Wire to operating condenser fan motor	
Wire to operating electric strip heater	
Any operating electric load	
Any operating electric load	

LAB 33.4 MEASURING OHMS

LABORATORY OBJECTIVE

You will demonstrate your ability to safely read the resistance of air conditioning components using a multimeter.

FUNDAMENTALS OF HVACR TEXT REFERENCE

Unit 33 Electrical Measuring and Test Instruments

LABORATORY NOTES

Whenever checking resistance, the power to the circuit being tested must be shutoff.
The ohmmeter uses a battery to furnish the current needed for resistance measurements.
The higher the resistance the lower the current flow.

REQUIRED TOOLS, EQUIPMENT, AND MATERIALS

Non-contact voltage detector
Multimeter
3 variable resistors

SAFETY REQUIREMENTS

A. Wear safety glasses
B. Never check resistance on an energized circuit.

PROCEDURE

1. Use a non-contact voltage detector to check the circuit before disconnecting any wires or ohming any part of the circuit.
2. Components in an electrical circuit should be isolated from the circuit before ohming. This is normally done by removing the wires from them.
3. Avoid touching the tips of the leads while checking resistance. Touching the meter leads while ohming devices with a high resistance will cause an inaccurate reading because your body has a measurable resistance to most digital ohm meters.
4. Pay attention to the scale, especially on auto ranging and auto selecting meters.
5. The Ω symbol stands for ohms. 3.5 Ω means 3.5 ohms, Figure 33-4-1.
   KΩ stands for thousand ohms. 3.5KΩ means 3500 ohms, Figure 33-4-2.
   MΩ stands for million ohms. 3.5 MΩ means 3500000 ohms, Figure 33-4-3.
   The V symbol stands for volts. If you see this, the circuit is energized!

Figure 33-4-1 3.5 Ohms   Figure 33-4-2 3500 Ohms   Figure 33-4-3 3.5 Million ohms

Record the readings from each component in the following table.
1. Record the number shown on the meter in the "Numeric Reading" column, the scale (Ω, KΩ, MΩ) in the "Scale" column, and the value in the "Actual Value" column.
2. For example, Figure 33-4-2 would be 3.5, KΩ, 3500 ohms.

Device	Numeric Reading	Scale	Actual Ω Reading
Resistor 1 Left to Middle			
Resistor 1 Right to Middle			
Resistor 1 End to End			
Resistor 2 Left to Middle			
Resistor 2 Right to Middle			
Resistor 2 End to End			
Resistor 3 Left to Middle			
Resistor 3 Right to Middle			
Resistor 3 End to End			
RBM 90-63 Relay Term 2 to 5			
RBM 90-64 Relay Term 2 to 5			
RBM 90-65 Relay Term 2 to 5			
RBM 90-66 Relay Term 2 to 5			
RBM 90-67 Relay Term 2 to 5			
RBM 90-68 Relay Term 2 to 5			
Transformer Yellow to Yellow			
Transformer Black to White			
Transformer Black to Red			
Transformer Black to Orange			
24-volt relay coil			
120-volt relay coil			
208/230-volt relay coil			
Electric strip heater			
Fuse			
Shaded pole motor			

LAB 33.5 CHECKING RESISTORS

LABORATORY OBJECTIVE

You will demonstrate your ability to safely read the resistance of fixed resistors.

FUNDAMENTALS OF HVACR TEXT REFERENCE

Unit 33 Electrical Measuring and Test Instruments

LABORATORY NOTES

You will determine the resistor rating using the color bands on the resistors. Then, you will use a multimeter to read the resistance of fixed resistors and compare the actual resistance to the rating.

REQUIRED TOOLS, EQUIPMENT, AND MATERIALS

Multimeter and fixed resistors.

SAFETY REQUIREMENTS

A. Wear safety glasses.
B. Never check resistance on an energized circuit.

PROCEDURE

1. Write down the color bands on the resistor.
2. Use the guide below to determine the rating of the resistor and record it in the data table.
3. Read the resistance of the resistor and compare to the rating.

Reading Resistor Color Bands

Resistors "resist" the flow of electrical current. Resistors are color coded to identify their resistance value. Each color is assigned a number. The table below lists the colors used and the number that each color represents.

Digits		Tolerances
Black	0	
Brown	1	1%
Red	2	2%
Orange	3	
Yellow	4	
Green	5	0.5%
Blue	6	0.25%
Violet	7	0.1%
Grey	8	0.05%
White	9	
Gold		5%
Silver		10%
None		20%

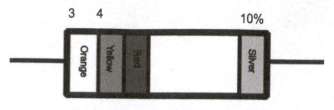

Four color band resistor 3400 ohms

Most resistors use a 4-band color code. The first two digits represent digits, the third band represents a power of ten multiplier, and the fourth band represents tolerance. The tolerance band is usually separated some from the other bands. A few resistors use a 5-band scheme that adds an extra digit band. They have three color bands, a multiplier band, and a tolerance band.

For example, a resistor with bands of **orange**, **yellow**, **red**, and **silver**: The first digit is 3 and the second digit is 4. The power of 10 is 2, for a multiplier of 10^2, which is 100. Silver indicates a 10% tolerance. $34 \times 10^2 = 3400$ ohms with a 10% tolerance. The 10% tolerance means that the actual value could be anywhere from 3060 to 3740 ohms.

Because resistors are not the exact value as indicated by the color bands, manufactures have included a tolerance color band to indicate the accuracy of the resistor. The tolerance band is usually gold or silver, but other colors are used for very close tolerance resistors. Some resistors may have no tolerance color. Gold band indicates the resistor is within 5% of what is indicated. Silver = 10% and None = 20%.

	Resistor 1	Resistor 2	Resistor 3	Resistor 4
Band 1 Color				
First Digit				
Band 2 Color				
Second Digit				
Band 3 Color				
Multiplier				
Band 4 Color				
Tolerance				
Rating				
Reading				

LAB 34.1 ELECTRICAL COMPONENT IDENTIFICATION

LABORATORY OBJECTIVE

The purpose of this lab is to identify common electrical components used in HVACR.

FUNDAMENTALS OF HVACR TEXT REFERENCE

Unit 34 Electrical Components

LABORATORY NOTES

You will examine the electrical components and identify each by name and function.

REQUIRED TOOLS, EQUIPMENT, AND MATERIALS

Safety glasses
Assortment of common electrical components

SAFETY REQUIREMENTS

Wear safety glasses

PROCEDURE

Examine each of the electrical components and write down their name and function.

#	Name	Function
Lab 34.1 Electrical Component Identification		
1		
2		
3		
4		
5		
6		
7		
8		
9		
10		

LAB 34.2 EXAMINING LOW-VOLTAGE THERMOSTATS

LABORATORY OBJECTIVE

The purpose of this lab is to learn the types of low-voltage thermostats and become familiar with their electrical details.

FUNDAMENTALS OF HVACR TEXT REFERENCE

Unit 34 Electrical Components

LABORATORY NOTES

You will examine an assortment of low-voltage thermostats and record their specifications.

REQUIRED TOOLS, EQUIPMENT, AND MATERIALS

Low-voltage thermostats
Multimeter
Small thermostat screwdriver, wire cutter-strippers

SAFETY REQUIREMENTS

Wear safety glasses

PROCEDURE

Examine four low-voltage thermostats and record the following specifications:
1. Voltage–What is the maximum voltage rating for the thermostat?
2. Amperage–What is the maximum amperage rating for the thermostat?
3. Operation–Is the thermostat a mechanical or digital?
4. Heat–Cool or Heat Pump? Is the thermostat designed to work with a furnace and air conditioner or a heat pump?
5. Stages–How many stages of heating and cooling does the thermostat have?
6. Terminal Identification–What are the terminal markings and what is each terminal's function?

Low-Voltage Thermostat Data				
	Thermostat 1	Thermostat 2	Thermostat 3	Thermostat 4
Volt Rating				
Amp Rating				
Operation				
Heat-Cool or Heat Pump				
Terminal				
Terminal				
Terminal				
Terminal				
Terminal				
Terminal				
Terminal				
Terminal				

LAB 34.3 PROGRAMING THERMOSTATS

LABORATORY OBJECTIVE

The purpose of this lab is to demonstrate your ability to program a programmable thermostat.

FUNDAMENTALS OF HVACR TEXT REFERENCE

Unit 34 Electrical Components

LABORATORY NOTES

Programmable thermostats are a popular energy management tool for the residential and light commercial markets. Programmable thermostats allow the customer to automatically change the setpoint temperatures for occupied times and set back or set up the unoccupied temperature depending on the heating or cooling mode. Reading the instruction manual is essential to understanding any particular thermostat. Many thermostats have common functions, but the function name may vary from brand to brand. Typically, most programmable thermostats have four daily programmable time periods (Waking, Daytime, Evening, and Sleeping). The four times are designed for two set back periods; one during the day when people are at work one during the night when people are sleeping. Some thermostats treat every day the same, so each of these time periods would occur at the same time each day and maintain the same temperature during that time. Others allow programming a different schedule every day of the week. However, the most common arrangement is to treat week days all the same, and treat Saturday and Sunday the same.

REQUIRED TOOLS, EQUIPMENT, AND MATERIALS

Programmable thermostat
Instruction manual

SAFETY REQUIREMENTS

Check all circuits for voltage before doing any service work.

PROCEDURE

Obtain a weekday/weekend (5+2) programmable thermostat with four daily setback periods. Most programmable thermostats today can be programmed while not on the system, operating on battery power. Install batteries and set the time and date. Complete the Thermostat Programming Check List and make sure to check each step off in the appropriate box as you finish it. This will help you to keep track of your progress. Show the instructor the program when you are done.

Thermostat Programming Check List (Weekday/Weekend)		
Prepare for Programming		
Step	Procedure	**Check**
1	Read and follow the manufacturer's instructions.	
2	Install batteries in thermostat.	
3	Set clock for correct time and date.	
Weekday Programming		
1	Set the weekday Morning schedule for a "typical" household 7:00 AM–8:00 AM, 70°F Heat setpoint, 75°F Cool setpoint	
2	Set weekday Away schedule for a "typical" household 8:00 AM–6:00 PM, 60°F Heat setpoint, 80°F Cool setpoint	
3	Set weekday Evening schedule for a "typical" household 6:00 PM–11:00 PM, 70°F Heat setpoint, 75°F Cool setpoint	
4	Set weekday Sleep schedule for a "typical" household 11:00 PM–7:00 AM, 65°F Heat setpoint, 75°F Cool setpoint	
Weekend Programming		
1	Set the Weekend Morning schedule for a "typical" household 7:00 AM–8:00 AM, 70°F Heat setpoint, 75°F Cool setpoint	
2	Set Weekend Daytime schedule for a "typical" household 8:00 AM–6:00 PM, 70°F Heat setpoint, 75°F Cool setpoint	
3	Set Weekend Evening schedule for a "typical" household 6:00 PM–11:00 PM, 70°F Heat setpoint, 75°F Cool setpoint	
4	Set Weekend Sleep schedule for a "typical" household 11:00 PM–7:00 AM, 65°F Heat setpoint, 75°F Cool setpoint	

LAB 34.4 EXAMINING LINE VOLTAGE THERMOSTATS

LABORATORY OBJECTIVE
The purpose of this lab is to learn the types of line voltage thermostats and become familiar with their electrical details.

FUNDAMENTALS OF HVACR TEXT REFERENCE
Unit 34 Electrical Components

LABORATORY NOTES
You will examine an assortment of line voltage thermostats, record their specifications, and test them with an ohmmeter.

REQUIRED TOOLS, EQUIPMENT, AND MATERIALS
Line voltage thermostats
Multimeter
6-in-1 screwdriver and wire cutter-strippers

SAFETY REQUIREMENTS
Wear safety glasses

PROCEDURE
Examine three-line voltage thermostats and record the following specifications:

1. Voltage–What is the maximum voltage rating for the thermostat?
2. Amperage–What is the maximum amperage rating for the thermostat?
3. Switching Action–Close on temperature rise, open on temperature rise, or both
4. Terminal Identification–What are the terminal markings and what is each terminal's function? (Where does power connect, where do the heating and cooling circuits connect?)
5. Temperature Range–At what temperatures can this thermostat function?
6. Check the thermostat with an ohm meter by adjusting the temperature while measuring the resistance across the contacts.

	Thermostat 1	Thermostat 2	Thermostat 3
Volt Rating			
Amp Rating			
Switching Action			
Temperature Range			
Terminal			
Terminal			
Terminal			

LAB 34.5 IDENTIFYING PRESSURE SWITCHES

LABORATORY OBJECTIVE
The purpose of this lab is to learn the types of pressure switches and become familiar with their electrical details.

FUNDAMENTALS OF HVACR TEXT REFERENCE
Unit 34 Electrical Components

LABORATORY NOTES
You will examine an assortment of pressure switches and record their specifications.

REQUIRED TOOLS, EQUIPMENT, AND MATERIALS
Pressure switches
Multimeter
Electrical hand tools

SAFETY REQUIREMENTS
Wear safety glasses

PROCEDURE
Examine four pressure switches and record the following specifications:

1. Voltage–What is the maximum voltage rating for the pressure switch?
2. Amperage–What is the maximum amperage rating for the pressure switch?
3. Switching Action–Close on rise, open on rise?
4. Pressure Range–At what pressures can this switch function?
5. Cut in pressure–What is the cut-in pressure setting?
6. Cut out setting–What is the cut-out setting?
7. Differential–What is the differential?

	Switch 1	Switch 2	Switch 3	Switch 4
Voltage				
Amperage				
Close or Open on Rise				
Range				
Cut-In				
Cut-Out				
Differential				

LAB 34.6 IDENTIFYING RELAY AND CONTACTOR PARTS

LABORATORY OBJECTIVE

The purpose of this lab is to learn how relays and contactors work.

FUNDAMENTALS OF HVACR TEXT REFERENCE

Unit 34 Electrical Components

LABORATORY NOTES

You will disassemble a contactor and a relay, identify their parts, and discuss their operation.

REQUIRED TOOLS, EQUIPMENT, AND MATERIALS

Electrical hand tools, relay to disassemble, contactor to disassemble

SAFETY REQUIREMENTS

Wear safety glasses

PROCEDURE

Step 1 – Inspect the sealed relay and identify parts.
    A.   Identify the coil. Where are the electrical connections to the coil?
    B.   Identify the armature. What type of armature does it have – swinging or sliding?
    C.   Identify the contacts. How many contacts does it have?
    D.   Are the contacts normally open or normally closed?
    E.   Where are the electrical connections for the contacts?

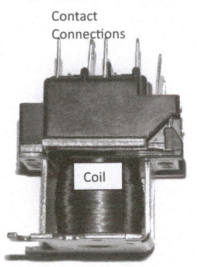

Contact
Connections

Coil

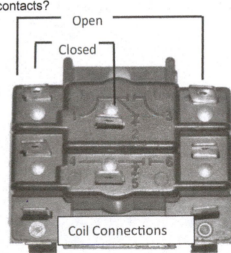

Open

Closed

Contact Set 1

Contact Set 2

Coil Connections

Step 2 – Disassemble a contactor and identify the parts.
   A. Identify the coil. Where are the electrical connections to the coil?
   B. Identify the armature. What type of armature does it have – swinging or sliding?
   C. Identify the contacts. How many contacts does it have?
   D. Are the contacts normally open or normally closed?
   E. Where are the electrical connections for the contacts?

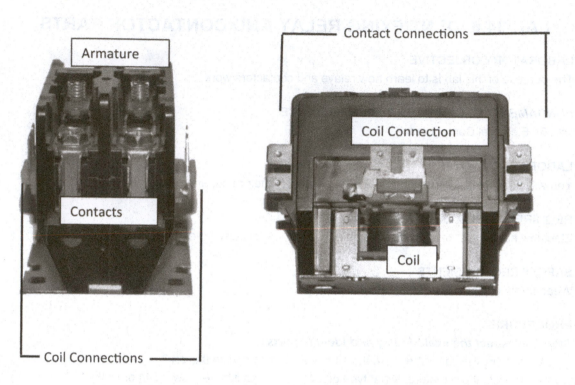

Step 3 – Reassemble the contactor and relay.

LAB 34.7 SETTING A LOW-PRESSURE SAFETY SWITCH

LABORATORY OBJECTIVE
The student will demonstrate how to properly set the cut-out and cut-in points on a low-pressure safety switch.

FUNDAMENTALS OF HVACR TEXT REFERENCE
Unit 34 Electrical Components

LABORATORY NOTES
For this lab you will set the cut-in and cut-out points of a low-pressure safety switch on an operating refrigeration system. Most refrigeration compressors are at least partially cooled by the refrigerant they pump. Without a low pressure cut-out, if the system loses charge, the compressor overheats from running continuously with a reduced flow of refrigerant. It is also possible for the suction pressure to drop into a vacuum, allowing air to enter the system at any leaks on the low side. Low-pressure switches protect against this as well. The low-pressure switch senses compressor suction pressure and opens on a drop in pressure. It is set to open at low pressures which can damage the compressor but remain closed at normal operating pressures. It is also known as a loss of charge or low-pressure cut-out switch.

REQUIRED TOOLS, EQUIPMENT, AND MATERIALS
Safety glasses and gloves
Operating refrigeration unit with low-pressure cut-out switch and king valve
Gauge manifold

SAFETY REQUIREMENTS
A. Always read the equipment manual to become familiarized with the refrigeration system and its accessory components prior to start up.
B. Wear safety goggles and gloves when working with refrigerants. Liquid refrigerant can cause frostbite when in contact with eyes and skin.
C. Use low loss hose fittings to reduce refrigerant spray when connecting and disconnecting refrigeration hoses.
D. Inspect all fittings before attaching hoses.

PROCEDURE
Step 1 – Determine Normal Operating Pressure
The operating manual with the refrigeration specifications should indicate what the expected suction pressure should be. If you have no information on the system, you can to calculate the expected suction pressure based upon the refrigerant type and box temperature of the system you are working on. The refrigerant temperature in the

evaporator should generally be no more than 15°F colder than the medium being cooled. For example, the evaporator saturation temperature for a 40°F walk-in cooler should be about 25°F. Once you know the evaporator saturation temperature you can determine the evaporator pressure from the P-T chart or phone app. A 0°F freezer charged with R-134a operating at a –15°F evaporator saturation temperature would have a pressure of 0 psig. If the space to be cooled was for vegetables to be kept at 40°F, then the corresponding refrigerant temperature would be 40°F – 15°F = 25°F which corresponds to an evaporator pressure of 22 psig.

Step 2 – Determine the Low-Pressure Cut-Out Setting

For low-pressure safety switches, the cut-out setting should be a pressure equivalent to 15°F to 20°F below the normal operating evaporator saturation temperature. For example, an air conditioning application which normally operates with a saturation temperature of 40°F would have a cut-out pressure equivalent to a saturation temperature of 20°F to 25°F. The cut-in setting should be high enough above the cut-out setting that the system will not continuously cycle on and off on the low-pressure switch. Common practice is to set the cut-in to a pressure equivalent to a normal operating evaporator saturation temperature. The Table 34-7-1 gives some suggested cut-in and cut-out pressures for different low-pressure safety switch applications.

Table 34-7-1 Typical Low-Pressure Safety Switch Settings				
Application	Refrigerant	Cut-Out	Cut-In	Differential
Medium Temp Refrigeration	134a	12 psig	26 psig	14 psig
Medium Temp Refrigeration	404A	44 psig	70 psig	26 psig
Medium Temp Refrigeration	450A	13 psig	21 psig	8 psig
Air Conditioning	32	80 psig	133 psig	53 psig
Air Conditioning	410A	80 psig	130 psig	50 psig
Air Conditioning	466A	84 psig	125 psig	41 psig

Step 3 – Adjust the Low-Pressure Cut-Out

The cut-out is not directly set on most low-pressure switches. Instead, they have two adjustments: a cut-In setting and a differential setting. The cut-out is the difference between the cut-in and the differential. For example, the cut-out pressure shown in Figure 34-7-1 is 35 psig. Cut-in 60 psig – differential 25 psig = 35 psig. Different terms are sometimes used instead of cut-in and cut-out. As shown in Figure 34-7-1, the cut-in is referred to as the "high event" and the cut-out is referred to as the "low event." Look at the label for the switching action to determine what happens on the high event. In the case of Figure 34-7-1 the switch closes on pressure rise. This tells you that the high event is the cut-in and the low event is the cut-out.

Figure 34-7-1 Low-pressure switch setting

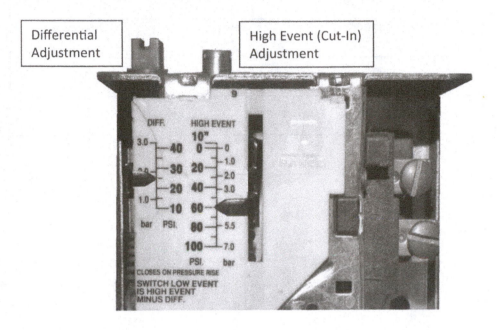

Step 4 – Check Operation

Connect a gauge manifold to the high and low sides of the system. Start the refrigeration system in the normal cooling mode and allow the pressures to stabilize. With the unit continuing to run, front-seat the king valve to shut off refrigerant flow to the evaporator. The suction pressure will begin to decrease. The compressor should stop when the suction pressure decreases to the cut-out setting. Observe and record the suction pressure when the compressor stops. Crack open the king valve. Observe and record the suction pressure when the compressor restarts.

Step 5 – Fine Tuning

Often the actual operating pressures vary from the settings. The pressure switch can be adjusted so that the operating cut-in and cut-out points match the desired pressures. The cut-out is more critical than the cut-in, but it is set by a combination of the cut-in and the differential, so it is easy for it to be a bit off. One way to get an accurate operating cut-out pressure is to first set the differential to the desired setting, which is the difference between the cut-in and the cut-out. Then set the set-in low enough to allow the unit to run. With the unit operating, run the king valve in towards the front-seat position, but do not seat it. You are trying to restrict the flow enough to reduce the suction pressure until the system operates at the pressure which you want the switch to open. Slowly increase the cut-in pressure adjustment until the system cuts off. Allow the system to run and stabilize prior to shutting down and then carefully disconnect the gauge manifold and any instrumentation that you attached to the unit.

Data Sheet 34-7-1 Low-Pressure Switch Setting and Operation			
	Recommended	Setting	Operation
Cut-In Pressure			
Differential			
Cut-Out Pressure			

LAB 34.8 SETTING A HIGH-PRESSURE CUT-OUT

LABORATORY OBJECTIVE
The purpose of this lab is to learn how to set the cut-out on a high-pressure switch used for compressor safety.

FUNDAMENTALS OF HVACR TEXT REFERENCE
Unit 34 Electrical Components

LABORATORY NOTES
You will set the high-pressure safety cut-out switch on the system assigned by the instructor and test the operation of the pressure switch.

REQUIRED TOOLS, EQUIPMENT, AND MATERIALS
Gauges, multimeter, electrical hand tools
Refrigeration system with high-pressure switch

SAFETY REQUIREMENTS
A. Wear safety goggles and gloves when working with refrigerants. Liquid refrigerant can cause frostbite when in contact with eyes and skin.
B. Use low loss hose fittings to avoid refrigerant spray when connecting and disconnecting refrigeration hoses.
C. Avoid touching electrical terminals on pressure switch.

PROCEDURE
Step 1 – Determine Normal Operating Pressure
The operating manual with the refrigeration specifications should indicate what the expected discharge pressure should be. If you have no information on the system, you can to calculate the expected discharge pressure based on the air or water flowing over the condenser and the normal temperature difference between the cooling medium and the refrigerant. Estimate the target condenser temperature by adding the estimated temperature difference between the air or water flowing over the condenser and the refrigerant. Use a pressure-temperature chart or a PT app to determine the target condenser pressure from the target condenser temperature. Typical differences are listed below.

A. 18 SEER Air Cooled Condenser Temp = Ambient Temp = 10°F
B. 16 SEER Air Cooled Condenser Temp = Ambient Temp + 15°F
C. 14 SEER Air Cooled Condenser Temp = Ambient Temp + 18°F
D. 13 SEER Air Cooled Condenser Temp = Ambient Temp + 20°F
E. Water Cooled Condenser Temperature = Inlet water temperature + 20°F

Step 2 – Determine the High-Pressure Cut-Out Setting

The high-pressure cut-out setting for air-cooled condensers is typically a pressure equivalent to 140°F to 150°F saturation temperature. The cut-in is typically a pressure equivalent to a saturation temperature of 120°F–130°F. For water-cooled condensers the cut-out setting is typically a pressure equivalent to a saturation temperature of 120°F to 130°F. The cut-in is typically a pressure equivalent to a saturation temperature of 95°F to 100°F.

Table 34-8-1 Typical High-Pressure Safety Switch Settings				
Application	Refrigerant	Cut-Out	Cut-In	Differential
Air Cooled Condenser	32	600 psig	400 psig	200 psig
Air Cooled Condenser	410A	650 psig	450 psig	200 psig
Air Cooled Condenser	454B	600 psig	400 psig	200 psig
Air Cooled Condenser	510A	240	140	100 psig
Water Cooled Condenser	32	450 psig	300 psig	150 psig
Water Cooled Condenser	410A	450 psig	300 psig	150 psig
Water Cooled Condenser	454B	400 psig	300 psig	100 psig
Water Cooled Condenser	510A	180 psig	120 psig	60 psig

Step 3 – Adjust the High-Pressure Cut-Out

Set the assigned high-pressure cut-out according to the settings determined in step 2. High-pressure cut-out switches open on rise of pressure, as seen in Figure 34-8-1. Adjustable pressure switches used as high-pressure cut-outs all have a cut-out setting, and some also have a differential setting. Switches that only have a cut-out setting have a fixed differential. The cut-in is not set directly, but is the difference between the cut-out and the differential. For example, the cut-out pressure shown in Figure 34-8-1 is 400 psig and the differential is 100 psig. This makes the cut-in 300 psig: 400 psig – 100 psig = 300 psig.

Figure 34-8-1 High pressure switch settings

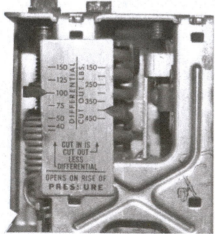

360

Step 4 – Check Operation

Connect a gauge manifold to the high and low sides of the system. Start the refrigeration system in the normal cooling mode and allow the pressures to stabilize. With the compressor continuing to run, shut off or restrict the condenser air or water to force an increase in discharge pressure. The discharge pressure will begin to decrease. The compressor should stop when the discharge pressure increases to the cut-out setting. Observe and record the discharge pressure when the compressor stops. Re-start the condenser air or water flow. The high side pressure should begin to drop. When the discharge pressure reaches the cut-in point the compressor should start back up. Some high-pressure switches have a manual reset. The reset on these will have to be pushed in order to close the high-pressure switch and allow the compressor to run.

High-Pressure Switch Setting and Operation			
	Recommended	Setting	Operation
Cut-Out Pressure			
Differential			
Cut-In Pressure			

LAB 35.1 ELECTRONIC COMPONENT IDENTIFICATION

LABORATORY OBJECTIVE

The purpose of this lab is to identify common electronic components used in HVACR.

FUNDAMENTALS OF HVACR **TEXT REFERENCE**

Unit 35 HVACR Electronic Controls

LABORATORY NOTES

You will examine the electronic components and identify each by name and function.

REQUIRED TOOLS, EQUIPMENT, AND MATERIALS

Safety glasses
Assortment of common electronic components

SAFETY REQUIREMENTS

Wear safety glasses

PROCEDURE

Examine each of the electronic components and write down their name and function.

#	Name	Function
Lab 35.1 Electronic Component Identification		
1		
2		
3		
4		
5		
6		
7		
8		
9		
10		

LAB 35.2 EXAMINING ELECTRONIC CONTROL BOARDS

LABORATORY OBJECTIVE

The purpose of this lab is to learn the function, inputs, and outputs of common electronic control boards used in HVACR.

***FUNDAMENTALS OF HVACR* TEXT REFERENCE**

Unit 35 HVACR Electronic Controls

LABORATORY NOTES

You will examine the electronic control boards and identify their function, inputs, and outputs.

REQUIRED TOOLS, EQUIPMENT, AND MATERIALS

Safety glasses
Non-communicating integrated furnace control board
Non-communicating heat pump defrost control (defrost only)
Non-communicating heat pump control board (more functions than just defrost)
Communicating control board, any type

SAFETY REQUIREMENTS

Wear safety glasses

PROCEDURE

Examine each of the electronic control boards, the matching unit diagrams, and manufacturer's literature for each control. Record each board's functions, inputs, and outputs in the following sections.

***Step 1* – Non-communicating Integrated Furnace Control Board**

Check off the board's functions is Table 35-2-1.

Table 35-2-1 Integrated Furnace Control Functions	
Functions	*Checkoff*
Inducer motor	
Ignition	
Draft assurance	
Flame safety	
Gas staging	
High-temperature limit safety	
Flame roll-out safety	

Indoor fan on/off control	
Indoor fan speed selection	
Indoor fan delay on and delay off	
Low-voltage fuse protection	
24-volt control terminals	
Error code reporting	

Locate the inputs on the non-communicating integrated furnace control board and check them off in Table 35-2-2.

Table 35-2-2 Integrated Furnace Control Inputs	
Functions	**Checkoff**
Line voltage	
24 volts from transformer secondary	
Flame sensor	
High limit switch	
Flame roll-out switches	
Draft switch	
G terminal – fan operation	
W terminal – heating operation	
Y terminal – cooling operation	

Locate the outputs on the non-communicating integrated furnace control board and check them off in Table 35-2-3.

Table 35-2-3 Integrated Furnace Control Outputs	
Functions	**Checkoff**
Line voltage to transformer primary	
24 volts to control system R and C	
Voltage to igniter	
Line voltage to draft inducer	
Line voltage to indoor fan	
24 volts to gas valve	
24 volts to terminals R and C	

Step 2 – Non-communicating Defrost Control Board

Check off the board's functions is Table 35-2-4.

Table 35-2-4 Defrost Control Functions	
Functions	*Checkoff*
Control reversing valve operation	
Control outdoor fan operation	
Energize auxiliary heat in defrost	
Measure ambient temperature	
Lockout compressor at low ambient	
Allow balance point adjustment	
Restrict auxiliary heat operation above balance point	
Off-cycle timer to prevent compressor short cycling	
Control compressor staging	

Locate the inputs on the non-communicating defrost control board and check them off in Table 35-2-5.

Table 35-2-5 Defrost Control Inputs	
Functions	*Checkoff*
Line voltage	
24 volts	
Coil temperature thermistor sensor	
Air temperature thermistor sensor	
Defrost thermostat	
Pressure differential switch	

Locate the outputs on the non-communicating defrost control board and check them off in Table 35-2-6.

Table 35-2-6 Defrost Control Outputs	
Functions	*Checkoff*
Voltage to reversing valve coil	
24 volts to bring on auxiliary heat	
Line voltage to outdoor fan	
24 volts to compressor contactor	
Voltage to compressor staging control	

Step 3 – Non-communicating Integrated Heat Pump Control Board

Check off the board's functions is Table 35-2-7.

Table 35-2-7 Heat Pump Integrated Control Functions	
Functions	*Checkoff*
Control reversing valve operation	
Control outdoor fan operation	
Energize auxiliary heat in defrost	
Measure ambient temperature	
Lockout compressor at low ambient	
Allow balance point adjustment	
Restrict auxiliary heat operation above balance point	
Off-cycle timer to prevent compressor short cycling	
Control compressor staging	

Locate the inputs on the non-communicating heat pump integrated control board and check them off in Table 35-2-8.

Table 35-2-8 Heat Pump Integrated Control Inputs	
Functions	*Checkoff*
Line voltage	
24 volts	
Coil temperature thermistor sensor	
Air temperature thermistor sensor	
Defrost thermostat	
Pressure differential switch	

Locate the outputs on the non-communicating integrated furnace control board and check them off in Table 35-2-9.

Table 35-2-9 Heat Pump Integrated Control Outputs	
Functions	*Checkoff*
Voltage to reversing valve coil	
24 volts to bring on auxiliary heat	
Line voltage to outdoor fan	
24 volts to compressor contactor	
Voltage to compressor staging control	

Step 4 – **Communicating Control Board**

Examine the communicating control, identify the communicating control terminals and list their function (control voltage or communications).

Table 35-2-10 Heat Pump Integrated Control Functions	
Functions	*Answer*
Does the board have legacy control connections?	
Communicating Control terminal1 label	
Communicating Control terminal1 function	
Communicating Control terminal2 label	
Communicating Control terminal1 function	
Communicating Control terminal3 label	
Communicating Control terminal1 function	
Communicating Control terminal4 label	
Communicating Control terminal1 function	

LAB 36.1 EXAMINE SHADED POLE MOTOR

LABORATORY OBJECTIVE

You will identify the parts of a shaded pole motor and explain its operation.

FUNDAMENTALS OF HVACR TEXT REFERENCE

Unit 36 Electric Motors

LABORATORY NOTES

You will disassemble the shaded pole motor assigned by the instructor, identify its parts, discuss its operation, and reassemble the motor. Refer to Figure 36-1-1 for help labeling parts.

REQUIRED TOOLS, EQUIPMENT, AND MATERIALS

Nut drivers
Shaded pole motor for disassembly

SAFETY REQUIREMENTS

Wear safety glasses

PROCEDURE

1. Use a marker to mark where the end bells align with the motor frame.
2. Loosen the nuts on the bolts that hold the motor together.
3. Disassemble the motor.
4. Identify the rotor and stator.
5. Identify the shaded poles.
6. Use the shaded poles to determine the direction of rotation.
7. Count the number of poles and calculate the motor speed (RPM).
8. Discuss your findings with the instructor.
9. Reassemble the motor.
10. Be careful to get the end bells aligned and check that the shaft turns freely after assembly.

Figure 36-1-1 Six pole shaded pole motor

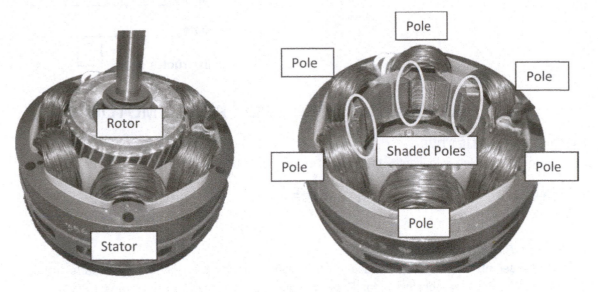

LAB 36.2 EXAMINE OPEN SPLIT PHASE MOTOR

LABORATORY OBJECTIVE

The purpose of this lab is to identify the parts of an open split phase motor and explain its operation.

FUNDAMENTALS OF HVACR TEXT REFERENCE

Unit 36 Electric Motors

LABORATORY NOTES

You will disassemble the open split phase motor assigned by the instructor, identify its parts, discuss its operation, and reassemble the motor. Refer to Figure 36-2-1 for help labeling parts.

REQUIRED TOOLS, EQUIPMENT, AND MATERIALS

Nut drivers

Open split phase motor for disassembly

SAFETY REQUIREMENTS

Wear safety glasses

PROCEDURE

1. Use a marker to mark where the end bells align with the motor frame.
2. Loosen the nuts on the bolts that hold the motor together.
3. Disassemble the motor.
4. Identify the rotor and stator.
5. Identify the start and run windings.
6. Identify the centrifugal switch.
7. Count the number of poles and calculate the motor speed (RPM).
8. Discuss your findings with the instructor.
9. Reassemble the motor.
10. Be careful to get the end bells aligned and check that the shaft turns freely after assembly.

Figure 36-2-1 Split phase motor parts

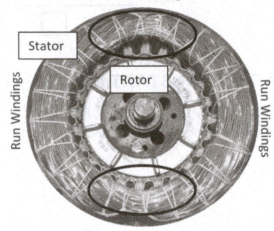

Start Windings

Stator

Rotor

Run Windings

Run Windings

Start Windings

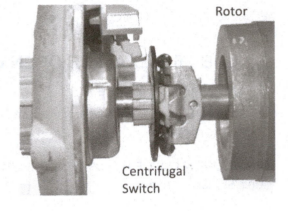

Rotor

Centrifugal
Switch

LAB 36.3 EXAMINE OPEN CAPACITOR START MOTOR

LABORATORY OBJECTIVE

The purpose of this lab is to identify the parts of an open capacitor start motor and explain its operation.

FUNDAMENTALS OF HVACR TEXT REFERENCE

Unit 36 Electric Motors

LABORATORY NOTES

You will disassemble the open capacitor start motor assigned by the instructor, identify its parts, discuss its operation, and reassemble the motor. Refer to Figure 36-3-1 for help labeling parts.

REQUIRED TOOLS, EQUIPMENT, AND MATERIALS

Nut drivers
Open capacitor start motor for disassembly

SAFETY REQUIREMENTS

Wear safety glasses

PROCEDURE

1. Use a marker to mark where the end bells align with the motor frame.
2. Loosen the nuts on the bolts that hold the motor together.
3. Disassemble the motor.
4. Identify the rotor and stator.
5. Identify the start and run windings.
6. Identify the centrifugal switch.
7. Examine how the start capacitor is wired to the centrifugal switch
8. Count the number of poles and calculate the motor speed (RPM).
9. Discuss your findings with the instructor.
10. Reassemble the motor.
11. Be careful to get the end bells aligned and check that the shaft turns freely after assembly.

Figure 36-3-1 Capacitor start motor

Start Capacitor

LAB 36.4 EXAMINE PERMANENT SPLIT CAPACITOR (PSC) MOTOR

LABORATORY OBJECTIVE
The purpose of this lab is to identify the parts of a PSC motor and explain its operation.

FUNDAMENTALS OF HVACR TEXT REFERENCE
Unit 36 Motors

LABORATORY NOTES
PSC motors have two windings: a run winding with more turns of larger gauge wire, and a start winding with fewer turns of smaller gauge wire. Unlike split phase or capacitor start motors, the start winding stays in the circuit during operation. For this reason, PSC motors do not need switches or relays to drop out the start winding. The difference between the two windings is less obvious than with split phase motors, Figure 36-4-1.

REQUIRED TOOLS, EQUIPMENT, AND MATERIALS
Nut drivers
Permanent split capacitor motor (PSC) for disassembly

SAFETY REQUIREMENTS
Wear safety glasses

PROCEDURE
1. Use a marker to mark where the end bells align with the motor frame.
2. Loosen the nuts on the bolts that hold the motor together.
3. Disassemble the motor.
4. Identify the rotor and stator.
5. Identify the start and run windings.
6. Count the number of poles and calculate the motor speed (RPM).
7. Discuss your findings with the instructor.
8. Reassemble the motor.
9. Be careful to get the end bells aligned and check that the shaft turns freely after assembly.

Figure   36-4-1 PSC motor

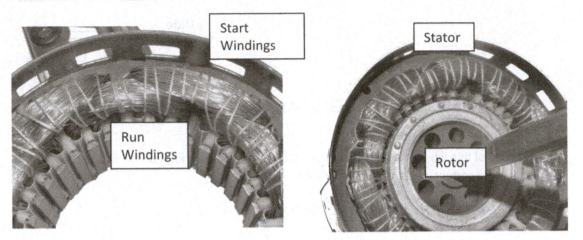

LAB 36.5 IDENTIFYING START AND RUN CAPACITORS

LABORATORY OBJECTIVE
The purpose of this lab is to learn how to distinguish between start and run capacitors and identify their important electrical characteristics.

FUNDAMENTALS OF HVACR TEXT REFERENCE
Unit 36 Electric Motors

LABORATORY NOTES
You will identify capacitors as either starting capacitors or running capacitors. You will also record their microfarad and voltage ratings.

REQUIRED TOOLS, EQUIPMENT, AND MATERIALS
Assortment of starting and running capacitors
20,000 ohm 2-watt bleed resistor

SAFETY REQUIREMENTS
A. Wear safety glasses
B. Capacitors can hold a ***potentially fatal*** electric charge.
C. Always discharge capacitors before touching their electrical terminals.
D. Better yet, do NOT touch the electrical terminals.

PROCEDURE
Examine the assortment of capacitors, separating the run capacitors from the starting capacitors. Discharge the capacitor using a bleed resistor or multimeter, Figure 36-5-1. Record the capacitor microfarad and voltage ratings on Data Sheet 36-5-1.

Figure 36-5-1 Discharging a run capacitor

Data Sheet 36-5-1

Capacitor Type	Voltage Rating	Microfarad Rating

LAB 36.6 EXAMINE ECM MOTOR

LABORATORY OBJECTIVE

You will identify the parts of an ECM motor and explain its operation.

***FUNDAMENTALS OF HVACR* TEXT REFERENCE**

Unit 36 Electric Motors

LABORATORY NOTES

You will disassemble the ECM motor assigned by the instructor, identify its parts, discuss its operation, and reassemble the motor.

REQUIRED TOOLS, EQUIPMENT, AND MATERIALS

Volt-Ohm meter and ammeter

Electrical hand tools

ECM motor for disassembly

SAFETY REQUIREMENTS

A. Wear safety glasses
B. The capacitors in the module can hold a 500-volt charge.
C. Wait at least 5 minutes after turning off power before removing the module to allow capacitors to discharge.

PROCEDURE

1. Use a marker to mark where the end bells align with the motor frame.
2. Loosen the nuts on the bolts that hold the motor together.
3. Disassemble the motor.
4. Identify the rotor, stator, and module.
5. Notice the difference between the rotor and the rotors of induction motors you examined earlier.
6. Ohm the windings.
7. Discuss your findings with the instructor.
8. Reassemble the motor.
9. Be careful to get the end bells aligned and check that the shaft turns freely after assembly.

Figure 36-6-1   ECM motor parts

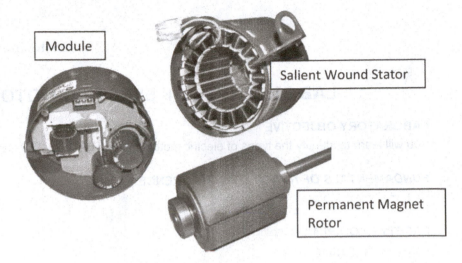

Module

Salient Wound Stator

Permanent Magnet Rotor

LAB 36.7 TYPES OF ELECTRIC MOTORS

LABORATORY OBJECTIVE

You will learn to identify the types of electric motors used in the HVACR field.

FUNDAMENTALS OF HVACR TEXT REFERENCE

Unit 36 Electric Motors

SAFETY REQUIREMENTS

Wear safety glasses

PROCEDURE

The Instructor will assign ten motors to identify according to type.

- SP–Shaded Pole
- Split Phase
- PSC–Permanent Split Capacitor
- CS–Capacitor Start
- CSR–Capacitor Start-Capacitor Run

Write down the type of motor in the table below. Use the other columns to check off motor characteristics to help you identify the motor type. You will be expected to explain to the instructor how you determined the motor types.

	Motor Type	Shaded Pole	Start Winding	Centrifugal Switch	Capacitors
1					
2					
3					
4					
5					
6					

7					
8					
9					
10					

LAB 37.1 DESIGN AND OPERATE RELAY CIRCUITS

LABORATORY OBJECTIVE
The purpose of this lab is to learn how to design and wire a simple relay circuit.

FUNDAMENTALS OF HVACR TEXT REFERENCE
Unit 37 Motor Controls
Unit 40 Control Systems

LABORATORY NOTES
You will design, wire, and operate a simple relay circuit. A toggle switch will control the coil on a 120-volt relay. And will control a 120-volt light.

REQUIRED TOOLS, EQUIPMENT, AND MATERIALS
Relay with 120-volt coil
Toggle switch
120-volt power cord
Electrical hand tools
120-volt light

SAFETY REQUIREMENTS
A. Wear safety glasses
B. Circuit should be unplugged during all wiring.
C. Do not energize circuit until given explicit instructions from instructor.

PROCEDURE
Step 1 – 120-volt Relay Coil
A. Draw a diagram with a toggle switch controlling a 120-volt relay and the normally open relay contacts operating a 120-volt light. There will be two circuits – one with the relay coil as the load, and one with the light as the load. The toggle switch will be in series with the relay coil to control it. The relay normally open contacts will be in series with the light to control it.
B. Wire the circuit you have drawn and have an instructor check it.
C. Operate your circuit.

382

Step 2 – 24-volt Relay Coil

D. Draw a diagram with a toggle switch controlling a 24-volt relay coil. The normally open relay contacts will operate a 120-volt light. There will be three circuits

- one 120-volt circuit to the transformer primary coil
- a second 120-volt circuit with the relay normally open contacts in series with a light
- and a 24-volt circuit from the transformer secondary coil through the toggle switch to the 24-volt relay coil.
- The transformer will provide 24 volts to operate the relay. The toggle switch will control the 24-volt relay coil. The relay normally open contacts will control the 120-volt light.

LAB 37.2 DESIGN AND OPERATE A 24-VOLT CONTACTOR CIRCUIT

LABORATORY OBJECTIVE
The purpose of this lab is to learn how to design and wire a simple contactor circuit.

FUNDAMENTALS OF HVACR TEXT REFERENCE
Unit 37 Motor Controls
Unit 40 Control Systems

LABORATORY NOTES
You will design, wire, and operate a simple contactor circuit. A toggle switch will control the coil on a 24-volt contactor. The normally open contactor contacts will control a 120-volt light.

REQUIRED TOOLS, EQUIPMENT, AND MATERIALS
Contactor with 24-volt coil
120-volt power cord
Electrical hand tools
Transformer
Toggle switch
120-volt light

SAFETY REQUIREMENTS
A. Wear safety glasses
B. Circuit should be unplugged during all wiring.
C. Do not energize circuit until given explicit instructions from instructor.

PROCEDURE
1. Draw a diagram with a toggle switch controlling a 24-volt contactor coil.
2. The normally open contactor contacts will operate a 120-volt light.
3. There will be three circuits
   a. one for the 120-volt primary transformer winding
   b. one with the 24-volt transformer secondary winding as the source with the contactor coil as the load,
   c. and one with the 120-volt light as the load.
4. The toggle switch will be in series with the contactor coil to control it. The contactor normally open contacts will be in series with the light to control.
5. Wire the circuit you have drawn and have an instructor check it.
6. Operate your circuit.

LAB 37.3 IDENTIFYING OVERLOADS

LABORATORY OBJECTIVE

The purpose of this lab is to learn the types of overloads and become familiar with their specifications.

FUNDAMENTALS OF HVACR TEXT REFERENCE

Unit 34 Electrical Components

Unit 37 Motor Controls

LABORATORY NOTES

You will examine an assortment of overloads and record their specifications.

REQUIRED TOOLS, EQUIPMENT, AND MATERIALS

Multimeter

Overloads

SAFETY REQUIREMENTS

Wear safety glasses

PROCEDURE

Examine an assortment of overloads and record the following specifications:

1. Type of overload–Magnetic, thermal, or current
2. Line duty or pilot duty
3. Rating–This can be different depending upon the type of overload.

   Some common overload ratings are

   a. Amperage
   b. Must hold current
   c. Must trip current
   d. Temperature

Type—Magnetic, thermal, or current	Line Duty or Pilot Duty	Rating	Rating

LAB 37.4 TROUBLESHOOTING THERMAL OVERLOADS

LABORATORY OBJECTIVE
The purpose of this lab is to learn to troubleshoot line duty thermal overloads.

FUNDAMENTALS OF HVACR TEXT REFERENCE
Unit 34 Electrical Components
Unit 37 Motor Controls

LABORATORY NOTES
You will check line duty thermal overloads using volt, ohms, and amps.

REQUIRED TOOLS, EQUIPMENT, AND MATERIALS
Multimeter
Ammeter
Electrical hand tools
Line duty thermal overloads

SAFETY REQUIREMENTS
A. Wear safety glasses
B. Circuit should be de-energized when using ohm meter.
C. Do not touch any electrical connections when operating unit.

PROCEDURE
1. On open overloads, physically look at the overload element and contacts and check to see if the contacts or element of the overload are burned out.
2. WITH POWER OFF check the contacts of the overload by ohming out the points. They should have a resistance of close to 0 ohms (closed).
3. If the contacts have a measurable resistance of more than a few tenths of an ohm, the overload is bad and should be changed.
4. If the overload is open, an infinite ohm reading, check:
   a. the voltage
   b. compressor windings
   c. run capacitor
   d. starting components
5. If all of the above components are good, attempt to start the motor after the overload closes and check the amp draw when the motor tries to start.
6. Compare the amp draw when the overload opens to its rating.
7. If amp draw is below the overload rating, change the overload.
8. If the operating amp draw is at or above the overload rating, it is just doing its job and is good. Further checks of the compressor and compressor circuit are required.
9. If the overload is bad, change with correct replacement.

LAB 37.5 PILOT DUTY OVERLOAD OPERATION

LABORATORY OBJECTIVE
The purpose of this lab is to learn how pilot duty overload relays work.

***FUNDAMENTALS OF HVACR* TEXT REFERENCE**
Unit 34 Electrical Components
Unit 37 Motor Controls

LABORATORY NOTES
You will build a circuit containing a pilot duty magnetic overload, a 24-volt contactor, and a 24-volt lock out relay to demonstrate how a pilot duty overload works. The heater will draw enough current to trip the overload and the lock-out relay will lock the circuit out after the overload trips.

REQUIRED TOOLS, EQUIPMENT, AND MATERIALS
Multimeter and ammeter
Pilot duty overloads, 24-volt contactor, 24- volt lock out relay, transformer, toggle switch
Electrical hand tools
Heat strips

SAFETY REQUIREMENTS
A. Wear Safety Glasses
B. All wiring should be done with the circuit not plugged in.
C. Do not energize circuit until given explicit instructions from the instructor.
D. Do not touch any electrical connections when operating unit.
E. Do not touch the electrical heaters during operation.

PROCEDURE

1. Use the diagram below to wire a pilot duty magnetic overload protecting a set of electric strip heaters controlled by a contactor.
2. Have an instructor check your wiring.
3. Turn the power on and operate the unit.

Figure 37-5-1

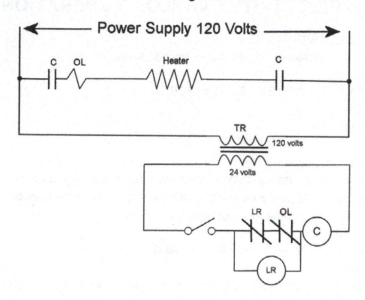

LAB 37.6 MAGNETIC STARTER FAMILIARIZATION

LABORATORY OBJECTIVE
The purpose of this lab is to learn how magnetic starters work.

FUNDAMENTALS OF HVACR TEXT REFERENCE
Unit 37 Motor Controls

LABORATORY NOTES
You will disassemble a magnetic starter, identify the parts, and discuss its operation.

Refer to Figure 37-6-1 for help.

REQUIRED TOOLS, EQUIPMENT, AND MATERIALS
Electrical hand tools

Magnetic starter to disassemble

SAFETY REQUIREMENTS
Wear safety glasses

PROCEDURE
1. Identify the coil.
2. Where are the electrical connections to the coil?
3. Identify the armature
4. What type of armature does it have – swinging or sliding?
5. Identify the contacts.
6. How many contacts does it have?
7. Are the contacts normally open or normally closed?
8. Where are the auxiliary contacts?
9. Where are the electrical connections for the contacts?
10. Identify the overloads.
11. Where are the electrical connections for the overloads?
12. Remove one of the overload heaters and discuss how to change them.
13. Where are the electrical connections for the contacts?
14. Reassemble the magnetic starter.

Figure 37-6-1

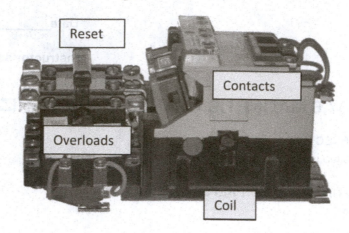

LAB 37.7 WIRE A MAGNETIC STARTER WITH A TOGGLE SWITCH

LABORATORY OBJECTIVE
The purpose of this lab is to learn how to design and wire a simple magnetic starter circuit.

FUNDAMENTALS OF HVACR TEXT REFERENCE
Unit 37 Motor Controls

LABORATORY NOTES
You will design, wire, and operate a basic magnetic starter circuit using a toggle switch will control the coil. The magnetic starter will control a 3 phase motor.

REQUIRED TOOLS, EQUIPMENT, AND MATERIALS
Electrical hand tools
Multimeter
Ammeter
Magnetic starter with 120-volt coil
Transformer 208 volt primary and 120 volt secondary
Toggle switch
3 phase motor

SAFETY REQUIREMENTS
A. Wear safety glasses
B. All wiring should be done with the circuit not plugged in.
C. Do not energize circuit until given explicit instructions from the instructor.
D. Do not touch any electrical connections when operating unit.
E. Do not touch the electrical heaters during operation.

PROCEDURE
1. Use Figure 37-7-1 to wire a toggle switch controlling the magnetic starter. Make sure the circuit to the coil passes through the normally closed overload contacts. The starter will control a three phase motor.
2. Have instructor check your wiring.
3. Operate your circuit.

Figure 37-7-1

Power Supply 208 Volts
3 Phase

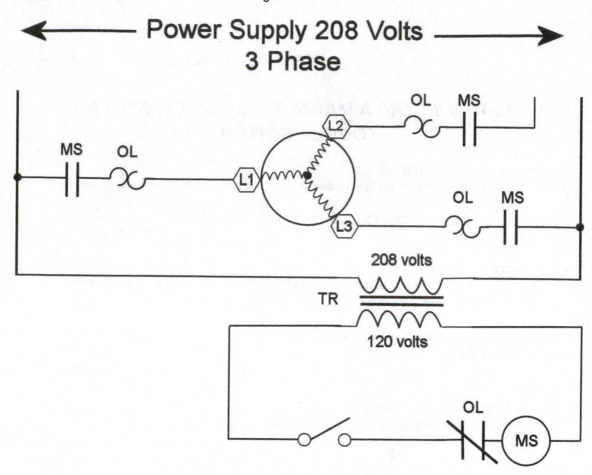

LAB 37.8 WIRE A MAGNETIC STARTER WITH A START-STOP SWITCH

LABORATORY OBJECTIVE

The purpose of this lab is to learn how to wire a magnetic starter circuit using a stop-start switch.

FUNDAMENTALS OF HVACR TEXT REFERENCE

Unit 37 Motor Controls

LABORATORY NOTES

You will use the diagram below to wire and operate a magnetic starter circuit using a stop-start switch. The magnetic starter will control a 3 phase motor.

REQUIRED TOOLS, EQUIPMENT, AND MATERIALS

Electrical hand tools, multimeter

Ammeter

Magnetic starter with 120-volt coil

Transformer 208 volt primary and 120 volt secondary

Stop-start switch

3 phase motor

SAFETY REQUIREMENTS

A. Wear safety glasses

B. All wiring should be done with the circuit not plugged in.

C. Do not energize circuit until given explicit instructions from the instructor.

D. Do not touch any electrical connections when operating unit.

E. Do not touch the electrical heaters during operation.

PROCEDURE

1. Wire the circuit using Figure 37-8-1.

2. Have an instructor check it.

3. Operate your circuit.

Figure 37-8-1

Power Supply 208 Volts
3 Phase

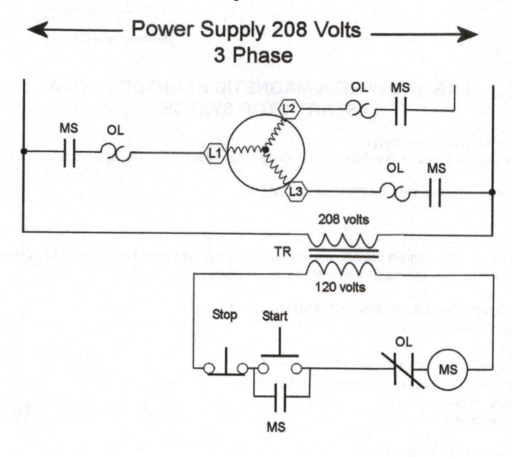

LAB 38.1 BELT DRIVES

LABORATORY OBJECTIVE
You will learn to adjust and correctly tension a belt drive.

FUNDAMENTALS OF HVACR TEXT REFERENCE
Unit 38 Motor Application and Troubleshooting

LABORATORY NOTES
You will use the diagram below to wire and operate a magnetic starter circuit using a stop-start switch. The magnetic starter will control a 3 phase motor.

REQUIRED TOOLS, EQUIPMENT, AND MATERIALS
6-in-1 screwdriver
Allen wrench set
Ammeter
Photo-tachometer
Belt drive blower with adjustable pulley

SAFETY REQUIREMENTS
A. Wear safety glasses
B. Lock out the power and check the voltage to ensure the power is off before beginning work.
C. Be cautious around rotating devices. You can be struck or caught in machinery.

PROCEDURE
Step 1 – Initial Readings
    A. Operate the belt drive blower assigned by the instructor.
    B. Use a digital photo-tachometer to measure the motor RPM and the blower RPM.
    C. Measure the motor amp draw.

Step 2 – Adjust Pulley
    A. Turn off the blower.
    B. Lock the power off.
    C. Loosen and remove the belt.
    D. Adjust the pulley out counterclockwise 1 turn. Make sure to position the set screw over a flat on the pulley before tightening the set screw, Figure 38-1-1.
    E. Reinstall and re-tension the belt.

Step 3 – Operate

A. Operate the blower and measure the motor amp draw.
B. Use a digital photo-tachometer to measure the motor RPM and the blower RPM.

Step 4 – Reset Pulley

A. Turn off the blower.
B. Lock the power off.
C. Loosen and remove the belt.
D. Adjust the pulley back in clockwise 1 turn. Make sure to position the set screw over a flat on the pulley before tightening the set screw, Figure 38-1-1.
E. Reinstall and re-tension the belt.

Step 5 – Operate and Check

A. Operate the blower and measure the motor amp draw.
B. Use a digital photo-tachometer to measure the motor RPM and the blower RPM.
C. Check the amp draw against the data plate to insure you are not leaving the blower in a condition that will cause excessive current draw.
D. If the Current draw is too high Use the wiring diagram on the motor and/or Figures
E. make sure the belt is not overtightened.
F. If the belt is properly tightened repeat step 2, adjusting the pulley out by 1/2 turn increments until the blower amp draw is at or lower than the data plate rating.

Figure 38-1-1 Adjustable pulley detail

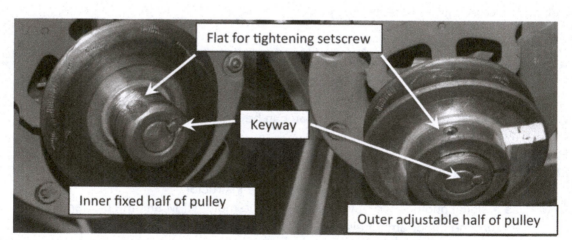

Flat for tightening setscrew

Keyway

Inner fixed half of pulley

Outer adjustable half of pulley

LAB 38.2 DIRECT DRIVE MOTOR APPLICATIONS

LABORATORY OBJECTIVE
You will learn to identify the types of direct drive electric motor applications in HVACR.

FUNDAMENTALS OF HVACR TEXT REFERENCE
Unit 38 Motor Application and Troubleshooting

LABORATORY NOTES
In a direct drive application the driven device turns exactly the same speed as the motor. The motor is connected to the device using a hub or coupling.

REQUIRED TOOLS, EQUIPMENT, AND MATERIALS
Non-contact voltage detector
6-in-1 screwdriver
Selection of direct drive motor applications

SAFETY REQUIREMENTS
A. Wear safety glasses
B. Be cautious around rotating devices. You can be struck or caught in machinery.
C. Do not touch live circuits.

PROCEDURE
Find three different direct drive electric motor applications in the shop.

| Elastomer Jaw Coupling | Blower Hub Coupling | Pump Spring Coupling |

Application	Connection: Hub, Spring, Rubber, Elastomer Jaw

LAB 38.3 CHECKING CAPACITORS USING A MULTIMETER

LABORATORY OBJECTIVE
The purpose of this lab is to learn how to check start and run capacitors using the capacitor test function on a multimeter.

FUNDAMENTALS OF HVACR TEXT REFERENCE
Unit 36 Electric Motors
Unit 38 Motor Application and Troubleshooting

LABORATORY NOTES
You will check both starting capacitors and running capacitors. You will also record their microfarad and voltage ratings and compare their rated values to the value displayed by the meter to determine if the capacitor is good or bad.

REQUIRED TOOLS, EQUIPMENT, AND MATERIAL
Multimeter with capacitor test function
20,000 ohm, 2-watt bleed resistor with insulated alligator clips
Assortment of starting and running capacitors

SAFETY REQUIREMENTS
A. Capacitors can hold a ***potentially fatal*** electric charge. Avoid touching the terminals.
B. Capacitors should be discharged before handling them. This can be done using a bleed resistor with insulated alligator clips or using the DC volt setting of your multimeter.

PROCEDURE
For each capacitor being tested, you should:
1. If the capacitor is in a circuit, ensure the capacitor is not energized using your multimeter on AC volts.
2. Discharge the capacitor by either
   A. Attaching a 20,000 ohm, 2-watt bleed resistor with insulated alligator clips (Figure 38-3-1) OR
   B. Using your multimeter set to DC volts.
3. Take a picture of the capacitor for later reference and remove the wires from the capacitor.
4. Set the multimeter to the capacitor test function, check its capacitance.
5. Compare the capacitor rating to the tested value and record in the table below.

Figure 38-3-1 Discharging a capacitor

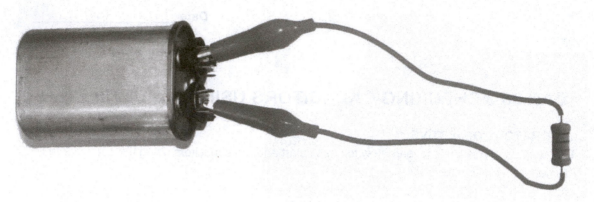

	Capacitor Type	Voltage Rating	Microfarad Rating	Microfarad Test	Condition
1					
2					
3					
4					
5					
6					
7					
8					
9					
10					

LAB 38.4 TESTING SINGLE PHASE COMPRESSOR WINDINGS

LABORATORY OBJECTIVE
The student will demonstrate how to properly test a single phase motor winding.

FUNDAMENTALS OF HVACR TEXT REFERENCE
Unit 38 Motor Application and Troubleshooting

LABORATORY NOTES
Terminals
Single phase compressor motors have two windings: Start and Run. One end of each winding is connected to a marked terminal: R for the Run winding an S for the Start winding. The other ends of the two windings are tied together and brought out to a third terminal. This terminal is referred to as Common because the two windings have it in common.

Windings
The start winding has a higher resistance than the run winding. For a good compressor, the sum of Run winding resistance (C to R) plus the Start winding resistance (C to S) should equal the total resistance read from S to R. A difference of a few tenths of an ohm is not unusual.

Shorted or Open Windings
A reading close to 0 ohms between any two terminals indicates a short. However, large motors (5 Hp and higher) may read very close to a short circuit because they have heavy copper windings to carry the motor current. A reading of infinity between any two terminals indicates an open winding.

Internal Overload
One caveat: if a compressor is hot and the internal overload opens it will read open from C to R and C to S, but S to R still indicates a resistance. In this case you need to wait for the compressor to cool before condemning it.

Grounded Windings
When read with a standard multimeter, any reading other than infinity between a terminal and ground indicates a grounded winding

REQUIRED TOOLS, EQUIPMENT, AND MATERIALS
Single phase compressor motor
6-in-1 screwdriver
20,000 ohm, 2-watt bleed resistor
Multimeter

SAFETY REQUIREMENTS

A. Turn the power off. Lock and tag out the power supply.
B. Confirm the power is secured by testing for zero voltage with your meter.
C. Capacitors should be discharged before you work on the system.

PROCEDURE

Step 1 – Preparing

A. Turn the power off and lock it out.
B. Verify with a voltmeter that the unit is not energized.
C. Remove a guard that encloses the terminals on a hermetic compressor. The terminals should be marked C, S, and R for Common, Start, and Run. These markings are sometimes on the terminal cover.
D. Refer to Figure 38-4-1 and read the winding resistances. The readings you should take are:

    a. C to S
    b. C to R
    c. S to R

Figure 38-4-1 Compressor terminals

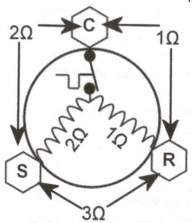

Step 2 – Identifying Terminals

If the terminals are not marked, they can be identified using the procedure shown in Figure 38-4-2. Measure the resistance across each pair of terminals and comparing the readings.

A. *Identify the Common Terminal first*

    The greatest resistance reading should be between the Start(S) and the Run(R)terminals. The remaining terminal not being touched is the Common terminal (C).

B. *Identify the Run Terminal*

The lowest reading should be between the Common(C) terminal and the Run(R) terminal. The terminal with the lowest reading between it and the C terminal which you just identified is the Run (R) terminal.

C. *Identify the Start Terminal*

The reading between the Common(C) terminal and the start(S) terminal should be higher than the reading between the Common(C) terminal and the Run (R) terminal, but lower than the reading between the Run(R) terminal and the Start(S) terminal. If you have already identified both the Common and Run terminals, then you know the remaining terminal is the start terminal.

Compressor Winding Readings			
Terminal Readings		**Readings to Ground**	
C to R		C to ground	
C to S		S to ground	
S to R		R to ground	

Figure 38-4-2 Ohming compressor terminals

Step 3 Check for a Grounded Winding

A. The meter must be able to read very high resistance, at least 100,000 ohms.
B. When testing for a grounded winding, one test lead is placed on the terminal and the other on the outer shell of the compressor. Be careful to make good contact with the motor shell as a coat of paint or a layer of dirt can hide a grounded winding. The copper lines going to the compressor make good contact points.
C. The temperature is important, so run the compressor for about five minutes before testing.
D. A reading of 1 mega ohm or higher indicates an ungrounded winding (Good). Most standard multimeters will read OL, Figure 38-4-3.
E. A reading less than a mega ohm indicates either possible contamination or a grounded winding, Figure 38-4-3.

Figure 38-4-3

Not Grounded - High Resistance

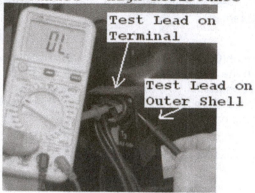

Test Lead on Terminal

Test Lead on Outer Shell

Grounded - Low Resistance

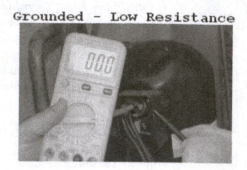

QUESTIONS
(Circle the letter that indicates the correct answer.)

1. Motor windings are insulated:
    a. with heavy copper wire.
    b. with red latex paint.
    c. with rubber.
    d. with a thin coat of varnish.

2. A shorted motor winding would indicate:
    a. an infinite resistance.
    b. a low resistance.
    c. no resistance.
    d. exactly 100 mega ohms.

3. An open winding would indicate:
    a. an infinite resistance.
    b. a low resistance.
    c. no resistance.
    d. exactly 100 mega ohms.

4. As motor windings get old, they turn from a tan color to black:
    a. True.
    b. False.

5. A resistance reading of OL between the S and R terminals indicates:
    a. An open winding.
    b. An open overload.
    c. A shorted winding.
    d. A grounded winding.

6. On a hot compressor you read OL between the C and R terminals indicates, OL between the C and S terminals, and 4 ohms between S and R terminals. Your readings most likely indicate:
   a. An open winding.
   b. An open overload.
   c. A shorted winding.
   d. A grounded winding.

7. A resistance reading of 0 ohms between the C and S terminals indicates:
   a. An open winding.
   b. An open overload.
   c. A shorted winding.
   d. A grounded winding.

8. A resistance reading of 500 ohms between the C terminal and the suction line indicates:
   a. An open winding.
   b. An open overload.
   c. A shorted winding.
   d. A grounded winding.

LAB 38.5 MOTOR INSULATION TESTING

LABORATORY OBJECTIVE
The student will demonstrate how to properly test for leakage resistance from motor windings.

FUNDAMENTALS OF HVACR TEXT REFERENCE
Unit 36 Electric Motors

LABORATORY NOTES
The wiring that is used for motor windings is insulated with a thin coat of varnish. As the motor is used and the windings are heated, this material will slowly carbonize. As it carbonizes, it will slightly change to a darker tan and eventually on to black. When it is completely carbonized, it does not provide any insulating capacity. An instrument referred to as a megger tests the degree to which carbonization has taken place within the motor winding insulation. Larger motors are checked for insulation breakdown by having megger readings taken on a regular basis in accordance with the preventative maintenance schedule. This provides a record of the motor condition over time and a lowering resistance indicates that the windings are becoming increasingly carbonized.

REQUIRED TOOLS, EQUIPMENT, AND MATERIALS
Motor
Megger

SAFETY REQUIREMENTS
A. Turn off and lock out power to the motor to be tested.
B. Confirm the power is secured by testing for 0 voltage with a multimeter.

PROCEDURE
Step 1 **Megger Familiarization**
    A. Meggers can test small amounts of current leakage at high voltage. 500 volts for 208 to 240-volt motors and 1000 volts for 480-volt motors.
    B. Meggers may be battery operated or have a built-in hand cranked generator.
    C. A megger can detect insulation faults that an ordinary multimeter cannot read.
    D. The megger shown in Figure 38-5-1 has a scale that reads from 20 to 1000 megohms.

Figure 38-5-1 Megohmmeter   Figure 38-5-2 Checking Compressor winding insulation

Step 2 Testing Motor Insulation

A. Turn off and lock out the power to the motor to be tested.

B. Confirm the power is secured by testing for 0 voltage with a multimeter.

C. Attach one lead of the Megger to the motor terminal and the other lead to ground. The motor frame or compressor piping are suitable, Figure 38-5-2.

D. Read and record the insulation resistance in Data Sheet 38-5-1.

E. Megohm readings are best used as part of a regular maintenance routine. Declining megohm readings can indicate the gradual failure of the winding insulation.

F. Some general guidelines are

    a. Below 30 megohms indicates extreme contamination or failed winding insulation.

    b. 30 to 100 megohms indicates contamination or degrading insulation.

    c. 100 megohms or greater indicates good winding insulation.

Data Sheet 38-5-1

Megohm Test	Megohm Reading
Common to ground	
Start to ground	
Run to ground	

407

LAB 38.6 WIRE AND OPERATE SHADED POLE MOTORS

LABORATORY OBJECTIVE
You will learn how to wire and operate shaded pole motors used in HVACR.

***FUNDAMENTALS OF HVACR* TEXT REFERENCE**
Unit 38 Motor Application and Troubleshooting

LABORATORY NOTES
You will wire shaded pole motors, operate them, and measure their amp draw.

REQUIRED TOOLS, EQUIPMENT, AND MATERIALS
Volt-Ohm meter
Ammeter
Electrical hand tools
Shaded pole motors

SAFETY REQUIREMENTS
A. Wear safety glasses
B. All wiring should be done with the motor not energized.
C. Motor should be physically secured before operating.
D. Do not energize the motor until given explicit instructions from the instructor.

PROCEDURE
1. Use the wiring diagram on the motor or Figure 38-6-1 to wire the shaded pole motors assigned by the instructor.
2. Be sure to check the motor nameplate voltage and compare it to the voltage you are connecting to the motor.
3. Operate the motor.
4. Use a clamp-on ammeter to read the motor current.
5. Write down your reading in the chart below.
6. Discus your results with the instructor.
7. Put up all the motors.

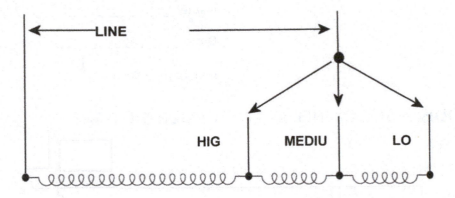

Figure 38-6-1   Multi-speed shaded pole motor

Voltage	Speeds	RPM	Nameplate FLA	Running Amps (for each speed)

LAB 38.7 TROUBLESHOOTING SHADED POLE MOTORS

LABORATORY OBJECTIVE
The purpose of this lab is to learn how to check shaded pole motors for faults.

FUNDAMENTALS OF HVACR TEXT REFERENCE
Unit 38 Motor Application and Troubleshooting

LABORATORY NOTES
You will check the shaded pole motors assigned by the instructor. You will wire and try to operate motors that ohm out correctly.

REQUIRED TOOLS, EQUIPMENT, AND MATERIALS
Volt-Ohm meter

Ammeter

Electrical hand tools

Shaded pole motors

SAFETY REQUIREMENTS
A. Wear safety glasses
B. All wiring should be done with the motor not energized.
C. Motor should be physically secured before operating.
D. Do not energize the motor until given explicit instructions from the instructor.

PROCEDURE
1. Check to see that the shaft turns easily by hand. If it does not turn easily by hand, the motor has a mechanical problem and should not be wired up.
2. Ohm the motor windings and record your readings.
3. A reading of infinity (OL) indicates an open motor winding.
4. A reading less than 1.0 ohm indicates a shorted winding.
5. A reading from the motor winding to the motor frame indicates a grounded winding.
6. If the motor ohms out correctly, wire and operate it.
7. Be sure to check the motor nameplate voltage and compare it to the voltage you are connecting to the motor.
8. Operate the motor.
9. Use a clamp-on ammeter to read the motor current.
10. Write down your analysis in the table below.
11. Discus your results with the instructor.
12. Put up all the motors.

SUMMARY

Voltage	Speeds	RPM	Nameplate FLA	Running Amps	Condition (Good, open, shorted, grounded, bad bearings)

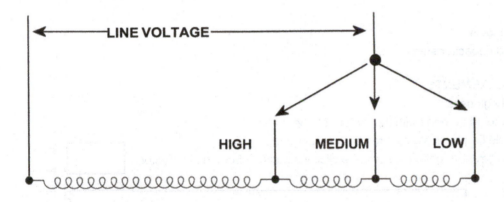

Figure 38-6-1 Multi-speed shaded pole motor

LAB 38.8 WIRE AND OPERATE OPEN SPLIT PHASE AND CAPACITOR START MOTORS

LABORATORY OBJECTIVE

You will learn how to wire and operate open split phase motors used in HVACR.

FUNDAMENTALS OF HVACR TEXT REFERENCE

Unit 38 Motor Application and Troubleshooting

LABORATORY NOTES

You will wire open split phase motors, operate them, and measure their amp draw.

REQUIRED TOOLS, EQUIPMENT, AND MATERIALS

Multimeter
Ammeter
Electrical hand tools
Split phase and capacitor start motors

SAFETY REQUIREMENTS

A. Wear safety glasses
B. All wiring should be done with the motor not energized.
B. Motor should be physically secured before operating.
C. Do not energize the motor until given explicit instructions from the instructor.

PROCEDURE

1. Use the wiring diagram on the motor and/or the diagrams in the book to wire the split phase motors assigned by the instructor. Figure 38-8-1 is a typical split phase connection diagram.
2. Check the motor nameplate voltage and compare it to the voltage you are connecting to the motor.
3. For dual voltage motors, check that the connections in the motor match the voltage. For example, by studying the diagram in Figure 38-8-1 you can see that the motor shown in Figure 38-8-2 is configured for low voltage because there is no connection on terminal 4.
4. Have the instructor check your wiring.
5. Operate the motor.
6. Use a clamp-on ammeter to read the motor current.
7. Write down your reading in Data Sheet 38-8-1.
8. Discus your results with the instructor.
9. Put up all the motors.

412

Data Sheet 38-8-1

Voltage	Speeds	RPM	Nameplate FLA	Running Amps

Figure 38-8-1 Split phase motor connection diagram

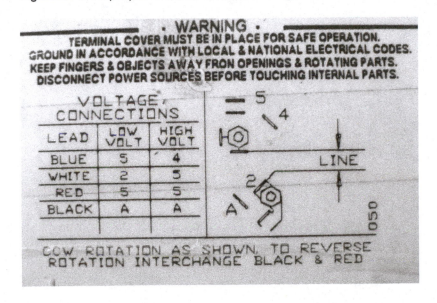

Figure 38-8-2 Split phase motor connections

414

LAB 38.9 TROUBLESHOOTING OPEN SPLIT PHASE & CAPACITOR START MOTORS

LABORATORY OBJECTIVE
The purpose of this lab is to learn how to check open split phase motors for faults.

FUNDAMENTALS OF HVACR TEXT REFERENCE
Unit 38 Motor Application and Troubleshooting

LABORATORY NOTES
You will check the open split phase motors assigned by the instructor. You will wire and try to operate motors that ohm out correctly.

REQUIRED TOOLS, EQUIPMENT, AND MATERIALS

Volt-Ohm meter   Open split phase motors

Ammeter   Open capacitor start motors

Electrical hand tools

SAFETY REQUIREMENTS
A. Wear safety glasses
B. All wiring should be done with the motor not energized.
C. Motor should be physically secured before operating.
D. Do not energize the motor until given explicit instructions from the instructor.

PROCEDURE
1. Check to see that the shaft turns easily by hand. If it does not turn easily by hand, the motor has a mechanical problem and should not be wired up.
2. Ohm the motor windings and record your readings.
   a. A reading of infinity (OL) indicates an open motor winding.
   b. A reading less than 1.0 ohm indicates a shorted winding.
   c. A reading from the motor winding to the motor frame indicates a grounded winding.
3. If the motor ohms out correctly, wire and operate it.
4. Be sure to check the motor nameplate voltage and compare it to the voltage you are connecting to the motor.
5. Operate the motor.
6. Use a clamp-on ammeter to read the motor current.
7. Write down your analysis in the table below and discus you results with the instructor.
8. Put up all the motors.

Voltage	Speeds	RPM	Nameplate FLA	Running Amps	Condition (Good, open, shorted, grounded, bad bearings)

LAB 38.10 WIRE AND OPERATE OPEN PSC MOTORS

LABORATORY OBJECTIVE
You will learn how to wire and operate open PSC motors used in HVACR.

FUNDAMENTALS OF HVACR TEXT REFERENCE
Unit 38 Motor Application and Troubleshooting

LABORATORY NOTES
You will wire open PSC motors, operate them, and measure their amp draw.

REQUIRED TOOLS EQUIPMENT AND MATERIALS
Volt-Ohm meter
Ammeter
Electrical hand tools
PSC motors

SAFETY REQUIREMENTS
A. Wear safety glasses
B. All wiring should be done with the motor not energized.
C. Motor should be physically secured before operating.
D. Do not energize the motor until given explicit instructions from the instructor.

PROCEDURE
1. Use the wiring diagram on the motor and/or Figures 38-10-1, 38-10-2, 38-10-3 to wire the PSC motors assigned by the instructor.
2. Be sure to check the motor nameplate voltage and compare it to the voltage you are connecting to the motor.
3. Operate the motor.
4. Use a clamp-on ammeter to read the motor current.
5. Write down your reading in the chart below.
6. Discus your results with the instructor.
7. Put up all the motors.

SUMMARY

Voltage	Speeds	RPM	Nameplate FLA	Running Amps (for each speed)

Figure 38-10-1 Single speed PSC motor with 1 capacitor wire

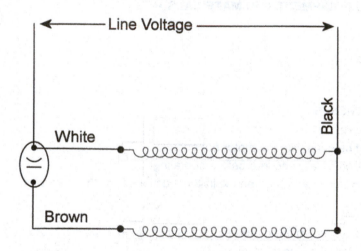

Figure 38-10-2 Single speed PSC motor with 2 capacitor wires

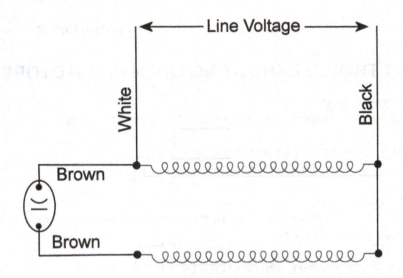

Figure 38-10-3 Three speed PSC motor with 1 capacitor wire

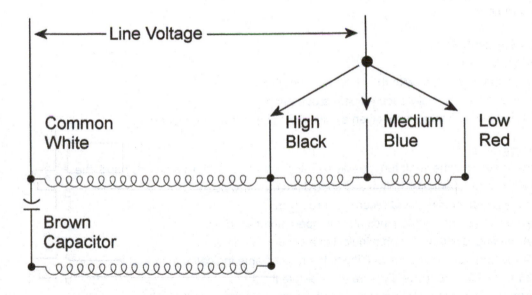

LAB 38.11 TROUBLESHOOTING OPEN PSC MOTORS

LABORATORY OBJECTIVE
The purpose of this lab is to learn how to check open PSC motors for faults.

FUNDAMENTALS OF HVACR TEXT REFERENCE
Unit 38 Motor Application and Troubleshooting

LABORATORY NOTES
You will check the open PSC motors assigned by the instructor. You will wire and try to operate motors that ohm out correctly.

REQUIRED TOOLS, EQUIPMENT, AND MATERIALS
Volt-Ohm meter
Electrical hand tools
Ammeter
Open PSC motors

SAFETY REQUIREMENTS
A. Wear safety glasses
B. All wiring should be done with the motor not energized.
C. Motor should be physically secured before operating.
D. Do not energize the motor until given explicit instructions from the instructor.

PROCEDURE
1. Check to see that the shaft turns easily by hand. If it does not turn easily by hand, the motor has a mechanical problem and should not be wired up.
2. Ohm the motor windings and record your readings.
    a. A reading of infinity (OL) indicates an open motor winding.
    b. A reading less than 1.0 ohm indicates a shorted winding.
    c. A reading from the motor winding to the motor frame indicates a grounded winding.
3. If the motor ohms out correctly, wire and operate it.
4. Be sure to check the motor nameplate voltage and compare it to the voltage you are connecting to the motor.
5. Operate the motor.
6. Use a clamp-on ammeter to read the motor current.
7. Write down your analysis in the table below.
8. Discus your results with the instructor.
9. Put up all the motors.

SUMMARY

Voltage	Speeds	RPM	Nameplate FLA	Running Amps	Condition (Good, open, shorted, grounded, bad bearings)

LAB 38.12 OHMING THREE PHASE MOTORS

LABORATORY OBJECTIVE
The purpose of this lab is to learn how to check three phase motors for faults using an ohm meter.

FUNDAMENTALS OF HVACR TEXT REFERENCE
Unit 38 Motor Application and Troubleshooting

LABORATORY NOTES
You will check the three phase motors assigned by the instructor.

REQUIRED TOOLS, EQUIPMENT, AND MATERIALS
Volt-Ohm meter
Electrical hand tools
Ammeter
Three phase motors

SAFETY REQUIREMENTS
A. Wear safety glasses
B. The motor should NOT be energized when checking winding resistance.

PROCEDURE
1. Check to see that the shaft turns easily by hand. If it does not turn easily by hand, the motor has a mechanical problem and should not be wired up.
2. Ohm the motor windings and record your readings on the data sheet.
   a. A reading of infinity (OL) indicates an open motor winding.
   b. A reading less than 1.0 ohm indicates a shorted winding.
   c. A reading from the motor winding to the motor frame indicates a grounded winding.

Single voltage three phase motors should have three equal readings.
Dual voltage motors will have two groups of readings. The reading in each group should be the same. Figure 38-12-1 shows a Wye configuration. Figure 38-12-2 shows a Delta configuration.

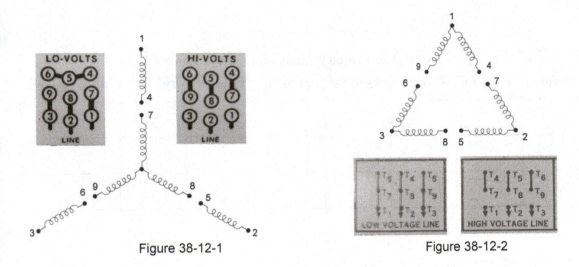

Figure 38-12-1   Figure 38-12-2

Lab 38.12 Data Sheet

Single Voltage Three Phase Motors						
Motor	Lead 1 to Lead 2	Lead 2 to Lead 3	Lead 3 to Lead 1	Lead 1 to Ground	Lead 2 to Ground	Lead 3 to Ground

Dual Voltage Three Phase Motors – Wye Wound							
Motor	T1 to T4	T2 to T5	T3 to T6	T7 to T8	T8 to T9	T9 to T7	All to Ground

Dual Voltage Three Phase Motors – Delta Wound										
Motor	T1 to T4	T1 to T9	T4 to T9	T2 to T7	T2 to T5	T5 to T7	T3 to T6	T3 to T8	T6 to T8	All to Ground

LAB 38.13 WIRE AND OPERATE THREE PHASE MOTORS

LABORATORY OBJECTIVE
You will learn how to wire and operate three phase motors used in HVACR.

FUNDAMENTALS OF HVACR TEXT REFERENCE
Unit 38 Motor Application and Troubleshooting

LABORATORY NOTES
You will wire three phase motors, operate them, and measure their amp draw.

REQUIRED TOOLS, EQUIPMENT, AND MATERIALS
Volt-Ohm meter and ammeter
Electrical hand tools
Three phase motors

SAFETY REQUIREMENTS
A. Wear safety glasses
B. All wiring should be done with the motor not energized.
C. Motor should be physically secured before operating.
D. Do not energize the motor until given explicit instructions from the instructor.

PROCEDURE
1. Use the wiring diagram on the motor and/or Figures 38-12-2 and 38-12-2 in Lab 38-12 to wire the three phase motors assigned by the instructor.
2. Be sure to check the motor nameplate voltage and compare it to the voltage you are connecting to the motor.
3. Operate the motor.
4. Use a clamp-on ammeter to read the motor current.
5. Write down your reading in the chart below.
6. Discus your results with the instructor.
7. Put up all the motors.

Voltage	RPM	Nameplate FLA	Running Amps

LAB 38.14 OPERATE ECM MOTOR

LABORATORY OBJECTIVE
You will describe the operation of an ECM blower motor and compare its operation to a traditional PSC blower motor.

FUNDAMENTALS OF HVACR TEXT REFERENCE
Unit 36 Electric Motors
Unit 38 Motor Application and Troubleshooting

LABORATORY NOTES
You will operate an ECM blower motor, measure its operating characteristics, and compare them to a traditional PSC blower motor.

REQUIRED TOOLS, EQUIPMENT, AND MATERIALS
Electrical hand tools, ECM motor trainer

SAFETY REQUIREMENTS
A. Wear safety glasses
B. All wiring should be done with the motor not energized.
C. Motor should be physically secured before operating.
D. Do not energize the motor until given explicit instructions from the instructor.

PROCEDURE
1. Connect the controls to the ECM trainer.
2. Have the instructor check your work.
3. Operate the blower and notice the blower RPM, amp draw, and watts.
4. Restrict the airflow and observe the changes in the RPM, amp draw, and watts.
5. Adjust the speed control and observe the changes in the RPM, amp draw, and watts.
6. Notice the difference between the rotor and the rotors of induction motors you examined earlier.
7. Remove the airflow restriction and observe the changes in the RPM, amp draw, and watts.

	RPM	AMPS	WATTS
Unrestricted			
50% Restricted			
75% Restricted			

Name_____

Date_____

Instructor's OK ☐

LAB 38.15 CHECK ECM MOTOR USING TECMATE

LABORATORY OBJECTIVE

You will learn how to check the operation of an ECM blower motor using the TECMATE tool.

***FUNDAMENTALS OF HVACR* TEXT REFERENCE**

Unit 36 Electric Motors

Unit 38 Motor Application and Troubleshooting

LABORATORY NOTES

You will connect a TECMATE to an ECM blower motor. You will then operate and test the motor using the TECMATE to control the motor.

REQUIRED TOOLS, EQUIPMENT, AND MATERIALS

Electrical hand tools

ECM motor trainer or blower with ECM motor

TECMATE

SAFETY REQUIREMENTS

A. Wear safety glasses

B. ECM motor modules receive power all the time! Even when NOT running.

C. Disconnect power to unit BEFORE removing power and control connections.

D. Connect TECMATE while power is turned off to unit.

E. Do not energize the motor until given explicit instructions from the instructor.

PROCEDURE

1. Turn power off to unit.
2. Remove the control plug from the ECM control module, Figure 38-15-1.
3. Connect the TECMATE to the control socket on the ECM control module, Figure 38-15-2.
4. Connect the two TECMATE alligator clips to 24 volts.
5. Have the instructor check your work.
6. Turn power on with the TECMATE switch in the OFF position.
7. Turn the TECMATE switch on. The blower should run.

Figure 38-15-1   Figure 38-15-2

Module Control Plug

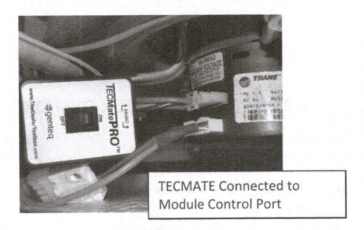

TECMATE Connected to
Module Control Port

LAB 38.16 SET BLOWER CFM ON ECM BLOWER BOARD

LABORATORY OBJECTIVE

You will learn how to adjust the airflow setting on blower control board for an ECM blower motor.

FUNDAMENTALS OF HVACR TEXT REFERENCE

Unit 36 Electric Motors

Unit 38 Motor Application and Troubleshooting

LABORATORY NOTES

Most systems with ECM blowers have settings allowing technicians to adjust blower CFM. These settings normally involve DIP switches or jumper wires, as in Figure 38-16-1. You will connect measure the airflow of an ECM equipped blower. Next, you will adjust the airflow using the blower control board settings. Then you will read the new CFM after the adjustment. Finally, you will return the adjustment to its original setting.

REQUIRED TOOLS, EQUIPMENT, AND MATERIALS

Electrical hand tools

Unit with ECM blower and ECM blower control board

SAFETY REQUIREMENTS

A. Wear safety glasses

B. Turn power off to unit BEFORE setting CFM switches or jumpers.

C. Do not energize the motor until given explicit instructions from the instructor.

PROCEDURE

1. Note the setting on the blower control board. Take a picture with your phone if possible.
2. Operate the blower and measure the CFM.
3. Shut off the blower and adjust the CFM on the blower control.
4. Operate the blower and measure the CFM.
5. Return the blower control to its original setting.

Figure 38-16-1 ECM blower settings

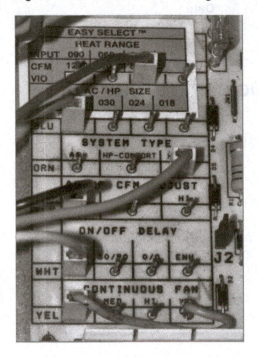

LAB 38.17 TROUBLESHOOTING CSR MOTORS

LABORATORY OBJECTIVE
The purpose of this lab is to learn how to check hermetic CSR motors for faults.

FUNDAMENTALS OF HVACR TEXT REFERENCE
Unit 38 Motor Application and Troubleshooting

LABORATORY NOTES
We will use an ammeter, volt meter, and an ohm meter to test CSR compressors and their starting components. You will determine if the starting components are good and if the motor is good, open, shorted, or grounded. If the motor is good, you will use the ohm readings to identify the Common, Start, and Run terminals.

REQUIRED TOOLS, EQUIPMENT, AND MATERIALS
Volt-Ohm meter   Electrical hand tools
Ammeter   Hermetic compressors with CSR motors

SAFETY REQUIREMENTS
A. Wear safety glasses
B. All wiring should be done with the motor not energized.
C. Motor should be physically secured before operating.
D. Do not energize the motor until given explicit instructions from the instructor.

PROCEDURE
1. First check to see that voltage is available to the unit.
2. Check the amp draw through the start capacitor when the compressor starts. The current should spike when the compressor starts and drop to 0 after it gets started.
3. If there is no amp draw thought the start capacitor when the compressor tries to start the problem is most likely with an open start capacitor or start relay.
4. If there is no current through the common wire with voltage available, you need to disconnect the power and ohm the compressor motor.
5. Take ohm readings between each combination of two terminals – there should be three readings
6. Write the readings down.
7. If any of the three readings is infinite, either the motor winding or internal overload is open.
8. If the reading from C to S and C to R are both infinite, but the reading from S to R is not, the internal overload is open. If the motor is hot, the overload may close and reset when the motor cools.
9. If any of the three readings is less than 0.3 ohms, the motor is shorted.
10. Read the resistance between each terminal and one of the copper lines on the compressor.
11. Any reading other than infinite indicates a grounded motor.

12. If the motor is not open, shorted, or grounded you can determine the C, S, and R terminals.

13. The highest reading will be between the start and ruin terminals. Therefore, the terminal that is NOT involved in the highest reading is common.

14. The lowest reading is between Common and run. The terminal that is involved in the lowest reading with the Common terminal is Run.

15. The start terminal will be the terminal left over after identifying both the common and run terminals.

16. Complete the table and be prepared to explain your results to the instructor.

The three most common terminal arrangements are shown in the boxes below.

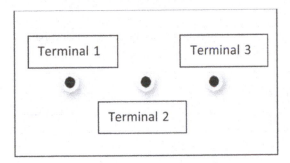

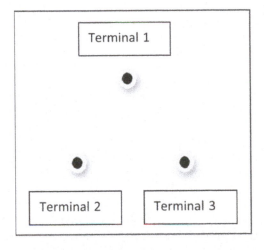
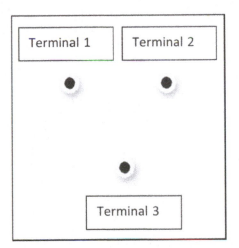

	Term 1 to Term 2	Term 1 to Term 3	Term 2 to Term 3	Terminal 1 to Ground	Terminal 2 to Ground	Terminal 3 to Ground	Condition Good Shorted Open Grounded
				Ohming Compressor Motors Resistance			
Comp 1							
Comp 2							
Comp 3							

	Terminal 1	Terminal 2	Terminal 3
	Determining Common Start and Run Terminal Designations (Common, Start, and Run)		
Comp 1			
Comp 2			
Comp 3			

LAB 38.18 TROUBLESHOOTING MOTORS

LABORATORY OBJECTIVE
The purpose of this lab is to apply your motor testing skills on a range of different types of electric motors.

FUNDAMENTALS OF HVACR TEXT REFERENCE
Unit 38 Motor Application and Troubleshooting

LABORATORY NOTES
You will test two of each type of motor that you have studied and list the specific fault of each motor. If the motor checks out, you will wire it, operate it, and record the amp reading.

REQUIRED TOOLS, EQUIPMENT, AND MATERIALS
Electrical hand tools
Multimeter and ammeter
Assortment of electric motors

SAFETY REQUIREMENTS
A. Wear safety glasses
B. All wiring should be done with the motor not energized.
C. Motor should be physically secured before operating.
D. Do not energize the motor until given explicit instructions from the instructor.

PROCEDURE
Check the motors assigned by the instructor and record the information requested on the chart. State the fault as bearing failure, winding open or shorted, starting component failure, etc. If the motor checks out, wire and operate it and record the amp reading. Figures 37-18-1, 37-18-2, 37-18-3, 37-18-4 can be used for reference.

Multi-speed PSC motor

Multi-speed shaded pole motor

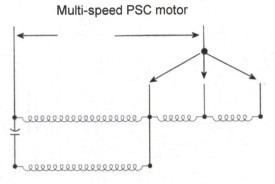

Figure 38-18-1

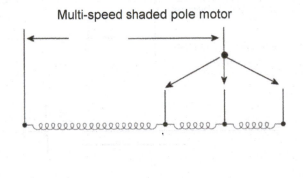

Figure 38-18-2

Single Speed, one capacitor wire PSC Motor

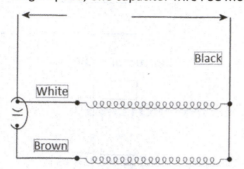

Figure 38-18-3

Single speed, two capacitor wire PSC motor

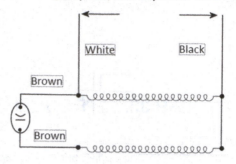

Figure 38-18-4

Motor	Voltage	Nameplate FLA	Running Amps	Condition (Good, open, shorted, grounded, bad bearings)
Shaded Pole				
Shaded Pole				
Split Phase (open)				
Capacitor Start (open)				
PSC (open)				
PSC (open)				
Three Phases				
Three Phases				
Capacitor Start (hermetic)				
Capacitor Start Run (hermetic)				

LAB 39.1 IDENTIFY ELECTRIC COMPONENT SYMBOLS

LABORATORY OBJECTIVE
You will identify ten common HVACR electrical components and draw the symbol that represents each of them.

FUNDAMENTALS OF HVACR TEXT REFERENCE
Unit 39 Electrical Diagrams

LABORATORY NOTES
In this lab exercise you will write down the name of the component and draw the symbol used to represent that component in electrical diagrams.

REQUIRED TOOLS, EQUIPMENT, AND MATERIALS
Pencil
Ten common HVACR electrical components

SAFETY REQUIREMENTS
Wear safety glasses

PROCEDURE

Name and draw the symbols for 10 electrical components provided by the instructor using the table below.

	Component Name	Symbol
1		
2		
3		
4		
5		
6		
7		

8		
9		
10		

LAB 39.2 IDENTIFY COMPONENTS ON AIR CONDITIONING DIAGRAM

LABORATORY OBJECTIVE
You will identify common HVACR electrical components on an air conditioning unit and locate their symbols on the unit electrical diagram.

***FUNDAMENTALS OF HVACR* TEXT REFERENCE**
Unit 34 Electrical Components
Unit 39 Electrical Diagrams

LABORATORY NOTES
In this lab exercise you will identify common electrical components on an air conditioner.
You will then identify the symbols in the unit diagram that represent the components.

REQUIRED TOOLS, EQUIPMENT, AND MATERIALS
Safety glasses
Non-contact voltage detector
6-in-1 screwdriver
Air conditioner with electrical components and an electrical diagram

SAFETY REQUIREMENTS
A. Turn off and lock out power to the unit before beginning work.
B. Check with the non-contact voltage detector to ensure that the panel is not energized.

PROCEDURE
    A. Turn off the power to the packaged air conditioner assigned by the instructor. Remove the electrical panel to the unit.
    B. Check with the non-contact voltage detector to ensure that the panel is not energized. DO NOT TOUCH ANYTHING YOU DON'T HAVE TO!
    C. You may use the unit component location diagram or pictorial diagram to locate the components. You should be able to locate the component on the diagram and on the unit.
    D. You must point out the components and their symbols to the instructor.
    E. Check them off in data sheet as you identify them.

Data Sheet for Air Conditioner Electric Components		
Component Name	Locate on Unit	Identify Symbol
Condenser fan motor (outdoor fan motor)		
Evaporator fan motor (indoor blower motor)		
Contactor		
Compressor		
Transformer		
Capacitor		
Indoor fan motor (blower)		
Relay		
Thermostat		

LAB 39.3 IDENTIFY COMPONENTS ON HEAT PUMP DIAGRAM

LABORATORY OBJECTIVE

You will identify common HVACR electrical components on a heat pump and locate
their symbols on the unit electrical diagram.

FUNDAMENTALS OF HVACR TEXT REFERENCE

Unit 34 Electrical Components

Unit 39 Electrical Diagrams

LABORATORY NOTES

In this lab exercise you will identify common electrical components on a heat pump.

You will then identify the symbols in the unit diagram that represent the components.

REQUIRED TOOLS, EQUIPMENT, AND MATERIALS

Safety glasses

Non-contact voltage detector

6-in-1 screwdriver

Heat pump with electrical components and an electrical diagram

SAFETY REQUIREMENTS

A. Turn off and lock out power to the unit before beginning work.

B. Check with the non-contact voltage detector to ensure that the panel is not energized.

PROCEDURE

    A. Turn off the power to the packaged heat pump assigned by the instructor.

    B. Remove the electrical panel to the unit.

    C. Check with the non-contact voltage detector to ensure that the panel is not energized.

    D. DO NOT TOUCH ANYTHING YOU DON'T HAVE TO!

    E. You may use the unit component location diagram or pictorial diagram to
       locate the components.

    F. You should be able to locate the component on the diagram and on the unit.

    G. You must point out the components and their symbols to the instructor.

    H. Check them off in data sheet as you identify them.

Data Sheet for Heat Pump Electrical Components		
Component Name	Locate on Unit	Identify Symbol
Outdoor fan motor		
Indoor blower motor		
Contactor		
Compressor		
Transformer		
Capacitor		
Indoor fan motor (blower)		
Relay		
Thermostat		
Reversing Valve		
Defrost control		
Defrost thermostat		
High-pressure switch		
Supplementary heater		
Thermal overload for electric heater		

LAB 39.4 CONNECTION WIRING DIAGRAMS

LABORATORY OBJECTIVE
The student will demonstrate how to properly draw a wiring schematic from a connection diagram.

FUNDAMENTALS OF HVACR TEXT REFERENCE
Unit 39 Electrical Diagrams

LABORATORY NOTES
The student may draw the wiring schematic from the connection diagram below or from an alternate connection diagram supplied by the Lab Instructor.

REQUIRED TOOLS, AND EQUIPMENT, AND MATERIALS
None

SAFETY REQUIREMENTS
None
The connection diagram shown in Figure 39-4-1 is for an air cooled condensing unit. The control system consists of a wiring panel enclosing the compressor starter, the start relay, the run capacitor, a thermostat and switch combination, a start capacitor, and a junction box.

External to the wiring panel are the fan motor, power supply, compressor motor, junction box, and high/low pressure control.

Figure 39-4-1

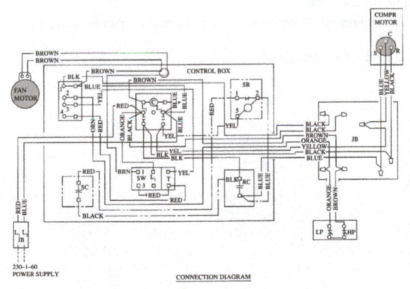

CONNECTION DIAGRAM

LEGEND FOR CONNECTION DIAGRAM

C – Contactor
RC – Run Capacitor
SC – Start Capacitor
SR – Start Relay
T – Thermostat
SW – Switch
HP – High-Pressure Switch
LP – Low-Pressure Switch
JB – Junction Box

———————————— Factory Wiring

– – – – – – – – Field Wiring

– . – . – . – . – . Alternate CSR Wiring

PROCEDURE

Step 1 **– Identify the Loads**

The first step in preparing the schematic is to locate the loads and determine the number of circuits. There are five loads and therefore five circuits.

    A. The five loads are:

      a)

      b)

      c)

      d)

      e)

Step 2 **– Identify the Switches**

The second step in preparing the schematic is to locate the switches in each circuit.

    A. The switches in each circuit are:

      a)

      b)

      c)

      d)

      e)

Step 3 – **Draw the Ladder**

The third step in preparing the schematic is to draw the ladder that represents the two legs of power. Draw two vertical lines several inches apart to represent L1 and L2.

Step 4 – **Draw the Circuits**

The last step is to draw the circuits between the two vertical lines. We will be drawing circuits from the top of the ladder to the bottom.

A. Draw the green test lamp circuit. Note that there are no controls or switches in this circuit.
B. Draw the compressor <u>motor </u>circuit, leaving the run and start capacitors out . Note that the contactor contacts are in series with the compressor motor.
C. Add the run capacitor to the compressor circuit.
D. Add the start capacitor and start relay to the compressor circuit.
E. Draw the switch and fan motor circuit.
F. Draw the switch and contactor coil circuit.

LAB 39.5 LADDER WIRING DIAGRAMS

LABORATORY OBJECTIVE
Students will demonstrate their understanding of a ladder diagram. They will draw lines representing wires between components according to the ladder diagram provided.

FUNDAMENTALS OF HVACR TEXT REFERENCE
Unit 39 Electrical Diagrams

LABORATORY NOTES
This lab is a chance to practice virtual wiring on air conditioning system components using a standard ladder diagram.

REQUIRED TOOLS, EQUIPMENT, AND MATERIALS
Pencil, pen, or markers. Colored pencils work well because they can be erased.

PROCEDURE
The ladder diagram shown in Figure 39-5-1 is for a packaged air conditioning unit. Figure 39-5-2 shows all the electrical components represented in the diagram. You are to "wire" the components according to the ladder diagram by drawing lines to represent wires. It may help to color code the lines. For example, one color for L1, another for L2, another for the R side of the transformer secondary, and another for the common side of the transformer secondary. You may want to use pencil so that erasures are possible.

LEGEND FOR LADDER DIAGRAM 39-5-1
COMP – Compressor
C – Contactor
IFR – Indoor Fan Relay
RC – Run Capacitor
OFM – Outdoor Fan Motor
IFM – Indoor Fan Motor
TR – Transformer
HP – High-Pressure Switch

Figure 39-5-1 Ladder diagram

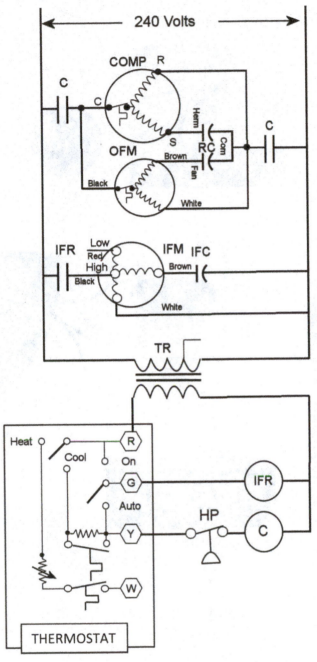

Figure 39-5-2 Air conditioning components

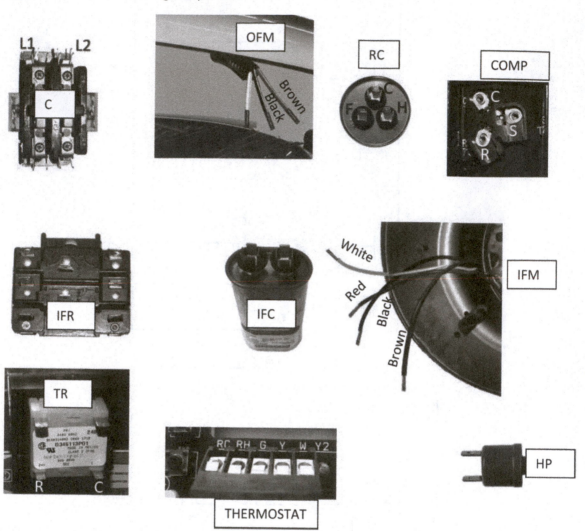

LAB 39.6 WIRING TRANSFORMERS

LABORATORY OBJECTIVE
You will demonstrate your understanding of transformer operation by safely wiring a transformer to control a contactor with a 24-volt coil.

FUNDAMENTALS OF HVACR TEXT REFERENCE
Unit 34 Electrical Components
Unit 39 Electrical Diagrams
Unit 40 Control Systems

LABORATORY NOTES
In this lab exercise you will construct a simple circuit with the SPST switch controlling the coil of a 24-volt contactor. A 110-volt primary to 24-volt secondary transformer will be the low voltage power supply.

REQUIRED TOOLS, EQUIPMENT, AND MATERIALS
Safety glasses
Wire and wire nuts
Wire cutter/stripper/crimper and electrical terminals
120-volt power cord
SPST switch
120-volt primary 24-volt secondary transformer
Contactor with 24-volt coil
Multimeter
6-in-1 screwdriver

SAFETY REQUIREMENTS
A. Wear safety glasses
B. Circuit should be unplugged during all wiring.
C. Do not energize circuit until given explicit instructions from instructor.

PROCEDURE
Step 1 – Wiring the Circuit
    A. Refer to Figure 39-6-1 to wire the circuit shown in Figure 39-6-2.
    B. The two wires from the power supply cord should feed the transformer 120-volt primary.
    C. One side of the transformer secondary should connect to the switch.
    D. The other side of the switch should connect to one side of the contactor coil.

E.   The other side of the contactor coil should connect to the remaining wire on the transformer secondary side.

F.   Have an instructor check your work before energizing it.

Step 2 – Operate the Circuit

Record the following voltage measurements on Data Sheet 39-6-1.

A.   Measure the transformer primary and secondary voltages.

B.   With the circuit energized and the switch off, measure the voltage across the switch.

C.   With the circuit energized and the switch off, measure the voltage at the contactor coil.

D.   With the circuit energized and the switch on, measure the voltage across the switch.

E.   With the circuit energized and the switch on, measure the voltage at the contactor coil.

Figure 39-6-1 Wiring 24-Volt contactor

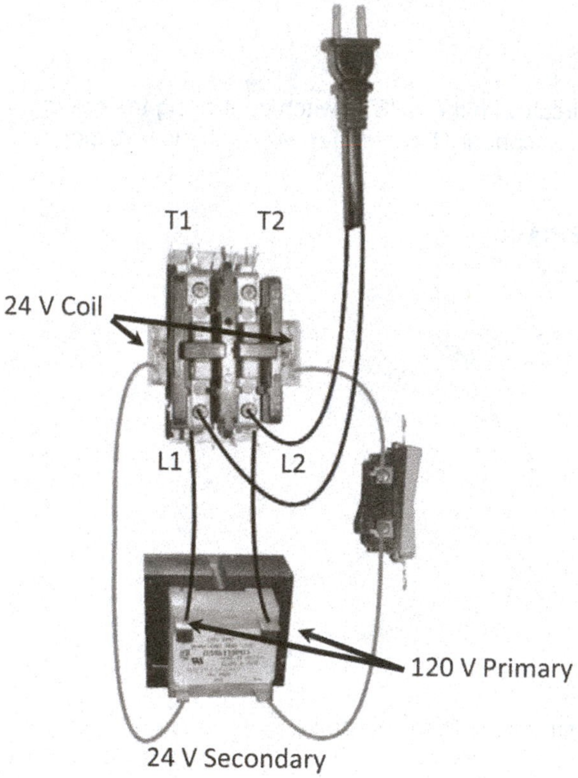

Figure 39-6-2 24-Volt contactor diagram

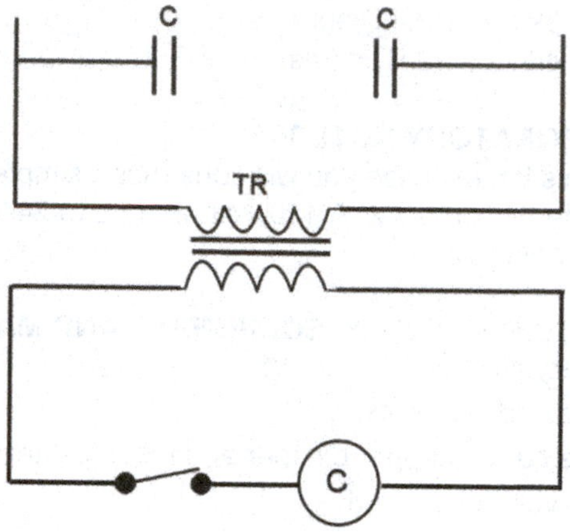

Data Sheet 39-6-1		
Location	Voltage Reading Switch OFF	Voltage Reading Switch ON
Primary Voltage		
Secondary Voltage		
Across the Switch		
Contactor Coil		

LAB 39.7 WIRING CONTACTORS

LABORATORY OBJECTIVE

You will demonstrate your understanding of contactor operation by safely wiring a 24-volt coil contactor to control a 120-volt light.

FUNDAMENTALS OF HVACR TEXT REFERENCE

Unit 34 Electrical Components
Unit 39 Electrical Diagrams
Unit 40 Control Systems

LABORATORY NOTES

In this lab exercise you will construct a simple circuit with the SPST switch controlling the coil of a 24-volt contactor. The contactor contacts will control the 120-volt light. A 110-volt primary to 24-volt secondary transformer will be the low voltage power supply.

REQUIRED TOOLS, EQUIPMENT, AND MATERIALS

Safety glasses	Wire
Multimeter	Electrical terminals
Wire stripper/crimper	Wire nuts
Wire cutter	SPST switch
6-in-1 screwdriver	120-volt primary 24-volt secondary transformer
120-volt power cord	Contactor with 24-volt coil
120-volt light	

SAFETY REQUIREMENTS

A. Wear safety glasses
B. Circuit should be unplugged during all wiring.
C. Do not energize circuit until given explicit instructions from instructor.

PROCEDURE

Step 1 – Wiring the Circuit

A. Refer to Figure 39-7-1 to wire the circuit shown in Figure 39-7-2.
B. The two wires from the power supply cord should wire to L1 and L2 of the contactor contacts.
C. The primary side of the transformer should also wire to L1 and L2 on the contactor.
D. The light should be wired to T1 and T2 of the contactor contacts.
E. One side of the transformer secondary should connect to the switch.
F. The other side of the switch should connect to one side of the contactor coil.
G. The other side of the contactor coil connects to the other side on the transformer secondary.
H. Have an instructor check your work before energizing it.

Step 2 – Operating the Circuit

A. Measure the transformer primary and secondary voltages.
B. With the circuit energized and the switch off, measure the voltage across the switch.
C. With the circuit energized and the switch off, measure the voltage at the contactor coil.
D. With the circuit energized and the switch off, measure the voltage between L1 and L2.
E. With the circuit energized and the switch off, measure the voltage between T1 and T2.
F. With the circuit energized and the switch off, measure the voltage between L1 and T1.
G. With the circuit energized and the switch on, measure the voltage across the switch.
H. With the circuit energized and the switch on, measure the voltage at the contactor coil.
I. With the circuit energized and the switch on, measure the voltage between L1 and L2.
J. With the circuit energized and the switch on, measure the voltage between T1 and T2.
K. With the circuit energized and the switch on, measure the voltage between L1 and T1.

Figure 39-7-1 Contactor controlling light   Figure 39-7-2 Contactor diagram

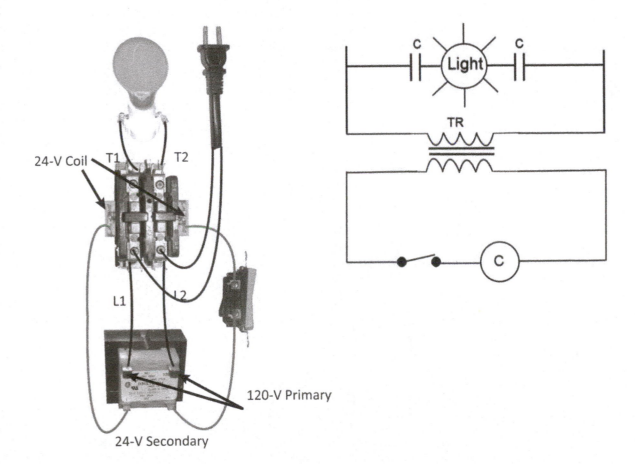

24-V Coil   T1   T2

L1   L2

120-V Primary

24-V Secondary

Data Sheet 39-7-1		
Location	Voltage Reading Switch OFF	Voltage Reading Switch ON
Primary Voltage		
Secondary Voltage		
Across the Switch		
Contactor Coil		
L1 and L2		
T1 and T2		
L1 and T1		
L2 and T2		

LAB 39.8 WIRING RELAYS

LABORATORY OBJECTIVE

You will demonstrate your understanding of relay operation by safely wiring a 24-volt coil relay to control two 120-volt lights; one light to operate only when the relay coil is not energized and the other light to operate only when the relay coil is energized.

FUNDAMENTALS OF HVACR TEXT REFERENCE

Unit 34 Electrical Components

Unit 39 Electrical Diagrams

Unit 40 Control Systems

LABORATORY NOTES

In this lab exercise you will construct a simple circuit with the SPST switch controlling the coil of a 24-volt relay. The NC relay contacts will control one 120-volt light; the NO contacts will control the other 120-volt light. A 110-volt primary to 24-volt secondary transformer will be the low voltage power supply.

REQUIRED TOOLS, EQUIPMENT, AND MATERIALS

Safety glasses

Wire and Wire nuts

Wire cutter/stripper/crimper and electrical terminals

Multimeter

SPST switch

6-in-1 screwdriver

120-volt primary 24-volt secondary transformer

Relay with 24-volt coil, 1 NO and 1 NC contacts

120-volt power cord and two 120-volt lights

SAFETY REQUIREMENTS

A. Wear safety glasses

B. Circuit should be unplugged during all wiring.

C. Do not energize circuit until given explicit instructions from instructor.

PROCEDURE

Step 1 – Wiring the Circuit

    A. Refer to Figure 39-8-1 to wire the circuit shown in Figure 39-8-2.

    B. Use a 110-volt to 24-volt transformer and a SPST switch to control a 24-volt relay coil.

    C. The relay contacts should be wired to control two 110-volt lights.

D. One light should come on when the relay is energized and the other should come on when the relay is de-energized.

E. The switch should be wired in series with the transformer secondary and the relay coil.

F. The relay normally open contacts should be wired in series with one light.

G. The relay normally closed contacts should be wired in series with the other light.

H. Have an instructor check your work before energizing it.

Step 2 – Operating the Circuit

A. With the switch off, plug in the circuit and notice which light comes on.

B. Turn the switch on and observe the change.

C. Explain the operation of relay normally pen and normally closed contacts.

Figure 39-8-1 Relay circuit   Figure 39-8-2 Relay circuit diagram

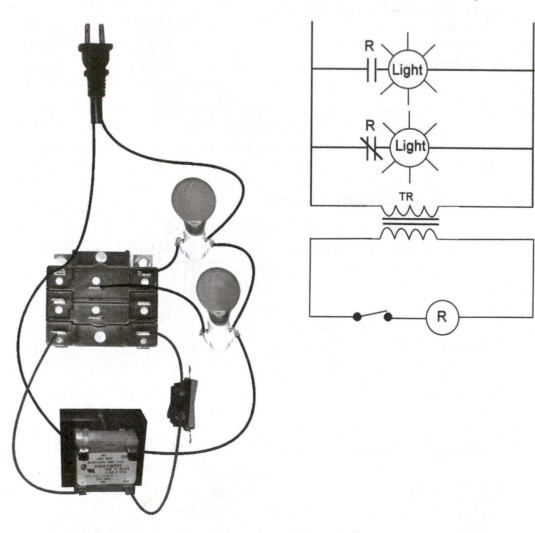

LAB 40.1 DRAW WINDOW UNIT DIAGRAM

LABORATORY OBJECTIVE
You will draw a schematic diagram for a window air conditioning unit.
Note: You must have completed this lab PRIOR to starting Lab 40.2, which is wiring this diagram.

FUNDAMENTALS OF HVACR TEXT REFERENCE
Unit 39 Electrical Diagrams
Unit 40 Control Systems

LABORATORY NOTES
You will draw a schematic diagram of a window air conditioning unit. This diagram will be used to wire the circuit in Lab 40.2.

REQUIRED TOOLS, EQUIPMENT, AND MATERIALS
Pencil
Paper

PROCEDURE
On a separate sheet of paper, draw a schematic diagram to meet the following specifications:

- UNIT TYPE: Window Air Conditioner
- COMPRESSOR TYPE: PSC Motor
- FAN MOTOR TYPE: PSC Single Speed Motor
- CONTROLS: Line Voltage
- SWITCH: Line Voltage Rotary Selector Switch to Control Fan Motor and Compressor
- THERMOSTAT: Line Voltage

LAB 40.2 WIRE WINDOW UNIT DIAGRAM

LABORATORY OBJECTIVE

You will wire the electrical components for a window air conditioning unit using the diagram you drew in Lab 40.1.

Note: You must have completed Lab 40.1 PRIOR to starting this lab.

FUNDAMENTALS OF HVACR TEXT REFERENCE

Unit 39 Electrical Diagrams

Unit 40 Control Systems

LABORATORY NOTES

You will wire the electrical components of a window air conditioning unit using the schematic diagram you drew in Lab 40.1.

REQUIRED TOOLS, EQUIPMENT, AND MATERIALS

Safety glasses	Wire
Multimeter	Wire connectors
Ammeter	Wire nuts
Wire cutter	Line voltage thermostat
Wire crimper/stripper	Line voltage rotary control switch
6-in-1 screwdriver	Light to represent compressor
120-volt power cord	Light to represent fan motor

SAFETY REQUIREMENTS

A. Wear safety glasses
B. Circuit should be unplugged during all wiring.
C. Do not energize circuit until given explicit instructions from instructor.

PROCEDURE

Build the diagram you drew in Lab 40.1.

Have instructor check you completed diagram BEFORE energizing it.

Energize and operate your diagram.

LAB 40.3 DRAW PACKAGED AC UNIT DIAGRAM

LABORATORY OBJECTIVE

You will draw a schematic diagram for a packaged central air conditioning unit.

Note: You must have completed this lab PRIOR to starting Lab 40.4, which is wiring this diagram.

FUNDAMENTALS OF HVACR TEXT REFERENCE

Unit 39 Electrical Diagrams

Unit 40 Control Systems

LABORATORY NOTES

You will draw a schematic diagram of a packaged air conditioning unit. This diagram will be used to wire the circuit in Lab 40.4.

REQUIRED TOOLS, EQUIPMENT, AND MATERIALS

Pencil

Paper

PROCEDURE

On a separate sheet of paper, draw a schematic diagram to meet the following specifications:

- UNIT TYPE: Packaged Unit
- COMP. TYPE: PSC Motor with Internal Overload
- CFM TYPE: PSC Motor
- IFM TYPE: Shaded Pole Motor
- CONTROL SYSTEM: 24-Volt Control System
- THERMOSTAT: 24-Volt Thermostat to Control Both Indoor Fan and Cooling
- PROTECTION: LP & HP Switch in Low-Voltage Circuit to Protect Compressor

LAB 40.4 WIRE PACKAGED AC DIAGRAM

LABORATORY OBJECTIVE
You will wire the electrical components for a window air conditioning unit using the diagram you drew in Lab 40.3.

Note: You must have completed Lab 40.3 PRIOR to starting this lab.

FUNDAMENTALS OF HVACR TEXT REFERENCE
Unit 39 Electrical Diagrams
Unit 40 Control Systems

LABORATORY NOTES
You will wire the electrical components of the packaged air conditioning unit using the schematic diagram you drew in Lab 40.3.

REQUIRED TOOLS, EQUIPMENT, AND MATERIALS
Safety glasses
Multimeter
Ammeter
Wire crimper/stripper
6-in-1 screwdriver
120-volt power cord
Wire
Three lights or motors to represent:
1. compressor
2. indoor fan motor
3. outdoor fan motor

Wire connectors
Wire nuts
Transformer
Low voltage heat-cool thermostat
24-volt contactor
24-volt relay

SAFETY REQUIREMENTS
A. Wear safety glasses
B. Circuit should be unplugged during all wiring.
C. Do not energize circuit until given explicit instructions from instructor.

PROCEDURE
Build the diagram you drew in Lab 40.3.
Have instructor check you completed diagram BEFORE energizing it.
Energize and operate your diagram.

LAB 40.5 DRAW 2-STAGE HEAT, 2-STAGE COOL DIAGRAM

(This lab is done at home, NOT during lab time.)

LABORATORY OBJECTIVE

You will draw a schematic diagram for a two-stage heating, two stage cooling air conditioning unit.

Note: You must have completed this lab PRIOR to starting Lab 40.6, which is wiring this diagram.

FUNDAMENTALS OF HVACR TEXT REFERENCE

Unit 39 Electrical Diagrams

Unit 40 Control Systems

LABORATORY NOTES

You will draw a schematic diagram of a two-stage heating, two stage cooling air conditioning unit. This diagram will be used to wire the circuit in Lab 40.6.

REQUIRED TOOLS, EQUIPMENT, AND MATERIALS

Pencil

Paper

PROCEDURE

On a separate sheet of paper, draw a schematic diagram to meet the following specifications:

- UNIT TYPE: 2-Stage Heat, 2-Stage Cool
- COMP. TYPE: Two PSC Compressor Motors with Internal Overload (one for each stage)
- CFM TYPE: PSC Motor
- IFM TYPE: PSC Motor
- COOL: One Compressor for Each Stage
- HEAT: Electric Strip Heat - 1 Heater for Each Stage
- CONTROL SYSTEM: 24-Volt Control System, 2-Stage Heat, 2-Stage Cool
- THERMOSTAT: 24-Volt Thermostat to Control Both Indoor Fan and Cooling
- PROTECTION: LP & HP Switch in Low-Voltage Circuit to Protect Each Compressor
- PROTECTION: Limit Switch in Series with Each Heat Strip to Protect from Overheating

LAB 40.6 WIRE 2-STAGE HEAT 2-STAGE COOL DIAGRAM

LABORATORY OBJECTIVE
You will wire the electrical components for a 2-stage heating, 2-stage cooling air conditioning unit using the diagram you drew in Lab 40.5.

Note: You must have completed Lab 40.5 PRIOR to starting this lab.

FUNDAMENTALS OF HVACR TEXT REFERENCE
Unit 39 Electrical Diagrams

Unit 40 Control Systems

LABORATORY NOTES
You will wire the electrical components of the packaged air conditioning unit using the schematic diagram you drew in Lab 40.5.

REQUIRED TOOLS, EQUIPMENT, AND MATERIALS
Safety glasses
Multimeter
Ammeter
Wire cutter
Wire crimper/stripper
6-in-1 screwdriver
120-volt power cord
Wire
Wire connectors
Wire nuts
Transformer
Low voltage heat-cool thermostat
Four 24-volt contactors
24-volt relay
Seven lights or small motors to represent: 2 compressors
2 condenser fan motors
2 outdoor fan motors
indoor fan motor

SAFETY REQUIREMENTS

A. Wear safety glasses

B. Circuit should be unplugged during all wiring.

C. Do not energize circuit until given explicit instructions from instructor.

PROCEDURE

Build the diagram you drew in Lab 40.5.

Have instructor check you completed diagram BEFORE energizing it.

Energize and operate your diagram.

LAB 40.7 WIRE A HEATING AND COOLING THERMOSTAT

LABORATORY OBJECTIVE

The purpose of this lab is to learn how to wire a low-voltage thermostat to a gas furnace and split system air conditioner.

FUNDAMENTALS OF HVACR TEXT REFERENCE

Unit 39 Electrical Diagrams

Unit 40 Control Systems

LABORATORY NOTES

You will wire a single stage heating-cooling thermostat to control a gas furnace with a split system air conditioner. You will need to wire the control wiring at the thermostat, furnace control panel, and condensing unit.

REQUIRED TOOLS, EQUIPMENT, AND MATERIALS

Low-voltage thermostat

Gas furnace

Split system air conditioner

Multimeter

Electrical hand tools

SAFETY REQUIREMENTS

A. Wear safety glasses

B. Power to unit should be turned off and locked out before beginning work.

C. Do not energize unit until given explicit instructions from instructor.

PROCEDURE

1. Read the installation instructions for the thermostat, furnace, and air conditioner.
2. Turn off and lock out the power to the furnace and air conditioning unit.
3. Check the voltage to the unit with a multimeter to be sure the power is off.
4. Wire the 5-wire at the thermostat and furnace.
5. Wire the 2-wire at the furnace and condensing unit.
6. Have an instructor check your wiring.
7. Turn the power on and operate the unit in both heating and cooling.

Figure 40-7-1 Typical heat-cool control wiring

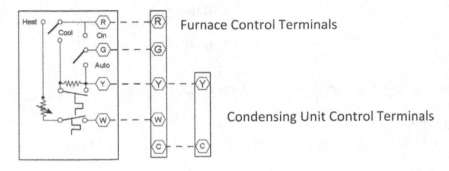

Furnace Control Terminals

Condensing Unit Control Terminals

LAB 40.8 WIRE A PACKAGED HEAT PUMP THERMOSTAT

LABORATORY OBJECTIVE
The purpose of this lab is to learn how to wire a low-voltage thermostat to a packaged heat pump.

FUNDAMENTALS OF HVACR TEXT REFERENCE
Unit 39 Electrical Diagrams

Unit 40 Control Systems

LABORATORY NOTES
You will wire a two-stage heat pump thermostat to control a packaged heat pump.

REQUIRED TOOLS, EQUIPMENT, AND MATERIALS
Low-voltage heat pump thermostat
Packaged heat pump
Multimeter
Electrical hand tools

SAFETY REQUIREMENTS
A. Wear safety glasses
B. Power to unit should be turned off and locked out before beginning work.
C. Do not energize unit until given explicit instructions from instructor.

PROCEDURE
1. Read the installation instructions for the thermostat and heat pump.
2. Turn off the power to the heat pump.
3. Check the voltage to the unit with a multimeter to be sure the power is off.
4. Wire the thermostat to control the unit.
5. Have an instructor check your wiring.
6. Turn the power on and operate the unit.

Figure 40-8-1 Typical heat pump control wiring

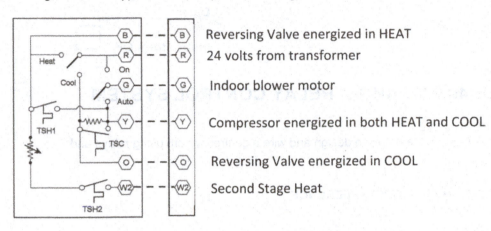

Reversing Valve energized in HEAT

24 volts from transformer

Indoor blower motor

Compressor energized in both HEAT and COOL

Reversing Valve energized in COOL

Second Stage Heat

Note: Terminals B and O are not usually both used.

LAB 40.9 WIRING A RELAY CONTROL SYSTEM

LABORATORY OBJECTIVE
The purpose of this lab is to learn how to design and wire a control system using relays and contactors.

FUNDAMENTALS OF HVACR TEXT REFERENCE
Unit 39 Electrical Diagrams
Unit 40 Control Systems

LABORATORY NOTES
You will Design a control system to control three lights. Each light will be controlled by a relay. Each relay will be controlled by its own toggle switch. All the lights will be 120-volt lights. The relays will each have a different coil voltage. One will be 24-volts, another 120-volts, and the other 230-volts.

REQUIRED TOOLS, EQUIPMENT, AND MATERIALS
24-volt relay
120-volt relay
230-volt relay
Transformer
3 toggle switches
3 120-volt lights
Multimeter
Electrical hand tools

SAFETY REQUIREMENTS
A. Wear safety glasses
B. Circuit should be unplugged and de-energized during all wiring.
D. Do not energize the circuit until given explicit instructions from instructor.

PROCEDURE

1. Gather and check the components.
    Note: It will save time to identify bad components BEFORE you wire the circuit.
2. Refer to Figure 40-9-1 to wire the control system.
3. Have the instructor check the wiring before energizing the circuit.
4. Operate the control system and troubleshoot any problems.

Figure 40-9-1 Relay control diagram

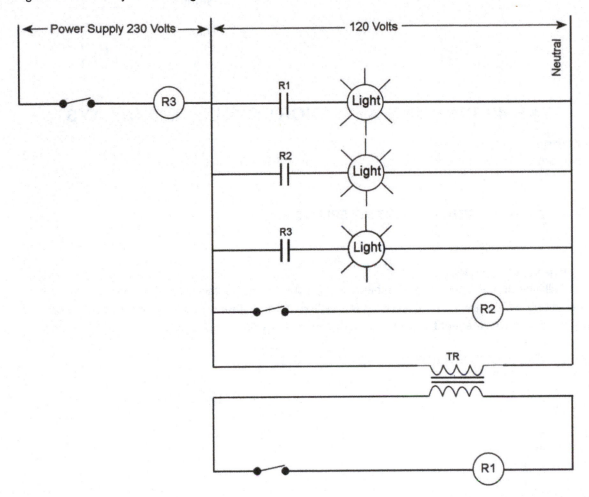

LAB 40.10 WIRING ANTI-SHORT CYCLE TIME DELAYS

LABORATORY OBJECTIVE
The purpose of this lab is to learn how to wire an anti-short cycle time delay to control a contactor.

FUNDAMENTALS OF HVACR TEXT REFERENCE
Unit 39 Electrical Diagrams
Unit 40 Control Systems

LABORATORY NOTES
You will wire an anti-short cycle time delay to control a 24-volt contactor coil. The contactor should energize when a thermostat calls for cooling. If the system is de-energized, the anti-short cycle timer should prevent the contactor from coming back on until after a delay of 3 to 5 minutes.

REQUIRED TOOLS, EQUIPMENT, AND MATERIALS
Low-voltage thermostat
Transformer
24-volt contactor
Time delay
Multimeter
Electrical hand tools

SAFETY REQUIREMENTS
A. Wear safety glasses
B. Circuit should be unplugged and de-energized during all wiring.
C. Do not energize the circuit until given explicit instructions from instructor.

PROCEDURE
1. Refer to Figure 40-10-1 to wire a delay on break time delay circuit.
2. Have the instructor check your wiring before energizing the circuit.
3. Operate the control system.
   A. There should be no delay when the switch is first closed. The contactor should energize and the light should come on.
   B. Open the switch and then close it again.
   C. This time there should be a delay. A delay on break timer does not start timing until after a break in the circuit.

Figure 40-10-1

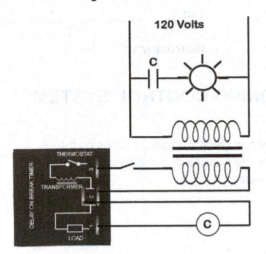

LAB 40.11 WIRING AIR CONDITIONING CONTROL SYSTEM

LABORATORY OBJECTIVE

You will draw a ladder diagram for an air conditioning system using the specifications on the next page. You will then select the components necessary to wire the air conditioning control system you drew. You will then wire and operate the system.

FUNDAMENTALS OF HVACR TEXT REFERENCE

Unit 39 Electrical Diagrams

Unit 40 Control Systems

LABORATORY NOTES

You will select the components necessary to construct the air conditioning control system that you drew. After building the system, you will operate it to verify that the controls work properly.

REQUIRED TOOLS, EQUIPMENT, AND MATERIALS

Safety glasses
Gloves
Electrical hand tools
Multimeter
Ammeter
Transformer
Low-voltage thermostat
Relays
Contactors
Motors
Lights to simulate loads
Pressure switches
Power wire
Control wire

SAFETY REQUIREMENTS

A. Wear safety glasses
B. Circuit should be unplugged and de-energized during all wiring.
C. Do not energize the circuit until given explicit instructions from instructor.

PROCEDURE

1. Draw a ladder type schematic diagram for an air conditioning control system using the specifications below.
2. Have the instructor check your design.
3. Select and check all necessary components.
4. Wire the control system.
5. Have the instructor check the wiring before energizing the circuit.
6. Operate the control system and troubleshoot any problems.

SPECIFICATIONS

- Unit Type: Air Cooled Packaged Unit
- Power Voltage: 230/1/60
- Control Voltage: 24-Volts
- Condenser Fan Motor: PSC Motor [230/1/60]
- Safety Controls: Low-Pressure Switch

  High-Pressure Switch

  Internal Overload for Compressor

- The condenser fan motor will be cycled by a close on rise pressure switch connected in series with the motor to control the head pressure.
- Power feeding the condenser fan motor will be from the load side of the compressor contactor
- Thermostat will be a three-wire low voltage cooling thermostat.

LAB 40.12 WIRING COMMERCIAL PACKAGED UNIT

LABORATORY OBJECTIVE

You will draw a ladder diagram for a commercial air conditioning system using the specifications on the next page. You will then select the components necessary to wire the commercial packaged unit control system you drew. You will then wire and operate the system.

FUNDAMENTALS OF HVACR TEXT REFERENCE

Unit 39 Electrical Diagrams

Unit 40 Control Systems

LABORATORY NOTES

You will select the components necessary to construct the commercial packaged unit control system that you drew. After building the system, you will operate it to verify that the controls work properly.

REQUIRED TOOLS, EQUIPMENT, AND MATERIALS

Safety glasses
Gloves
Electrical hand tools
Multimeter
Ammeter
Transformer
Low-voltage thermostat
Relays
Contactors
Motors
Lights to simulate loads
Pressure switches
Power wire
Control wire

SAFETY REQUIREMENTS

A. Wear safety glasses
B. Circuit should be unplugged and de-energized during all wiring.
C. Do not energize the circuit until given explicit instructions from instructor.

PROCEDURE

1. Draw a ladder type schematic diagram for the commercial packaged unit control system using the specifications below.
2. Have the instructor check your design.
3. Select and check all necessary components.
4. Wire the control system.
5. Have the instructor check the wiring before energizing the circuit.
6. Operate the control system and troubleshoot any problems.

SPECIFICATIONS

- Unit Type: 10 Ton Condensing Unit with Two-Stage Cooling
- Power Voltage: 208/3/60
- Control Voltage: 24-Volts
- Each condenser fan and compressor to have its own contactor
- Control relay to operate compressor contactor and condenser fan contactor
- Coil for the compressor contactor and condenser fan contactor to be 208 Volts
- First Stage Cool: Controls Compressor, Condenser Fans, and Indoor Fan
- Second Stage Cool: Controls Solenoid Valve for #2 Refrigerant Circuit
- Indoor fan motor controlled by thermostat
- Magnetic starter controlling each compressor
- Compressor Voltage: 208/3/60
- Condenser Fan Motors (2): 203/1/60 with Overload Protection
- Second CFM controlled by HPS to maintain constant head pressure
- Compressor protected by current type pilot duty overload
- Crankcase heater controlled thorough a control relay to energize heater on off cycle
- CFM motors controlled through control relay to operate when CR energized and CFM thermostat
- LP & HP switch for safety

LAB 40.13 WIRING TWO-STAGE AIR CONDITIONING SYSTEM

LABORATORY OBJECTIVE

You will draw a ladder diagram for a two-stage air conditioning system using the specifications listed on the next page. You will then select the components necessary to wire the two-stage air conditioning control system you drew. You will then wire and operate the system.

FUNDAMENTALS OF HVACR TEXT REFERENCE

Unit 39 Electrical Diagrams

Unit 40 Control Systems

LABORATORY NOTES

You will select the components necessary to construct the two-stage commercial air conditioning control system that you drew. After building the system, you will operate it to verify that the controls work properly.

REQUIRED TOOLS, EQUIPMENT, AND MATERIALS

Safety glasses

Gloves

Electrical hand tools

Multimeter

Ammeter

Transformer

Low-voltage thermostat

Relays

Contactors

Motors

Lights to simulate loads

Pressure switches

Power wire

Control wire

SAFETY REQUIREMENTS

A. Wear safety glasses
B. Circuit should be unplugged and de-energized during all wiring.
C. Do not energize the circuit until given explicit instructions from instructor.

PROCEDURE

1. Draw a ladder type schematic diagram for the two-stage air conditioning control system using the specifications below.
2. Have the instructor check your design.
3. Select and check all necessary components.
4. Wire the control system.
5. Have the instructor check the wiring before energizing the circuit.
6. Operate the control system and troubleshoot any problems.

SPECIFICATIONS

- Unit Type: Roof Top Two-Stage Cool and Two-Stage Heating
- Power Voltage: 230/1/60
- Control Voltage: 24-Volts
- Compressors {2} Type: 3 Phase Motors
- Condenser Fan Motors (2) Type: PSC Motors
- Indoor Fan Motor: PSC Motor
- Two-Stage Heating (2) & Cooling Low-Voltage Thermostat
- Electric Heaters 230/1/60
- 1st Compressor: 1st Stage Cool
- 2nd Compressor: 2nd Stage Cool
- 1st Heater: 1st Stage Heat
- 2nd Heater: 2nd Stage Heat
- Pressure switch in series with #2 CFM to control head pressure
- Safety Controls:
  a. Low-pressure switch
  b. High-pressure switch
  c. Pilot duty current overload relays
  d. Limit switch to interrupt power if temperature rises above 180°F

Note: Rating of transformer must be considered in this diagram because of the Number of 24-volt loads. You may use any type pilot duty system that you desire.

LAB 41.1 INSTALLING A COMMUNICATING THERMOSTAT

LABORATORY OBJECTIVE

Learn how to install a communicating thermostat.

FUNDAMENTALS OF HVACR TEXT REFERENCE

Unit 41 Communicating Control Systems

LABORATORY NOTES

You will install a communicating thermostat on a residential communicating control system. You will then start up the system and verify that all system components are operating correctly.

REQUIRED TOOLS, EQUIPMENT, AND MATERIALS

Safety glasses
Electrical hand tools
Multimeter
Ammeter
Communicating thermostat
Residential system with communicating controls
Control wire of the type specified by the installation instructions

SAFETY REQUIREMENTS

Wear safety glasses

PROCEDURE

Step 1 – Wiring Thermostat

    A. Read the installation instructions for the communicating system and thermostat.
    B. Pull the proper control wire for the system according to the instructions.
    C. Mount the thermostat.
    D. Make connections at the equipment and thermostat according to the instructions.
    E. Have instructor check your work.

Step 2 – Operating Thermostat

    A. Turn the system on.
    B. The system should go through an initial system set up if this system has never been energized.
    C. Verify that the thermostat found all the system components.
    D. If the thermostat has trouble finding all the components you may need to go into the service screen and instruct it to redo the initial setup. This occurs mainly when a thermostat is moved from one system to another.

LAB 41.2 ACCESS COMMUNICATING THERMOSTAT SERVICE SCREENS

LABORATORY OBJECTIVE
Learn how to access the service and installation screens on a communicating thermostat.

FUNDAMENTALS OF HVACR TEXT REFERENCE
Unit 41 Communicating Control Systems

LABORATORY NOTES
You will access the installation and service screens on a communicating thermostat, list any current faults, and list all previous faults.

REQUIRED TOOLS, EQUIPMENT, AND MATERIALS
Safety glasses
Communicating thermostat
Residential system with communicating controls

SAFETY REQUIREMENTS
Wear safety glasses

PROCEDURE
   A.   Read the installation instructions for the communicating system and thermostat.
   B.   Use the manufacturer's instructions to access the installation, maintenance, and service screens.
   C.   Note any current system faults.
   D.   List the fault history.

LAB 41.3 TROUBLESHOOTING COMMUNICATING CONTROLS

LABORATORY OBJECTIVE
Learn how to troubleshoot a communicating control system.

FUNDAMENTALS OF HVACR TEXT REFERENCE
Unit 41 Communicating Control Systems

LABORATORY NOTES
You will troubleshoot a communicating control system. Problems with communicating control systems are often the result of either incorrect installation or poor connections. The first place to start when troubleshooting communicating controls is to check to ensure all components are communicating. A communication error usually can be caused by a component not being energized, a poor electrical connection, changing components without running the installation software, or a failed board. Next, look at current faults and fault history on the thermostat. Communicating systems keep track of system failures.

REQUIRED TOOLS, EQUIPMENT, AND MATERIALS
Safety glasses
Communicating thermostat
Residential system with communicating controls

SAFETY REQUIREMENTS
Wear safety glasses

PROCEDURE
    A. Read the installation instructions for the communicating system and thermostat.
    B. Check to see that all installed communicating components are energized.
    C. Check to see that all installed communicating components are communicating.
        a. Check wiring with manufacturer's schematic
        b. Check connections
    D. Use the manufacturer's instructions to access the installation, maintenance, and service screens.

LAB 42.1 WIRE DISCONNECT SWITCH AND POWER SUPPLY

LABORATORY OBJECTIVE
You will size and install the power wiring and disconnect switch to an air conditioning system

***FUNDAMENTALS OF HVACR* TEXT REFERENCE**
Unit 42 Electrical Installation
Unit 89 Installation Techniques

LABORATORY NOTES
You will use the unit minimum circuit ampacity to size the conductor and disconnect switch. You will use the maximum overcurrent protection to size the fuse or circuit breaker. You will then install the disconnect switch, run conduit between the disconnect and the unit, and run power wire from the disconnect to the unit.

REQUIRED TOOLS, EQUIPMENT, AND MATERIALS

Safety glasses	Wire cutter
Multimeter	Wire crimper/stripper
Ammeter, wire	6-in-1 screwdriver
Conduit	Disconnect switch
Conduit connections	Air conditioning unit

SAFETY REQUIREMENTS
A. Wear safety glasses
B. Lock out power to circuit before beginning.
C. Check all circuits before beginning energizing.

PROCEDURE

***Step 1* – Sizing Wire, Conduit, Overcurrent Protection, and Disconnect**
    A. Find the minimum circuit ampacity and maximum overcurrent protection on the data plate.
    B. Refer to Tables 42-1-1, 42-1-2, 42-1-3 and the National Electric Code to size the wire and conduit.
    C. The circuit breaker or fuse should be no larger than the maximum overcurrent protection.
    D. The disconnect switch must be rated for no less than 115% of the minimum circuit ampacity.
        In most cases, a disconnect switch that will hold the maximum allowable fuse size is adequate.

Step 2 – Wiring

A. Lock out power to the circuit which will feed the disconnect witch.
B. Mount the disconnect switch.
C. Run conduit to the entry point of the disconnect switch.
D. Run wire from the power panel to the disconnect switch.
E. Run conduit between the disconnect switch and the unit.
F. Pull the power wire between the disconnect switch and the unit.
G. Have the instructor check your work.

Step 3 – Checkout

A. With the disconnect off but power to it, check the voltage into the top of the disconnect.
B. Compare the voltage to the minimum and maximum allowable voltages printed on the nameplate.

    a. If the voltage is outside of the allowable voltage range LEAVE THE DISCONNECT SWITCH OFF! You need to determine why the voltage does not match the specifications before proceeding any further.

    b. If the voltage is within the allowable voltage range, proceed to the next step.

C. If the voltage to the disconnect switch is within the manufacturer's specifications, make sure the system controls are set to OFF and turn the disconnect on.
D. Check voltage to the unit.
E. If the voltage to the unit is within the allowable voltage range, turn the unit on.
F. Check the voltage to the unit with the unit operating.
G. The voltage should still be within the allowable voltage range.
H. The difference between the voltage at the unit with the unit off and the voltage to the unit with the unit operating is the voltage drop through the circuit.

Table 42-1-1 Maximum Copper Wire Ampacity				
Wire Gauge	60°C Copper Wire Ampacity	75°C Copper Wire Ampacity	90°C Copper Wire Ampacity	* Maximum Overcurrent Protection
14*	15	20	25	15
12*	20	25	30	20
10*	30	35	40	30
8	40	50	60	NA
6	55	65	75	NA
* Indicates wire gauges whose overcurrent protection is limited to the values in the Maximum Overcurrent Protection column				

Table 42-1-2 Ambient Temperature Correction Factors (30°C)			
Ambient Temp	**60°C Wire**	**75°C Wire**	**90°C Wire**
50°F or less	1.29	1.2	1.15
51°F–59°F	1.22	1.15	1.12
60°F–68°F	1.15	1.11	1.08
69°F–77°F	1.08	1.05	1.04
78°F–86°F	1.0	1.0	1.0
87°F–95°F	0.91	0.94	0.96
96°F–104°F	0.82	0.88	0.91
105°F–113°F	0.71	0.82	0.87
114°F–122°F	0.58	0.75	0.82
123°F–131°F	NR	0.67	0.76
132°F–140°F	NR	0.58	0.71

Table 42-1-3 Factors for Four or More Current-Carrying Wires	
# of Wires	75°C Wire
4–6	0.80
7–9	0.70
10–20	0.50

LAB 43.1 TESTING TRANSFORMERS

LABORATORY OBJECTIVE
The purpose of this lab is to learn how to check transformers for faults.

FUNDAMENTALS OF HVACR TEXT REFERENCE
Unit 43 Electrical Troubleshooting

LABORATORY NOTES
You will check the resistance of the primary and secondary windings on five transformers assigned by the instructor. You will wire the transformers that ohm out correctly and check their voltage output.

REQUIRED TOOLS, EQUIPMENT, AND MATERIALS
Transformers
Volt-Ohm meter
Electrical hand tools

SAFETY REQUIREMENTS
A. Wear safety glasses
B. Ohm Meters should never be used on energized circuits.
C. Do not touch BOTH transformer leads when checking ohms.
D. Do not touch any electrical connections while transformers are energized.

PROCEDURE
1. Make sure the transformer is disconnected from any power source.
2. Check the resistance of the primary and secondary windings on the transformers.
   a. A reading of infinity (OL) indicates an open winding.
   b. A reading less than 1.0 ohm on the primary indicates a shorted winding.
   c. A reading less than 0.7 ohms on the secondary indicates a shorted winding.
3. Record the readings in the data table.
4. If both windings are good, apply the correct voltage to the primary winding and read the voltage on the secondary winding.
5. Record the applied primary voltage and induced secondary voltage in the data table.

Transformer VA Rating	Primary Resistance	Secondary Resistance	Primary Voltage	Secondary Voltage

LAB 43.2 CHECKING LOW-VOLTAGE THERMOSTATS

LABORATORY OBJECTIVE
The purpose of this lab is to learn how to test low-voltage thermostats using an ohm meter.

FUNDAMENTALS OF HVACR TEXT REFERENCE
Unit 43 Electrical Troubleshooting

LABORATORY NOTES
You will wire a low-voltage thermostat to a short set of control wires and test the thermostat operation with an ohm meter.

REQUIRED TOOLS, EQUIPMENT, AND MATERIALS
Low-voltage thermostats
Multimeter
Electrical hand tools

SAFETY REQUIREMENTS
A. Wear safety glasses
B. Ohm Meters should never be used on energized circuits.
C. When ohming components on a system, turn power off and check with a meter BEFORE using ohm meter.

PROCEDURE
You are going to set the thermostat to its various settings, check for continuity between R and the other terminals, and record your results in Data Sheet 43-2-1.

Step 1 – Preparation
A. Connect a set of 18-gauge, solid control wires to the terminals on the thermostat or thermostat sub base.
B. Connect the thermostat to the sub base.

Step 2 – Fan Check
A. Set the fan switch to "AUTO" and the system switch to "OFF"
B. Check resistance between the red wire and all the others. It should be infinite (OL)
C. Set the fan switch to "ON"
D. Check the resistance between R and G – it should read shorted (less than 1 ohm)

Step 3 – Heating Cycle Check
A. Set the fan switch back to "AUTO" and the system switch to "HEAT" and put the setpoint above the room temperature so there is no call for heating.
B. Check resistance between the red wire and all the other wires.
   a. The wire on the "B" terminal should have continuity to the red wire.
   b. All the other wires should read infinite (OL) to the red wire.

C. Leaving the fan switch set to "AUTO" and the system switch set to "HEAT," run the temperature setting up above room temperature.
D. Check the resistance between the red wire and all the other wires.
   a. The wires on the "B" and "W" terminals should have continuity to the red wire.
   b. Some thermostats will also create a circuit to the wire on the "G" terminal, others will not.

Step 4 – Cooling Cycle Check

1. Leave the fan switch on "AUTO," move the system switch to "COOL," and put the setpoint above the room temperature so there is no call for cooling.
2. Check resistance between the red wire and all the other wires.
   The wire on the "O" terminal should have continuity to the red wire.
   All the other wires should read infinite (OL) to the red wire.
3. Leaving the fan switch set to "AUTO" and the system switch set to "COOL," run the temperature setting down below room temperature, creating a call for cooling.
4. Check the resistance between the red wire and all the other wires.
   The wires on the "O", "Y", and "G" terminals should have continuity to the red wire.

Terminal	System OFF Fan AUTO	System OFF Fan ON	System HEAT Fan AUTO Setpoint below room temp	System HEAT Fan AUTO Setpoint above room temp	System COOL Fan AUTO Setpoint above room temp	System COOL Fan AUTO Setpoint below room temp
G						
Y						
W						
O						
B						

LAB 43.3 INSTALL REPLACEMENT THERMOSTAT

LABORATORY OBJECTIVE
The purpose of this lab is to demonstrate your ability to install a replacement thermostat.

FUNDAMENTALS OF HVACR TEXT REFERENCE
Unit 56 Gas Furnace Controls

LABORATORY NOTES
Many thermostats are multiday programmable, and may even be multi-zone programmable. Older thermostats may have a mercury switch and the proper disposal for this type is at a toxic waste collection site. The Thermostat Recycling Corporation collects mercury bulb thermostats.

REQUIRED TOOLS, EQUIPMENT, AND MATERIALS
Operating HVAC unit
Multimeter, clamp-on ammeter, thermometer
Thermostat
Thermostat screwdriver

SAFETY REQUIREMENTS
A. Check all circuits for voltage before doing any service work.
B. Stand on dry nonconductive surfaces when working on live circuits.
C. Never bypass any electrical protective devices.

PROCEDURE
Step 1 – System Data Collection
   A. The first step is to collect data from the system to ensure the replacement thermostat works. Examine the system and complete Data Sheet 43-3-1 System Data.
   B. Examine the existing thermostat and complete Data Sheet 43-3-2.

Data Sheet 43-3-1 System Data	
Unit Description (Furnace, Air Conditioner, Heat Pump, etc.)	
Fuel Type (Electric, gas, oil, etc.)	
Ignition System (Standing Pilot, Intermittent Pilot, Direct Spark, Hot Surface)	
Control Transformer VA Rating and Location	
Fan Relay Type and Number	

Heating Circuit Device and Rated Amperage Draw	
Cooling Capacity	
Number of Stages of Heat	
Number of Stages of Cooling	
Other Functions to Control or Supervise	

Data Sheet 43-3-2 Thermostat Data	
Make:	
Model Number:	
Operation	(circle one) Mechanical / Digital / Communicating
Stages of Cooling:	
Stages of Heat:	
Heating Cycles per Hour (Digital Thermostats)	
Cooling Cycles per Hour (Digital Thermostats)	
System Switching (circle all that apply)	HEAT OFF AUTO COOL
Fan Switching (circle all that apply)	ON AUTO
Heating Cycle Humidity Control	
Cooling Cycle Humidity Control	
Gauge and Number of Control Wires	
Colors of Control Wires (Circle)	White Red Green Blue Yellow Orange Black Brown

Step 2 – Remove Old Thermostat

A. Remove the old thermostat from the subbase.
B. Take a picture of the subbase wiring with your phone.
C. Complete Data Sheet 43-3-3, including
   a. the gauge and number of thermostat control wires
   b. the colors of the thermostat wires, including unused wires
   c. the terminals the wires are on
   d. the color of any unused wires
D. Remove the old sub-base.

Step 3 – Install New Thermostat

A. Mount the new sub-base on the wall. Make sure it is mounted level.
B. Connect the thermostat wires to the terminals on the new sub-base.
C. Take a picture with your phone of the sub-base and wire connections.
D. Record the terminals that the wires are connected to in Data Sheet 43-3-4.
E. Mount the new thermostat on the new sub-base.

Data Sheet 43-3-3 Sub-base Data

Number of control wires			
Terminal	Color Wire	Terminal	Color Wire
R		G	
RC		Y	
RH		Y1	
C		Y2	
X		W	
B		W1	
O		W2	
Unused color1			
Unused color2			
Unused color3			
Unused color4			
Unused color5			

Data Sheet 43-3-4 New Sub-base Wiring

Terminal	Color Wire	Terminal	Color Wire

Step 4 – Check New Thermostat Operation

A. Install batteries in the thermostat if required.
B. Obtain an accurate room temperature at the thermostat location and compare this reading to the thermostat temperature reading. This is best performed before handling the thermostat a great deal, or after it has been left alone for a time.
C. Set clock if required.
D. Set heating and cooling cycles per hour.
E. Use the thermostat instructions to program if programmable.
F. Set the thermostat to the OFF position and turn the system power on.
G. Put the fan setting to the "ON" position. Observe fan turn on.
H. Put the fan setting to the "AUTO" position. Observe fan turn off.
I. Set the MODE to HEAT, adjust the setpoint above room temperature, and observe heat operation.
J. Adjust the setpoint below the room temperature and observe the heat cycling off.
K. Set the MODE to COOL, adjust the setpoint below room temperature and observe cooling operation.
L. Adjust the setpoint above room temperature and observe cooling cycling off.

LAB 43.4 CHECKING RELAYS AND CONTACTORS

LABORATORY OBJECTIVE
The purpose of this lab is to learn how to test relays and contactors.

FUNDAMENTALS OF HVACR TEXT REFERENCE
Unit 43 Electrical Troubleshooting

LABORATORY NOTES
You will test an assortment of relays and contactors by checking the resistance of the coil and contacts. If the relay or contactor ohms correctly, you will apply the correct voltage to the coil and check to see that the contacts open and close by ohming them.

REQUIRED TOOLS, EQUIPMENT, AND MATERIALS
Relays
Contactors
Transformer
Multimeter
Electrical hand tools

SAFETY REQUIREMENTS
A. Wear safety glasses
B. Ohm Meters should never be used on energized circuits.
C. When ohming components on a system, turn power off and check with a meter BEFORE using ohm meter.

PROCEDURE

1. Check the resistance of the coil. The resistance should be measurable – not 0 and not infinite.

2. The coil condition listed as Open, Shorted, or Good. A coil resistance measurement of less than 1 ohm indicates a shorted coil. A measurement of OL (infinite) indicates an open coil. A very rough estimate for coil resistance is 1 ohm per volt coil rating.

3. Ohm all the contacts. With no power to the coil, normally open contacts should be infinite (OL) and normally closed should be shorted (less than 1 ohm)

4. The normal position of all the contacts should be listed and the reading for each set of contacts should be listed.

5. If the coil is good, energize the coil with the correct voltage and check the resistance of the contacts. With the coil energized, normally open contacts should read shorted (less than 1) and normally closed contacts should read infinite (OL).

6. Record the resistance reading of the contacts with the coil energized.

Relay Data					
	Relay 1	Relay 2	Relay 3	Relay 4	Relay 5
Coil Voltage					
Coil Resistance					
Coil Condition					
Contacts Set 1 Normal					
Contacts Set 1 Coil Energized					
Contacts Set 2 Normal					
Contacts Set 2 Coil Energized					
Contacts Set 3 Normal					
Contacts Set 3 Coil Energized					

Contactor Data					
	Contactor 1	Contactor 2	Contactor 3	Contactor 4	Contactor 5
Coil Voltage					
Coil Resistance					
Coil Condition					
Contacts Set 1 Normal					
Contacts Set 1 Coil Energized					
Contacts Set 2 Normal					
Contacts Set 2 Coil Energized					
Contacts Set 3 Normal					
Contacts Set 3 Coil Energized					

LAB 43.5 CHECKING PRESSURE SWITCHES

LABORATORY OBJECTIVE

The student will demonstrate how to properly test pressure switches for correct operation.

FUNDAMENTALS OF HVACR TEXT REFERENCE

Unit 34 Electrical Components

Unit 43 Electrical Troubleshooting

LABORATORY NOTES

For this lab you will check pressure switches and determine if they are operating properly. You will use a nitrogen cylinder and a pressure regulator to control the pressure to the switch. You will test the switch contacts for proper operation using a multimeter.

REQUIRED TOOLS, EQUIPMENT, AND MATERIALS

Safety glasses and gloves

Nitrogen cylinder and nitrogen regulator

Gauge manifold

Multimeter

SAFETY REQUIREMENTS

A. Never use nitrogen without a pressure regulator. The cylinder pressure can be over 2000 psig.

B. Wear safety glasses

PROCEDURE

You are going to impose a controlled pressure on the pressure switch and check the resistance of the pressure switch contacts at both the high and low event setpoints. If the switch contacts read anything other than 0 ohms (or a few tenths) or infinite ohms (OL) the contacts are bad.

Step 1 – Determine the Switch Settings

Record the cut-in, cut-out, and differential settings on Data Sheet 43-5-1 for each pressure switch assigned by the instructor. Refer to the information below for help in determining these settings.

Low Pressure

The cut-out is not directly set on most low-pressure switches. Instead, they have two adjustments: a cut-In setting and a differential setting. The cut-out is the difference between the cut-in and the differential. For example, the cut-out pressure shown in Figure 43-5-1 is 35 psig.

Cut-in 60 psig – differential 25 psig = 35 psig. Different terms are sometimes used instead of cut-in and cut-out. As shown in Figure 43-5-1, the cut-in is referred to as the "high event" and the cut-out is referred to as the "low event." Look at the label for the switching action to determine what happens on the high event. In the case of Figure 43-5-1 the switch closes on pressure rise. This tells you that the high event is the cut-in and the low event is the cut-out.

High Pressure

Set the assigned high-pressure cut-out according to the settings determined in step 2. High-pressure cut-out switches open on rise of pressure, as seen in Figure 43-5-2. Adjustable pressure switches used as high-pressure cut-outs all have a cut-out setting, and some also have a differential setting. Switches that only have a cut-out setting have a fixed differential. The cut-in is not set directly, but is the difference between the cut-out and the differential. For example, the cut-out pressure shown in Figure 43-5-2 is 400 psig and the differential is 100 psig. This makes the cut-in 300 psig: 400 psig – 100 psig = 300 psig.

Figure 43-5-1 Low-pressure switch settings

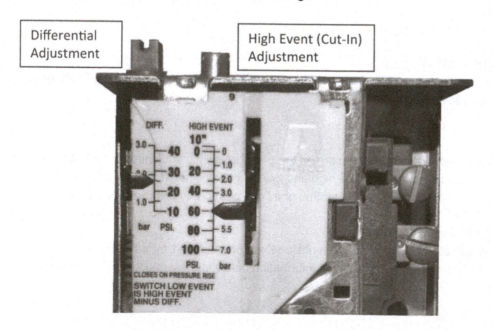

494

Figure 43-5-2 High-pressure switch settings

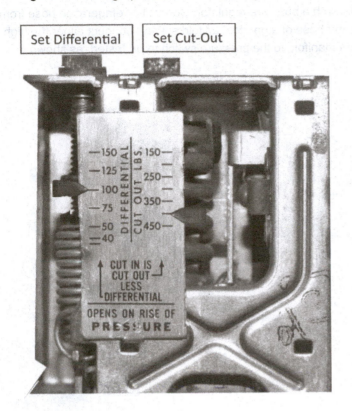

Step 2 – Connect Nitrogen

You will need a nitrogen cylinder with a pressure regulator. Connect a refrigeration hose from the nitrogen regulator to the center hose of a gauge manifold set and connect either the high side or the low side of the gauge manifold to the pressure switch to be tested, as shown in Figure 43-5-3.

Figure 43-5-3 Testing pressure switch

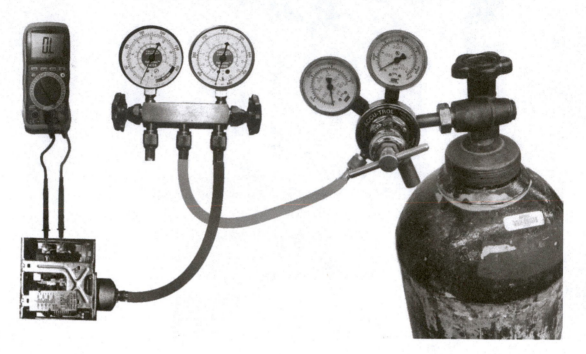

Step 4 – Check High Event Operation

The following procedure assumes that the both the cut-in and cut-out setpoints are above 0 psig.

Start with the pressure at 0 psig. The meter should be reading 0 ohms (or a few tenths) if the switching action is open on rise, and infinite (OL) if the switching action is close on rise. Open the gauge manifold and slowly adjust the nitrogen regulator in until the pressure indicated on the manifold is at or above the high event on the pressure switch. Often you will hear an audible click when the switch changes position. Record the pressure when the switch changed position on Data Sheet 43-5-1. The meter should be reading infinite (OL) ohms if the switching action is open on rise, and 0 ohms (or a few tenths) if the switching action is close on rise. If the switch does not change position after the pressure is significantly above the high event, the switch is bad. Before condemning the switch double check your pressure reading, the switch setpoints, and make sure the meter leads are making good contact.

Step 5 – Check Low Event Operation

Close the valve on the nitrogen cylinder. Use the open side of the gauge manifold to slowly release pressure from the pressure switch until the pressure is at or below the low event setting. The switch should change position. Record the pressure when the switch changed position on Data Sheet 43-5-1. The meter should be reading 0 ohms (or a few tenths) if the switching action is open on rise, and infinite (OL) if the switching action is close on rise. If the switch does not change position after the pressure is significantly below the low event, the switch is bad. Before condemning the switch double check your pressure reading, the switch setpoints, and make sure the meter leads are making good contact.

Data Sheet 43-5-1 Pressure Switch Setting and Operation			
		Setting	Operation
Switch 1	Cut-In Pressure		
	Differential		
	Cut-Out Pressure		
Switch 2	Cut-In Pressure		
	Differential		
	Cut-Out Pressure		
Switch 3	Cut-In Pressure		
	Differential		
	Cut-Out Pressure		
Switch 4	Cut-In Pressure		
	Differential		
	Cut-Out Pressure		

LAB 43.6 TROUBLESHOOTING SCENARIO

LABORATORY OBJECTIVE
Given a system with a problem, you will identify the problem, its root cause, and recommend corrective action.

FUNDAMENTALS OF HVACR TEXT REFERENCE
Unit 43 Electrical Troubleshooting

REQUIRED TOOLS, EQUIPMENT, AND MATERIALS
Safety glasses
Hand tools
Manometer
Multimeter
Refrigeration gauges
System with a problem

SAFETY REQUIREMENTS
A. Wear Safety Glasses
B. Check all circuits for voltage before doing any service work.
C. Stand on dry non-conductive surfaces when working on live circuits.
D. Never bypass any electrical protective devices.

PROCEDURE
Troubleshoot the system assigned by the instructor. Be sure to be complete in your description of the problem. You should include:

What is the unit is doing wrong?

What component or condition is causing this?

What tests did you perform that told you this?

How would you correct this?

LAB 44.1 USING SLING PSYCHROMETER

LABORATORY OBJECTIVE
You will demonstrate your ability to use a sling psychrometer to take wet bulb readings.

FUNDAMENTALS OF HVACR TEXT REFERENCE
Unit 44 Comfort and Psychrometrics
Unit 78 Testing and Balancing Air Systems

LABORATORY NOTES
You will take wet bulb and dry bulb temperature readings in the lab, outside, in an air conditioning return air stream, and in an air conditioning supply air stream. You will use these readings and the scale on the psychrometer to determine the relative humidity in each of these locations.

REQUIRED TOOLS, EQUIPMENT, AND MATERIALS
Sling psychrometer
Psychrometric chart
Operating air conditioner

SAFETY REQUIREMENTS
A. Wear safety glasses
B. Exercise caution around moving fan blades

PROCEDURE
1. Wet the sock on the wet bulb of the psychrometer with a few drops of room temperature water.
2. Spin the psychrometer in the air or hold in an airstream for approximately one minute.
3. *Note:* It is not necessary to spin the psychrometer when reading a moving air stream.
4. Read the wet bulb temperature first, and then the dry bulb temperature.
5. Use a psychrometric chart to determine the relative humidity and dew point.
6. Repeat to complete table.

Sling Psychrometer Readings				
	DRY BULB	WET BULB	%HUMID	DEW POINT
Lab				
Outside				
AC Return Air				
AC Supply Air				

LAB 44.2 USING ELECTRONIC HYGROMETER (DIGITAL PSYCHROMETER)

LABORATORY OBJECTIVE
You will demonstrate your ability to use a digital psychrometer to take wet bulb readings.

FUNDAMENTALS OF HVACR TEXT REFERENCE
Unit 44 Comfort and Psychrometrics
Unit 78 Testing and Balancing Air Systems

LABORATORY NOTES
You will take wet bulb and dry bulb temperature readings in the lab, outside, in an air conditioning return air stream, and in an air conditioning supply air stream. You will use the digital psychrometer to read the dry bulb, wet bulb, relative humidity, and dew point in each of these locations.

REQUIRED TOOLS, EQUIPMENT, AND MATERIALS
Digital psychrometer
Psychrometric chart
Operating air conditioner

SAFETY REQUIREMENTS
A. Wear safety glasses
B. Exercise caution around moving fan blades

PROCEDURE
1. Turn on the electronic hygrometer and note the location of the dry bulb, wet bulb, and relative humidity.
2. Open the sensing probe area with a gentle twist.
3. Read the dry bulb temperature, wet bulb temperature, and relative humidity.
4. Push the select button to read the dew point.
5. Repeat to complete table.

Table 44-2-1 Digital Psychrometer Readings				
	DRY BULB	WET BULB	%HUMID	DEW POINT
Lab				
Outside				
AC Return Air				
AC Supply Air				

LAB 45.1 AC INDUCTION MOTOR BLOWER PROPERTIES

LABORATORY OBJECTIVE
You will demonstrate the effect airflow restriction has on a forward curved centrifugal blower with an AC induction motor.

FUNDAMENTALS OF HVACR TEXT REFERENCE
Unit 45 Fans and Airflow

LABORATORY NOTES
You will operate the blower and read the motor amp draw. You will then restrict first the intake, and then the exhaust and measure the change in the blower motor current.

REQUIRED TOOLS, EQUIPMENT, AND MATERIALS
Clamp ammeter
Centrifugal blower with PSC motor
Piece of cardboard, wood, or metal to block airflow
Flow hood or anemometer
Digital phototachometer

SAFETY REQUIREMENTS
A. Safety glasses
B. Be careful around rotating equipment. It is possible to get hands caught in moving parts.

PROCEDURE
1. Operate the blower with no restriction and measure the blower motor amp draw, RPM, and CFM.
2. Record on Table 45-1-1.
3. Restrict the blower intake and operate the blower.
4. Measure the amp draw, blower RPM, and blower CFM.
5. Record on Table 45-1-1.
6. What happens to the blower motor amp draw when the intake is restricted?
7. What happens to the blower RPM when the intake is restricted?
8. What happens to the blower CFM when the intake is restricted?
9. Restrict the blower exhaust and operate the blower.
10. Measure the amp draw, blower RPM, and blower CFM.
11. Record on Table 45-1-1.
12. What happens to the blower motor amp draw when the exhaust is restricted?
13. What happens to the blower RPM when the exhaust is restricted?
14. What happens to the blower CFM when the exhaust is restricted?

Table 45-1-1 PSC Motor Blower Characteristics			
	Amps	RPM	CFM
Normal			
Intake restricted			
Exhaust restricted			

504

LAB 45.2 ECM MOTOR BLOWER PROPERTIES

LABORATORY OBJECTIVE
You will demonstrate the effect airflow restriction has on a forward curved centrifugal blower with an ECM permanent magnet motor.

FUNDAMENTALS OF HVACR TEXT REFERENCE
Unit 45 Fans and Airflow

LABORATORY NOTES
You will operate the blower and read the motor amp draw. You will then restrict first the intake, and then the exhaust and measure the change in the blower motor current.

REQUIRED TOOLS, EQUIPMENT, AND MATERIALS
Clamp ammeter
Centrifugal blower with ECM motor
Piece of cardboard, wood, or metal to block airflow
Flow hood or anemometer
Digital phototachometer

SAFETY REQUIREMENTS
A. Safety glasses
B. Be careful around rotating equipment. It is possible to get hands caught in moving parts.

PROCEDURE
1. Operate the blower with no restriction and measure the blower motor amp draw, RPM, and CFM.
2. Record on Table 45-2-1.
3. Restrict the blower intake and operate the blower.
4. Measure the amp draw, blower RPM, and blower CFM.
5. Record on Table 45-2-1.
6. What happens to the blower motor amp draw when the intake is restricted?
7. What happens to the blower RPM when the intake is restricted?
8. What happens to the blower CFM when the intake is restricted?
9. Restrict the blower exhaust and operate the blower.
10. Measure the amp draw, blower RPM, and blower CFM.
11. Record on Table 45-2-1.
12. What happens to the blower motor amp draw when the exhaust is restricted?
13. What happens to the blower RPM when the exhaust is restricted?
14. What happens to the blower CFM when the exhaust is restricted?

Table 45-2-1 ECM Motor Blower Characteristics			
	Amps	RPM	CFM
Normal			
Intake restricted			
Exhaust restricted			

LAB 46.1 CHANGING PANEL FILTERS

LABORATORY OBJECTIVE
You will demonstrate your ability to change the disposable panel filters in an air conditioning system.

FUNDAMENTALS OF HVACR TEXT REFERENCE
Unit 46 Indoor Air Quality and Filtration

LABORATORY NOTES
You will remove the existing disposable panel filter and replace it with a clean filter, making sure that the filter is oriented properly.

REQUIRED TOOLS, EQUIPMENT, AND MATERIALS
6-in-1 screwdriver 5/16" nut driver

Respirator (if allergic to dust)

Air conditioner with disposable panel filter

Clean disposable panel filter

SAFETY REQUIREMENTS
A. Wear safety glasses
B. Turn power off and lock out before changing filter
C. Test equipment with non-contact voltage detector to ensure power is off before touching unit.

PROCEDURE
1. Turn off and lock out power to the unit.
2. Test cabinet with non-contact voltage detector to ensure unit is safe to touch
3. Open the access door or filter access panel.
4. Remove the existing filter.
5. Replace the filter with a new, clean filter of the same dimensions.
6. Pay attention to the arrows on the filter indicating the proper airflow direction.
7. The arrows should point toward the unit.
8. Replace the access panel and turn the unit back on.

LAB 46.2 SERVICING MEDIA FILTERS

LABORATORY OBJECTIVE
You will demonstrate your ability to change the media in a high efficiency media filter.

FUNDAMENTALS OF HVACR TEXT REFERENCE
Unit 46 Indoor Air Quality and Filtration

LABORATORY NOTES
You will remove the existing filter media replace it with clean filter media, making sure that the filter is oriented properly.

REQUIRED TOOLS EQUIPMENT AND MATERIALS
6-in-1 screwdriver and 5/16" nut driver

Respirator (if allergic to dust)

Air conditioner with high efficiency media filter

New replacement filter media material

SAFETY REQUIREMENTS
A. Wear safety glasses
B. Turn power off and lock out before changing filter
C. Test equipment with non-contact voltage detector to ensure power is off before touching unit.

PROCEDURE
1. Turn off and lock out the power to the unit.
2. Test the cabinet with a non-contact voltage detector to ensure it is safe to touch.
3. Open the filter access panel.
4. Remove the filter box.
5. Study the way the media is in the box.
6. Carefully unsnap the combs one at a time by tilting them at an angle until they unsnap and then lifting them out.
7. Unsnap the edge holding the filter media in.
8. Remove the old media and place it in a plastic bag.
9. Lay the new media in place.
10. Snap the edges down to hold it in place.
11. Install the first comb being careful to match up the comb teeth with the pleats in the media.

12. Continue to install the combs until they are all in place.
13. Reinstall the filter into the cabinet—paying attention to the airflow directional arrows. The arrows should point toward the unit.
14. Replace the access panel and turn the unit back on.

LAB 46.3 ELECTRONIC AIR CLEANERS

LABORATORY OBJECTIVE
You will demonstrate your ability to use a high voltage probe to measure the voltage output of an electronic air cleaner power supply.

FUNDAMENTALS OF HVACR TEXT REFERENCE
Unit 46 Indoor Air Quality and Filtration

LABORATORY NOTES
You will use a high voltage probe to measure the voltage output of the charging section of the air cleaner.

REQUIRED TOOLS, EQUIPMENT, AND MATERIALS
6-in-1 screwdriver and 5/16" nut driver

Air conditioner with electronic air cleaner

High voltage probe

Digital multimeter with correct impedance to work with high voltage probe

Source of dry, compressed air

SAFETY REQUIREMENTS
A. Wear safety glasses
B. CAUTION! The power supply produces 6,000 volts!
C. Do NOT try to measure the voltage with your standard volt meter! Your meter will be fried and you along with it. A high voltage adapter is required!
D. Do NOT touch ANYTHING on the power cell!

PROCEDURE
1. Turn on the power supply with the fan off. Does the indicator light come on?
2. Now, turn on the fan. What reaction do you get from the indicator light?
3. Turn off the power supply.
4. With the power off and locked out, remove the electronic cells.
5. Examine the cell and identify the ionizer wires and plates and the collector plates.
   You will require instructor help for the next step.
   The cell produces 6,000 volts! Do NOT touch ANYTHING on the power cell!
6. Take the power supply to the bench and connect 120 volts to the incoming voltage section with an extension cord.
7. Check the voltage output using a high-voltage probe.

The cell produces 6,000 volts! Do NOT try to measure the voltage with your standard volt meter! Your meter will be fried and you along with it. A high voltage adapter is required!

8. What voltage is read on the high voltage probe?

9. Use the air hose to blow air across the airflow sensor.

10. Read the voltage on the high voltage probe while blowing air across the air flow sensor.

11. What voltage is read on the high voltage probe when air is blowing across the sensor?

LAB 47.1 IDENTIFY DEHUMIDIFIER COMPONENTS

LABORATORY OBJECTIVE
You will inspect a dehumidifier, identify its components, and explain its operation.

FUNDAMENTALS OF HVACR TEXT REFERENCE
Unit 47 Ventilation and Dehumidification

LABORATORY NOTES
You will examine a dehumidifier with the panels removed and identify its components. You will then describe the operation of the dehumidifier.

REQUIRED TOOLS, EQUIPMENT, AND MATERIALS
Safety glasses
Gloves
Hand tools
Dehumidifier

SAFETY REQUIREMENTS
A. Wear safety glasses and gloves
B. The dehumidifier should NOT be plugged in while you are examining its components.

PROCEDURE
1. Remove any panels on dehumidifier so you can see its components.
2. Identify the following components:
   a. Compressor
   b. Condenser
   c. Metering device
   d. Fan
3. Describe the refrigeration cycle.

4. Show the way that air flows through the unit.

LAB 47.2 DEHUMIDIFIER OPERATION

LABORATORY OBJECTIVE
Observe the effect on the air passing through a dehumidifier.

FUNDAMENTALS OF HVACR TEXT REFERENCE
Unit 47 Ventilation and Dehumidification

LABORATORY NOTES
You will operate a dehumidifier, measure the dry bulb and wet bulb temperatures,
and determine the relative humidity of the air entering and leaving the dehumidifier

REQUIRED TOOLS, EQUIPMENT, AND MATERIALS
Safety glasses
Gloves
Psychrometer
Dehumidifier

SAFETY REQUIREMENTS
A. Wear safety glasses and gloves
B. The dehumidifier should be plugged into a grounded outlet.
C. If an extension cord is used, the extension cord should be grounded and sized for the
   dehumidifier.

PROCEDURE
1. Plug in the dehumidifier and set its control to operate.
2. Let the dehumidifier run for at least 10 minutes.
3. Take the following readings:
   a. Dry bulb temperature of the entering air
   b. Wet bulb temperature of the entering air
   c. Dry bulb temperature of the leaving air
   d. Wet bulb temperature of the leaving air
4. Use a psychrometric chart to determine the following:
   a. Dew point of the entering air.
   b. Dew point of the leaving air
   c. Relative humidity of the entering air
   d. Relative humidity of the leaving air
   (*Note:* If you are using a digital psychrometer, it will be able to do the wet bulb and dew
   point calculation for you)

LAB 48.1 IDENTIFY ROOM AIR CONDITIONER COMPONENTS

LABORATORY OBJECTIVE

You will inspect a room air conditioner, identify its components, and explain its operation.

***FUNDAMENTALS OF HVACR* TEXT REFERENCE**

Unit 48 Room Air Conditioners

LABORATORY NOTES

You will examine a room air conditioner with the cover removed and identify its components. You will then describe the operation of the room air conditioner.

REQUIRED TOOLS, EQUIPMENT, AND MATERIALS

Safety glasses

Gloves

Hand tools

Room air conditioner

SAFETY REQUIREMENTS

A. Wear safety glasses and gloves

B. The air conditioner should NOT be plugged in while you are examining its components.

PROCEDURE

1. Remove the cover on the air conditioner and refer to Figures 48-1-1 and 48-1-2 to identify components.
2. Identify the following components:
   a. Compressor
   b. Condenser
   c. Metering device
   d. Fan

Figure 48-1-1 Capillary Tube Metering Device

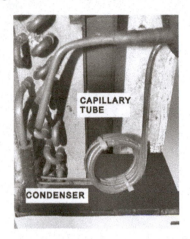

Figure 48-1-2 Window Unit Components

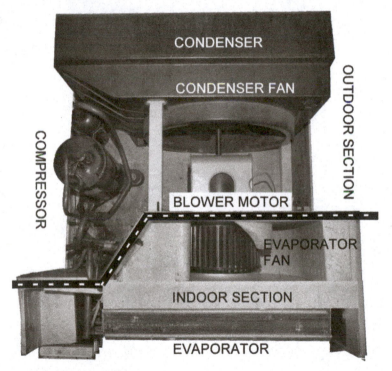

3. Describe the refrigeration cycle.

4. Describe the path the airflow takes over the evaporator.

5. Describe the path the airflow takes over the condenser.

6. Describe how condensate from the evaporator is removed.

LAB 48.2 DESCRIBE ROOM AIR CONDITIONER OPERATION

LABORATORY OBJECTIVE
Observe and describe room air conditioning system operation.

***FUNDAMENTALS OF HVACR* TEXT REFERENCE**
Unit 48 Room Air Conditioners

LABORATORY NOTES
You will operate a room air conditioning system, measure its operational characteristics, and describe its operation.

REQUIRED TOOLS, EQUIPMENT, AND MATERIALS
Safety glasses and gloves
Digital psychrometer
Psychrometric chart
Room air conditioner

SAFETY REQUIREMENTS
A. Wear safety glasses and gloves
B. The unit should be properly grounded
C. Keep hands away from rotating fans

PROCEDURE
1. Plug in the unit and operate it long enough for the temperatures to stabilize.
2. Record the following measurements in table 48-2-1:
    a. Return air dry bulb
    b. Return air wet bulb
    c. Return air relative humidity
    d. Supply air dry bulb
    e. Supply air wet bulb
    f. Supply air relative humidity
    g. Ambient air dry bulb temperature entering condenser
    h. Ambient air wet bulb temperature entering condenser
    i. Ambient air relative humidity
    j. Condenser discharge air dry bulb temperature
    k. Condenser discharge air wet bulb temperature
    l. Condenser discharge air relative humidity
3. Use a psychrometric chart to determine the absolute humidity and enthalpy of the return air, supply air, ambient air entering the condenser, and condenser discharge air.

4. Describe the changes in the air flowing over the evaporator.

5. Describe the changes in the air flowing over the condenser.

Table 48-2-1

CHARACTERISTIC	
Return Air DB	
Return Air WB	
Return Air RH	
Return Air Grains of Moisture	
Return Air Enthalpy	
Supply Air DB	
Supply Air WB	
Supply Air RH	
Supply Air Grains of Moisture	
Supply Air Enthalpy	
Ambient Air DB	
Ambient Air WB	
Ambient Air RH	
Ambient Air Grains of Moisture	
Ambient Air Enthalpy	
Condenser Discharge Air DB	
Condenser Discharge Air WB	
Condenser Discharge Air RH	
Discharge Air Grains of Moisture	
Discharge Air Enthalpy	

LAB 49.1 TYPES OF AIR CONDITIONING SYSTEMS

LABORATORY OBJECTIVE
Recognize different types of air conditioning systems.

FUNDAMENTALS OF HVACR TEXT REFERENCE
Unit 49 Residential Air Conditioning

LABORATORY NOTES
You will examine four air conditioning units. For each unit you will describe type of air conditioning system and its operation.

REQUIRED TOOLS EQUIPMENT AND MATERIALS
Safety glasses
Gloves
Hand tools
Four air conditioning systems

SAFETY REQUIREMENTS
A. Wear safety glasses and gloves
B. The units should be off with the power locked out while you are examining them.

PROCEDURE
1. For each unit:
2. Remove any unit panels necessary to see the unit components.
3. Identify the type of system.
4. For refrigeration cycle units, identify the following components:
    a. Compressor
    b. Condenser
    c. Metering device
    d. Evaporator
5. Complete Table 49-1-1.
6. Describe the unit operation.

Table 49-1-1 Air Conditioning System Details
Circle the best description

	Unit 1	Unit 2	Unit 3	Unit 4
Type of Unit	Packaged Split System Mini-split System PTAC Window Unit Evaporative	Packaged Split System Mini-split System PTAC Window Unit Evaporative	Packaged Split System Mini-split System PTAC Window Unit Evaporative	Packaged Split System Mini-split System PTAC Window Unit Evaporative
Where is this unit typically located (you may select more than one)	Outside Basement Crawlspace Attic Window Wall Conditioned Room	Outside Basement Crawlspace Attic Window Wall Conditioned Room	Outside Basement Crawlspace Attic Window Wall Conditioned Room	Outside Basement Crawlspace Attic Window Wall Conditioned Room
Compressor	Reciprocating Scroll Rotary	Reciprocating Scroll Rotary	Reciprocating Scroll Rotary	Reciprocating Scroll Rotary
Metering Device	Capillary Tube Orifice TEV EEV	Capillary Tube Orifice TEV EEV	Capillary Tube Orifice TEV EEV	Capillary Tube Orifice TEV EEV
Evaporator	Cu tubing/Al fin Al tubing/Al fin Microchannel Spiney fin	Cu tubing/Al fin Al tubing/Al fin Microchannel Spiney fin	Cu tubing/Al fin Al tubing/Al fin Microchannel Spiney fin	Cu tubing/Al fin Al tubing/Al fin Microchannel Spiney fin

LAB 49.2 AIR CONDITIONING SYSTEM OPERATION

LABORATORY OBJECTIVE
Demonstrate the effect of airflow changes on air conditioning system operation.

FUNDAMENTALS OF HVACR TEXT REFERENCE
Unit 44 Comfort and Psychometrics
Unit 49 Residential Air Conditioning

LABORATORY NOTES
You will operate an air conditioning system and measure its operational characteristics. You will then vary the airflow and recheck the operational characteristics. Finally, you will compare the operational characteristics at different airflow rates.

REQUIRED TOOLS, EQUIPMENT, AND MATERIALS

Safety glasses and gloves	Digital temperature clamp
Refrigeration gauges	Psychrometer
Air conditioning system	Amp meter

SAFETY REQUIREMENTS
A. Wear Safety Glasses and Gloves
B. The unit should be properly grounded
C. Keep hands away from rotating fans

PROCEDURE
1. Connect refrigeration gauges to the unit.
2. Operate the unit long enough for the pressures to stabilize.
3. Record the following measurements on the NORMAL column of table 49-2-1:
   a. Return Air Dry Bulb
   b. Supply Air Dry Bulb
   c. Supply Air Wet Bulb
   d. Suction Line Temperature
   e. Liquid Line Temperature
   f. Suction Pressure (Low Side)
   g. Liquid Pressure (High Side)
4. Turn the unit off and adjust the evaporator blower to deliver less air. This can be done either by moving a speed tap for PSC blowers or making an adjustment on the ECM blower board for ECM motors.
5. Operate long enough for the pressures to stabilize.
6. Repeat the measurements and record in the LOW column of Table 49-2-1.

7. Turn the unit off and adjust the evaporator blower to deliver more air. This can be done either by moving a sped tap for PSC blowers or making an adjustment on the ECM blower board for ECM motors.
8. Operate long enough for the pressures to stabilize.
9. Repeat the measurements and record in HIGH column of Table 49-2-1.
10. Turn the unit off and put the evaporator blower back to the original setting.

Table 49-2-1

CHARACTERISTIC	LOW CFM	NORMAL CFM	HIGH CFM
Suction Pressure			
Suction Saturation Temp (SST)			
Suction Line Temp (SLT)			
Superheat (SLT-SST)			
Liquid Pressure			
Liquid Saturation Temp (LST)			
Liquid Line Temp (LLT)			
Subcooling (LST-LLT)			
Return Air DB			
Return Air WB			
Return Air RH			
Supply Air DB			
Supply Air WB			
Supply Air RH			

LAB 50.1 MOUNTING A MINI-SPLIT WALL MOUNT UNIT

LABORATORY OBJECTIVE
You will mount a mini-split indoor wall mount unit. This includes cutting hole through the wall; installing the wall mount plate on the wall; running the refrigerant lines, wiring, and drain through the hole; and hanging the unit.

FUNDAMENTALS OF HVACR TEXT REFERENCE
Unit 50 Mini-Split, Multi-split, and Variable Refrigerant Flow Systems

LABORATORY NOTES
A mini-split system consists of two parts—an indoor unit that mounts in the conditioned space and an outdoor condensing unit. You will determine where the indoor wall mount unit will be installed, cut the hole through the wall, and mount the indoor unit.

REQUIRED TOOLS, EQUIPMENT, AND MATERIALS
Safety glasses
Gloves
Hand tools
Level
Extension cord
Fiberglass ladder
Mini-split system air conditioning unit
Wall mounting bracket for system being installed
Hole Saw
Mighty Bracket™
Drill

SAFETY REQUIREMENTS
A. Wear safety glasses and gloves
B. Extension cord should be rated for the amp draw of the drill being used.
C. Drill should be grounded
D. Ladder should be fiberglass (NOT METAL).

PROCEDURE
1. Hold the wall mount bracket against the wall while checking it with a level.
2. Mark the mounting holes and the hole for the lines, wiring, and drain.
3. Check to see if the mounting holes are over a stud area. If not, wall anchors will be needed. Note that you may want to check for studs in the wall prior to locating the unit to make installation easier.

4. Do not actually mount the plate until AFTER cutting the hole through the wall.
5. Use the hole saw to cut the hole in the wall. The hole should be at least 3" in diameter and angle down towards the other side of the wall to facilitate drainage.
6. Mount the wall bracket and check it with a level.
7. Where needed, select wall anchors rated for the weight of the indoor unit.
8. Hang the Mighty Bracket™ on the wall bracket.
9. Place the indoor unit on the Mighty Bracket™ leaving room between the unit and the wall.
10. Run the refrigerant lines, wiring and drain through the hole in the wall.
    Note: Pay attention to the drain line. It MUST slope down.
11. Connect the refrigerant lines and wiring to the unit.
12. Hang the unit on the wall bracket and remove the Mighty Bracket™

LAB 50.2 WIRING A MINI-SPLIT SYSTEM

LABORATORY OBJECTIVE
You will wire a mini-split air conditioning unit, including:

- Sizing the power wire to the outdoor unit.
- Sizing the circuit breaker to outdoor unit.
- Sizing the disconnect switch to the outdoor unit.
- Sizing the wire from the outdoor unit to indoor unit.
- Connecting the wire from the disconnect to the outdoor unit.
- Connecting the wire from the outdoor unit to the indoor unit.

FUNDAMENTALS OF HVACR TEXT REFERENCE
Unit 35 HVACR Electronic Controls
Unit 50 Mini-Split, Multi split, and Variable Refrigerant Flow Systems

LABORATORY NOTES
A mini-split system consists of two parts—an indoor unit that mounts in the conditioned space and an outdoor condensing unit. Power is supplied to the outdoor unit from a disconnect switch. Power for the indoor unit comes from the outdoor unit. Many smaller mini-split systems use 120 volt power supplies. You will size the power wire and disconnect for the condenser based on the unit minimum circuit ampacity on the outdoor unit data plate. The over-current protection will be sized by the maximum over-current protection on the outdoor unit data plate. You will wire the power wire from the disconnect to the outdoor unit. You will wire the control and power wire from the outdoor unit to the indoor unit. After completing the installation, you will operate the unit and check the voltage and amp draw.

REQUIRED TOOLS, EQUIPMENT, AND MATERIALS
Safety glasses

Gloves

Electrical hand tools

Multimeter

Amp meter

Mini-Split system air conditioning unit

Power wire

4 conductor power/control wire matching manufacturer's specifications

SAFETY REQUIREMENTS

A. Wear safety glasses
B. Turn off and lock out power before opening panel or doing any wiring.
C. Check power supply with non-contact voltage detector or meter BEFORE touching anything or doing any wiring
D. Disconnect switch and unit should both be grounded.

PROCEDURE

Step 1 – Sizing

1. Find and record the minimum circuit ampacity (MCA) and the maximum overcurrent protection on the unit data plate of the outdoor unit (power for both parts is supplied to the outdoor unit).
2. Use the National Electrical Code Wire sizing charts or the manufacturer's installation literature to look up the correct wire size for the system.
3. Size the disconnect switch by multiplying 1.15 x MCA.
4. The fuse or circuit breaker for the system should be no larger than the Maximum Overcurrent Protection on the outdoor unit data plate.

Step 2 – Wiring Disconnect and Outdoor Unit

1. Turn off and lock out the power supply before doing ANY wiring.
2. Use a non-contact voltage detector or meter to check the power supply to insure it is off.
3. Mount the disconnect switch within sight of the outdoor unit.
4. Run weatherproof conduit between the disconnect switch and the outdoor unit.
5. Run the power wire between the disconnect switch and the unit. Single-phase systems require 3 wires—2 power wires and a ground.

Step 3 – Wiring to Indoor Unit

1. Run the power/control wire between the outdoor unit and the indoor unit.
2. Make sure to use the type of wire recommended by the manufacturer.
3. Follow the unit diagram to connect the power and control wires.
4. Have the instructor check your work.

Step 4 – Operational Check

1. Before starting the unit, check the voltage at the disconnect switch and compare it to the unit nameplate to make sure that the correct voltage is supplied to the disconnect switch.
2. If the voltage going to the disconnect switch is correct turn the disconnect switch on.
3. Start the unit.
4. Check and record the operating voltage and current for the system.

LAB 50.3 MINI-SPLIT REFRIGERATION PIPING

LABORATORY OBJECTIVE
You will install the refrigerant lines for a mini-split air conditioning unit.

FUNDAMENTALS OF HVACR TEXT REFERENCE
Unit 50 Mini-Split, Multi split, and Variable Refrigerant Flow Systems

LABORATORY NOTES
A mini-split system consists of two parts—an indoor unit that mounts in the conditioned space and an outdoor condensing unit. They are connected by a small, saturated liquid line and a larger vapor line. Both lines must be insulated because they are both cold during operation. The lines are connected to the unit with flare connections. You will install the refrigeration lines, including flaring and leak testing.

REQUIRED TOOLS, EQUIPMENT, AND MATERIALS
Safety glasses and gloves
Flaring tool, flare nuts, and torque wrench for flare nuts
Copper tubing and pipe insulation
Hand tools
Mighty Bracket™
Drill, extension cord and hole saw
Minisplit system
Air-conditioning unit
Nitrogen cylinder with pressure regulator

SAFETY REQUIREMENTS
A. Wear safety glasses and gloves
B. Extension cord should be rated for the amp draw of the drill being used.
C. Drill should be grounded
D. Ladder should be fiberglass (NOT METAL).

PROCEDURE
This lab continues the installation after the wall bracket has been mounted
Step 1 – Flare tubing
1. Slide the flare nuts over the copper tubing ends.
2. Flare the copper tubing using a flaring tool with a 45°chamfer, Figure 50-3-1.

3. Hand tighten flare plugs in each nut to keep debris out while physically running the lines.

Step 2 – Running the lines
1. Hang the Mighty Bracket™ on the wall bracket.
2. Place the indoor unit on the Mighty Bracket™ leaving room between the unit and the wall.
3. Run the refrigerant lines, wiring and drain through the hole in the wall.
4. Apply Nylog™ to the flares, connect them to the indoor unit, and hand-tighten.
5. Tighten the flare nuts to the manufacturer's specified torque using a flare-nut torque wrench.
6. Apply Nylog™ to the flares, connect them to the outdoor unit, and hand-tighten.
7. Tighten the flare nuts to the manufacturer's specified torque using a flare-nut torque wrench.

Step 3 – Leak Test
1. Connect a nitrogen cylinder with a regulator to the unit charging valve at the outdoor unit.
2. Pressurize the lines with a minimum of 100 psig nitrogen pressure.
3. Test each of the four flare connections for leaks using soap bubbles or an ultrasonic leak detector.

Step 4 – Finishing up
After you have verified that the line connections do not leak you can complete installation of the indoor unit, including wiring and running the drain. Then hang the indoor unit on the wall bracket. Outside, release the nitrogen charge from the lines and prepare to evacuate them.

Figure 50-3-1 A 45° Flaring Tool

LAB 50.4 EVACUATE AND CHARGE A MINI-SPLIT SYSTEM

LABORATORY OBJECTIVE
You will demonstrate your ability to evacuate and charge a mini-split system on a new installation.

FUNDAMENTALS OF HVACR TEXT REFERENCE
Unit 28 Refrigerant System Evacuation
Unit 29 Refrigerant System Charging
Unit 50 Mini-split, Multisplit and Variable Refrigerant Flow Systems

LABORATORY NOTES
You will pull a deep vacuum on the lines and coil of a mini-split system air conditioner, open the charging valves, and adjust the refrigerant charge if necessary.

REQUIRED TOOLS, EQUIPMENT, AND MATERIALS
Schrader core removal tool and charging hose
Vacuum pump, vacuum gauge, and 1/2-inch vacuum hose
Extension cord
Mini-split system

SAFETY REQUIREMENTS
Wear safety goggles and gloves when working with refrigerants. Liquid refrigerant can cause frostbite when in contact with eyes or skin.

PROCEDURE
Step 1 – Leak Test
1. Connect a nitrogen cylinder with a regulator to the service valve on the gas line service valve at the outdoor unit. Typically, there is only one service access on a mini-split system.
2. Pressurize the lines with a minimum of 100 psig nitrogen pressure.
3. Test each of the four flare connections for leaks using soap bubbles or an ultrasonic leak detector.
4. Release the nitrogen from the lines and coil once you have verified that they do not leak.

Step 2 – Evacuation
1. Use a core removal tool to remove the Schrader core and connect a vacuum gauge.
2. Connect a vacuum pump using a large diameter 1/2-inch hose.
3. Evacuate the unit to below 500 microns.
4. Close the shutoff valve on the core tool to isolate the system from the vacuum pump.

Step 3 – Opening Charging Valves

1. Open the liquid and gas charging valves to allow the refrigerant into the lines and coil. The charge shipped with the unit is typically good for lines of 25 to 30 feet.
2. Refrigerant will need to be added for longer lines. Manufacturers provide tables indicating the amount of refrigerant that should be added per foot of line above a standard length.

LAB 51.1 WIRE DISCONNECT SWITCH AND POWER SUPPLY

LABORATORY OBJECTIVE
You will size and install the power wiring and disconnect switch to an air conditioning system

FUNDAMENTALS OF HVACR TEXT REFERENCE
Unit 42 Electrical Installation
Unit 51 Installing Residential Air Conditioning

LABORATORY NOTES
You will use the unit minimum circuit ampacity to size the conductor and disconnect switch. You will use the maximum overcurrent protection to size the fuse or circuit breaker. You will then install the disconnect switch, run conduit between the disconnect and the unit, and run power wire from the disconnect to the unit.

REQUIRED TOOLS, EQUIPMENT, AND MATERIALS

Safety glasses	Wire cutter
Multimeter	Wire crimper/stripper
Ammeter wire	6-in-1 screwdriver
Conduit	Disconnect switch
Conduit connections	Air conditioning unit

SAFETY REQUIREMENTS
A. Wear safety glasses.
B. Lock out power to circuit before beginning.
C. Check all circuits before beginning energizing.

PROCEDURE
Step 1 – **Sizing Wire, Conduit, Overcurrent Protection, and Disconnect**
  A. Find the minimum circuit ampacity and maximum overcurrent protection on the data plate.
  B. Refer to Tables 51-1-1, 51-1-2, 51-1-3 and the National Electric Code to size the wire and conduit.
  C. The circuit breaker or fuse should be no larger than the maximum overcurrent protection.
  D. The disconnect switch must be rated for no less than 115% of the minimum circuit ampacity. In most cases, a disconnect switch that will hold the maximum allowable fuse size is adequate.

Step 2 – Wiring
A. Lock out power to the circuit which will feed the disconnect witch.
B. Mount the disconnect switch.
C. Run conduit to the entry point of the disconnect switch.
D. Run wire from the power panel to the disconnect switch.
E. Run conduit between the disconnect switch and the unit.
F. Pull the power wire between the disconnect switch and the unit.
G. Have the instructor check your work.

Step 3 – Checkout
A. With the disconnect off but power to it, check the voltage into the top of the disconnect.
B. Compare the voltage to the minimum and maximum allowable voltages printed on the nameplate.
   a. If the voltage is outside of the allowable voltage range LEAVE THE DISCONNECT SWITCH OFF! You need to determine why the voltage does not match the specifications before proceeding any further.
   b. If the voltage is within the allowable voltage range, proceed to the next step.
C. If the voltage to the disconnect switch is within the manufacturer's specifications, make sure the system controls are set to OFF and turn the disconnect on.
D. Check voltage to the unit.
E. If the voltage to the unit is within the allowable voltage range, turn the unit on.
F. Check the voltage to the unit with the unit operating.
G. The voltage should still be within the allowable voltage range.
H. The difference between the voltage at the unit with the unit off and the voltage to the unit with the unit operating is the voltage drop through the circuit.

Table 51-1-1 Maximum Copper Wire Ampacity				
Wire Gauge	60°C Copper Wire Ampacity	75°C Copper Wire Ampacity	90°C Copper Wire Ampacity	* Maximum Overcurrent Protection
14*	15	20	25	15
12*	20	25	30	20
10*	30	35	40	30
8	40	50	60	NA
6	55	65	75	NA
* Indicates wire gauges whose overcurrent protection is limited to the values in the maximum overcurrent protection column				

Table 51-1-2 Ambient Temperature Correction Factors (30°C)			
Ambient Temp	60°C Wire	75°C Wire	90°C Wire
50°F or less	1.29	1.2	1.15
51°F–59°F	1.22	1.15	1.12
60°F–68°F	1.15	1.11	1.08
69°F–77°F	1.08	1.05	1.04
78°F–86°F	1.0	1.0	1.0
87°F–95°F	0.91	0.94	0.96
96°F–104°F	0.82	0.88	0.91
105°F–113°F	0.71	0.82	0.87
114°F–122°F	0.58	0.75	0.82
123°F–131°F	NR	0.67	0.76
132°F–140°F	NR	0.58	0.71

Table 51-1-3 Factors for Four or More Current-Carrying Wires	
# of Wires	75°C Wire
4–6	0.80
7–9	0.70
10–20	0.50

LAB 51.2 WIRING PACKAGED UNIT CONTROLS

LABORATORY OBJECTIVE
You will wire a packaged air conditioning unit, including installing the thermostat and connecting the control wiring.

FUNDAMENTALS OF HVACR TEXT REFERENCE
Unit 40 Control Systems

Unit 51 Residential Split-System Air-Conditioning Installations

LABORATORY NOTES
You will mount the thermostat, connect the control wiring from the thermostat to the unit and operate the unit.

REQUIRED TOOLS, EQUIPMENT, AND MATERIALS
Safety glasses

Electrical hand tools

Multimeter

Packaged air-conditioning

Unit Low voltage thermostat

Control Wire

SAFETY REQUIREMENTS
A. Wear safety glasses
B. Turn off and lock out power before opening panel or doing any wiring.
C. Check power supply with non-contact voltage detector or meter BEFORE touching anything or doing any wiring
D. Disconnect switch and unit should both be grounded.

PROCEDURE
1. Examine the unit wiring diagram and installation instructions to determine the type of thermostat and number of control wires needed.
2. Mount the thermostat 5 feet off the floor on an inside wall out of drafts and sunlight.
3. Run 18-gauge solid thermostat wire between the thermostat and the unit.
   Do not use wire smaller than 18 gauge. Do not use stranded wire.
4. Follow the unit diagram to wire the control wires.
5. Have the instructor check your work.
6. Before starting the unit, check the voltage at the disconnect and compare it to the unit nameplate to make sure it is the correct voltage for the unit.
7. Start the unit and check its operation.

LAB 51.3 WIRING A SPLIT SYSTEM

LABORATORY OBJECTIVE

You will wire a split system air conditioning unit, including:

- Sizing the power wire to blower coil
- Sizing the power wire to condenser
- Sizing the disconnect switch to blower
- Sizing the disconnect switch to condenser
- Sizing the circuit breaker to blower
- Sizing the circuit breaker to condenser
- Connecting the power wire
- Installing the thermostat
- Connecting the control wiring

FUNDAMENTALS OF HVACR TEXT REFERENCE

Unit 35 HVACR Electronic Controls

Unit 51 Residential Split-System Air-Conditioning Installations

LABORATORY NOTES

A split system consists of two parts—an indoor blower coil and an outdoor condensing unit. Each has its own disconnect and power supply. You will need to size the power wire, disconnect, and breaker size for each unit separately. You will size the power wire and disconnect for both the blower and condenser based on the unit minimum circuit ampacity on the unit data plates. The overcurrent protection will be sized by the maximum overcurrent protection on the unit data plates. You will wire the power wire from each disconnect to each unit. You will wire the control wire from the thermostat to the blower and from the blower to the condenser. After completing the installation, you will operate the unit and check the voltage and amp draw.

REQUIRED TOOLS, EQUIPMENT, AND MATERIALS

Safety glasses	Split system air-conditioning unit
Gloves	Low voltage thermostat
Electrical hand tools	Power wire
Multimeter	Control wire
Amp meter	

SAFETY REQUIREMENTS

A. Wear safety glasses
B. Turn off and lock out power before opening panel or doing any wiring.
C. Check power supply with non-contact voltage detector or meter BEFORE touching anything or doing any wiring
D. Disconnect switch and unit should both be grounded.

PROCEDURE

1. Find and record the minimum circuit ampacity (MCA) and the maximum overcurrent protection on the unit data plates of both units (indoor blower and outdoor condensing unit).

2. Use the National Electrical Code wire sizing charts or the manufacturer's installation literature to look up the correct wire size for both units.

3. Size the disconnect switch for each unit by multiplying 1.15 x MCA.

4. The fuse or circuit breaker for each unit should be no larger than the maximum overcurrent protection.

5. Turn off and lock out the power supply before doing ANY wiring.

6. Use a non-contact voltage detector or meter to check the power supply to insure it is off.

7. Mount each disconnect switch (one for each unit) within sight of the unit it controls. The disconnect switches can be mounted on the units, but they should not be mounted on service access panels.

8. Run weatherproof conduit between each disconnect switch and the unit it serves.

9. Run the power wire between each disconnect switch and the unit it serves. Single-phase systems require 3 wires–2 power wires and a ground. Three phase systems require four wires–3 power wires and a ground.

10. Mount the thermostat 5 feet off the floor on an inside wall out of drafts and sunlight.

11. Run 18-4 solid thermostat wire between the thermostat and the blower coil. Do not use wire smaller than 18 gauge. Do not use stranded wire.

12. Run 18-2 solid thermostat wire between the blower coil and the condensing unit.

13. Follow the unit diagram to wire the control wires.

14. Before starting the unit, check the voltage at each disconnect switch and compare it to the unit nameplate to make sure that the correct voltage is supplied to each unit.
    The voltage requirements may not be the same for both units. Furnaces normally operate on 120 volts while condensing units operate on 240 volts. Be careful to put the correct voltage to each unit.

15. Have the instructor check your work.

16. Start the unit.

17. Check and record the operating current for both the indoor and outdoor units.

LAB 51.4 EVACUATE AND CHARGE A SPLIT SYSTEM AIR CONDITIONER

LABORATORY OBJECTIVE
You will demonstrate your ability to evacuate and charge a split system heat pump.

FUNDAMENTALS OF HVACR TEXT REFERENCE
Unit 28 Refrigerant System Evacuation
Unit 29 Refrigerant System Charging
Unit 51 Residential Split-System Air-Conditioning Installations

LABORATORY NOTES
You will pull a deep vacuum on a split system air conditioner down to 500 microns and weight in the correct refrigerant charge in liquid form into the high side of the system.

REQUIRED TOOLS, EQUIPMENT, AND MATERIALS
Hand tools
Refrigeration gauges
Temperature tester
Split system heat pump
Vacuum pump
Extension cord
Vacuum gauge
Refrigerant scale

SAFETY REQUIREMENTS
Wear safety goggles and gloves when working with refrigerants. Liquid refrigerant can cause frostbite when in contact with eyes or skin.

PROCEDURE
1. Connect your gauges to the system.
2. Connect the vacuum gauge and vacuum pump.
3. Evacuate the system down to 500 microns on the vacuum gauge.
   Note: On a new split system you are only evacuating the lines and evaporator coil. With a leak free system, evacuating just the lines and coil should not take very long.
4. While the vacuum is pulling, calculate any additional required charge.
   For most systems, the amount of refrigerant shipped with the condensing unit is adequate for 15 feet of line. You will need to add for every to add for every foot over 15 feet of line.
   Refer to Table 51-4-1 for additional charge per foot of line.

5. Once the vacuum is achieved, close the refrigeration gauges and turn off the vacuum pump.

6. On new split systems, weigh in the amount of refrigerant needed for additional line length and then open the unit installation valves to let the refrigerant that is in the condenser into the rest of the system.

7. If the system is a dry system or an older system (not newly installed), you need to weigh in the full charge, which is the factory charge plus the line length allowance.

Table 51-4-1

Liquid Line Diameter (in)	Ounces per Foot of Liquid Line	
	R22	R410A
$\frac{1}{4}$	0.23	0.19
$\frac{5}{16}$	0.40	0.33
$\frac{3}{8}$	0.62	0.51
$\frac{1}{2}$	1.12	1.01
$\frac{5}{8}$	1.81	1.64

LAB 51.5 AIR CONDITIONING SYSTEM START AND CHECK

LABORATORY OBJECTIVE
You will demonstrate your ability to perform an initial unit start and check procedure.

FUNDAMENTALS OF HVACR **TEXT REFERENCE**
Unit 51 Residential Split-System Air-Conditioning Installations

LABORATORY NOTES
You will operate an air conditioning system, perform an initial unit start and check procedure, and record operational data on the worksheet.

REQUIRED TOOLS, EQUIPMENT, AND MATERIALS
Safety glasses and gloves
Multimeter
Ammeter
Hand tools
Refrigeration gauges
Temperature tester
Air-conditioner

SAFETY REQUIREMENTS
Wear safety glasses and gloves

PROCEDURE
Step 1 – Prestart Inspection
1. Measure and record the outdoor ambient temperature, the indoor temperature, and the indoor wet bulb temperature.
2. Check to see that all packing materials and shipping supports have been removed.
3. Spin the condenser and evaporator fan blades by hand to verify that they are not hanging or seized.
4. Check to see that an air filter is installed.
5. Pour water in the evaporator condensate drain and verify that the drain work properly
6. Connect refrigeration gauges and verify that the unit has a refrigerant charge.

Step 2 – Initial Power-Up

1. Power should be applied to the unit for 24 hours before turning it on.
   (This requirement is waived for this lab)
2. Check the incoming voltage to the unit with the thermostat set to off and the disconnect switch on. The voltage should be between the minimum and maximum operating voltages stated on the unit data plate.
3. If the voltage is NOT within the acceptable operating range, turn the disconnect switch off and determine why.
4. Set the thermostat to "Cool" and run the temperature down to call for cooling.
5. After the unit starts, recheck the voltage to the unit.
6. It should still be within the acceptable operating voltages listed on the unit data plate.
7. If the voltage is NOT within the acceptable operating range, the voltage drop through the power wire is unacceptably high—turn off the power!
8. Check the amp draw of the power wire feeding the unit and compare it to the unit data plate.
   The amp draw should be lower than the minimum circuit ampacity stated on the data plate.
9. Check the amp draw of the compressor, condenser fan motor, and evaporator fan motor.
10. Compare to the Compressor RLA and fan motor FLA ratings on the data plate.
11. The readings most likely will not be exactly what is on the data plate, but they should be in the same range.

Step 3 – Operational Checks

1. If the system has ductwork, read the static pressure difference between the return and supply duct.
2. Compare this total external static reading to the manufacturer's specification.
3. If you do not have a specification, assume 0.5" wc to be "normal". Anything over 0.7" wc is usually too high.
4. Measure the system airflow using a flow-hood or other airflow measuring instrument.
5. Compare the actual airflow reading to the manufacturer's specification.
6. If you do not have a manufacturer's specification, the airflow should be approximately 400 CFM per ton.
7. After the unit has operated long enough for the temperatures and pressures to stabilize, record the system suction and discharge pressures.
8. Compare the pressures to the manufacturer's specifications.
9. Read the temperature of the return air and supply air.
10. Calculate the temperature drop across the coil (return air temperature–supply air temperature).
11. Compare the coil temperature drop to the manufacturer's specifications. This varies depending upon the conditions and airflow.
12. Normal temperature drops can be anywhere between 10°F to 20°F.

Table 51-5-1

Prestart Check List					
Electrical Prestart Checks		**Refrigeration Prestart Checks**		**Mechanical Checks**	
Minimum Supply		Outdoor Ambient Temperature		Shipping Materials	
Maximum Supply		Indoor Dry Bulb Temperature		Condenser Fan Spins	
Compressor RLA		Indoor Wet Bulb Temperature		Evaporator Fan Spins	
Condenser Fan FLA		Type of Refrigerant		Air Filter in Place	
Evaporator Fan FLA		Equalized Refrigerant Pressure		Evaporator Drain Works	
MCA (Min Cir Amps)					
Wire sized to MCA					
Max Fuse Size					
Actual Fuse Size					

Initial Power-Up	
Voltage at unit with unit off	
Voltage at unit while running	
Amp draw on power wire	
Compressor amp draw	
Condenser fan motor amp draw	
Evaporator fan motor amp draw	

Refrigeration Operational Checks		**Airflow Operational Checks**	
Suction Pressure		Total External Static Specified	
Evaporator Saturation		Measured total external static	
Condenser Pressure		Airflow CFM Specified	
Condenser Saturation		Measured CFM	
Suction Line Temperature		Return air temperature	
Suction Superheat		Supply air temperature	
Liquid Line Temperature		Evaporator Temp Difference	
Liquid Subcooling			
Discharge line temperature			

LAB 52.1 DUCT COMPONENTS

LABORATORY OBJECTIVE
You will demonstrate your ability to identify common duct system components.

***FUNDAMENTALS OF HVACR* TEXT REFERENCE**
Unit 52 Duct Installation

LABORATORY NOTES
You will identify common duct system components and explain their purpose in the system.
SAFETY NOTE: Sheet metal duct can cause serious cuts and lacerations. Always wear gloves and handle carefully!

REQUIRED TOOLS, EQUIPMENT, AND MATERIALS
gloves
An assortment of duct components and an assembled duct system

SAFETY REQUIREMENTS
A. Wear safety glasses
B. Wear gloves when handling sheet metal
C. The edges of sheet metal duct components can cause severe cuts

PROCEDURE
Examine the duct system components assigned by the instructor. Be prepared to name each of the components and discuss their function. Check off each item in Data Sheet 52-1-1 as you examine it.

Data Sheet 52-1-1 Duct Component Identification	
Component	**Description of Use**
Round Metal Duct	
Rectangular Metal Duct	
S-lock	
Drive Cleat	
Plenum	
Transition	
End Cap	
Takeoff	
Wye	
Boot	
Mastic	
Flex Duct	
Duct Strap	

LAB 53.1 AIR CONDITIONING SYSTEM STARTUP

LABORATORY OBJECTIVE
You will demonstrate your ability to perform an initial unit start and check procedure.

***FUNDAMENTALS OF HVACR* TEXT REFERENCE**
Unit 53 Troubleshooting Air-Conditioning Systems

LABORATORY NOTES
You will operate an air conditioning system, perform an initial unit start and check procedure, and record operational data on the worksheet.

REQUIRED TOOLS, EQUIPMENT, AND MATERIALS
Safety glasses and gloves
Multimeter
Ammeter
Hand tools
Refrigeration gauges
Temperature tester
Air-conditioner

SAFETY REQUIREMENTS
Wear safety glasses and gloves

PROCEDURE
Step 1 – Prestart Inspection
1. Measure and record the outdoor ambient temperature, the indoor temperature, and the indoor wet bulb temperature.
2. Check to see that all packing materials and shipping supports have been removed.
3. Spin the condenser and evaporator fan blades by hand to verify that they are not hanging or seized.
4. Check to see that an air filter is installed.
5. Pour water in the evaporator condensate drain and verify that the drain work properly
6. Connect refrigeration gauges and verify that the unit has a refrigerant charge.

Step 2 – Initial Power-Up
1. Power should be applied to the unit for 24 hours before turning it on.
2. (This requirement is waived for this lab)
3. Check the incoming voltage to the unit with the thermostat set to off and the disconnect switch on. The voltage should be between the minimum and maximum operating voltages stated on the unit data plate.

4. If the voltage is NOT within the acceptable operating range, turn the disconnect switch off and determine why.
5. Set the thermostat to "Cool" and run the temperature down to call for cooling.
6. After the unit starts, recheck the voltage to the unit.
7. It should still be within the acceptable operating voltages listed on the unit data plate.
8. If the voltage is NOT within the acceptable operating range, the voltage drop through the power wire is unacceptably high–turn off the power!
9. Check the amp draw of the power wire feeding the unit and compare it to the unit data plate.
10. The amp draw should be lower than the minimum circuit ampacity stated on the data plate.
11. Check the amp draw of the compressor, condenser fan motor, and evaporator fan motor.
12. Compare to the Compressor RLA and fan motor FLA ratings on the data plate.
13. The readings most likely will not be exactly what is on the data plate, but they should be in the same range.

Step 3 – Operational Checks

1. If the system has ductwork, read the static pressure difference between the return and supply duct.
2. Compare this total external static reading to the manufacturer's specification.
3. If you do not have a specification, assume 0.5" wc to be "normal". Anything over 0.7" wc is usually too high.
4. Measure the system airflow using a flow-hood or other airflow measuring instrument.
5. Compare the actual airflow reading to the manufacturer's specification.
6. If you do not have a manufacturer's specification, the airflow should be approximately 400 CFM per ton.
7. After the unit has operated long enough for the temperatures and pressures to stabilize, record the system suction and discharge pressures.
8. Compare the pressures to the manufacturer's specifications.
9. Read the temperature of the return air and supply air.
10. Calculate the temperature drop across the coil (return air temperature–supply air temperature)
11. Compare the coil temperature drop to the manufacturer's specifications. This varies depending upon the conditions and airflow.
12. Normal temperature drops can be anywhere between 10°F to 20°F.

Prestart Check List					
Electrical Prestart Checks		**Refrigeration Prestart Checks**		**Mechanical Checks**	
Minimum Supply Voltage		Outdoor Ambient Temperature		Shipping Materials Removed	
Maximum Supply Voltage		Indoor Dry Bulb Temperature		Condenser Fan Spins	
Compressor RLA		Indoor Wet Bulb Temperature		Evaporator Fan Spins	
Condenser Fan FLA		Type of Refrigerant		Air Filter in Place	
Evaporator Fan FLA		Equalized Refrigerant Pressure		Evaporator Drain Works	
MCA (Min Cir Amps)					
Wire sized to MCA					
Max Fuse Size					
Actual Fuse Size Installed					
Initial Power-Up					
Voltage at unit with unit off					
Voltage at unit while running					
Amp draw on power wire					
Compressor amp draw					
Condenser fan motor amp draw					
Evaporator fan motor amp draw					
Refrigeration Operational Checks		**Airflow Operational Checks**			
Suction Pressure		Total External Static Specified			
Evaporator Saturation		Measured total external static measured			
Condenser Pressure		Airflow CFM Specified			
Condenser Saturation		Measured CFM			
Suction Line Temperature		Return air temperature			
Suction Superheat		Supply air temperature			
Liquid Line Temperature		Evaporator Temp Difference			
Liquid Subcooling					
Discharge line temperature					

LAB 53.2 TROUBLESHOOTING SCENARIO

LABORATORY OBJECTIVE
Given a system with a problem, you will identify the problem, its root cause, and recommend corrective action.

FUNDAMENTALS OF HVACR TEXT REFERENCE
Unit 53 Troubleshooting Air-Conditioning Systems

REQUIRED MATERIALS PROVIDED BY STUDENT
Safety glasses
Refrigeration gauges
Multimeter
Hand tools
Manometer
Air-conditioning system with a problem

SAFETY REQUIREMENTS
A. Wear safety glasses
B. Do not touch energized electrical components and circuits
C. Use non-contact voltage detector or meter to see if components and circuits are energized
D. Turn power off to unit and lock it out before replacing components or repairing circuits

PROCEDURE
Troubleshoot the system assigned by the instructor. Be sure to be complete in your description of the problem. You should include:

- What the unit is doing wrong?
- What component or condition is causing this?
- What tests did you perform that told you this?
- How would you correct this?

LAB 54.1 PRINCIPLES OF COMBUSTION

LABORATORY OBJECTIVE
You will demonstrate the effect of the correct fuel-air mixture for combustion.

FUNDAMENTALS OF HVACR TEXT REFERENCE
Unit 54 Principles of Combustion and Safety

LABORATORY NOTES
You will operate an atmospheric gas burner, observe the flames and measure the CO concentration in the flue gas. You will then close the primary air shutters, observe the flames, and measure the CO concentration. Finally, you will summarize the effect of primary air on proper gas combustion.

REQUIRED TOOLS, EQUIPMENT, AND MATERIALS
Safety glasses
Operating gas furnace with primary air shutters
Meter with CO measurement capability

SAFETY REQUIREMENTS
Wear safety glasses
Always familiarize yourself with the equipment and operating manuals prior to starting up any system.

PROCEDURE
1. Examine the furnace burners.
    a. Where does the primary air enter?
    b. Where does the secondary air enter?
2. Operate a furnace that has atmospheric burners with a primary air adjustment.
3. Observe the flames.
4. Measure the CO content of the flue gas.
5. Close the primary air adjustment all the way and observe the flames.
6. Measure the CO content of the flue gas.
    a. How did reducing the primary air affect the CO content of the flue gas?
7. Open the primary air shutter until all the yellow flame tips disappear.
8. Measure the CO content of the flue gas.

LAB 54.2 GAS SAFETY INSPECTION

LABORATORY OBJECTIVE
You will describe typical gas safety hazards.

FUNDAMENTALS OF HVACR TEXT REFERENCE
Unit 54 Principles of Combustion and Safety

LABORATORY NOTES
You will describe common gas safety hazards to the instructor using a gas furnace to show where to look for the hazard.

REQUIRED TOOLS, EQUIPMENT, AND MATERIALS
Safety glasses
Gas furnace

SAFETY REQUIREMENTS
Wear safety glasses

PROCEDURE
Describe at least five gas-related safety hazards, recommend corrective action for each safety hazard and ways to avoid these hazards. If the hazard does not actually exist in the shop, you can describe the hazard and how it would affect gas safety. Hazards should include one from each of the following categories:

- Combustion air
- Venting
- Gas leaks
- Clearance
- Flammable material storage

LAB 54.3 SERVICING GAS BURNERS

LABORATORY OBJECTIVE
You will adjust the primary air on an atmospheric gas burner to achieve a clean–burning, efficient flame.

FUNDAMENTALS OF HVACR TEXT REFERENCE
Unit 54 Principles of Combustion and Safety

LABORATORY NOTES
You will operate an atmospheric gas burner, adjust the primary air shutters and observe the effect the primary air adjustment has on the flames.

REQUIRED TOOLS, EQUIPMENT, AND MATERIALS
Safety glasses
Hand tools
Operating gas furnace with primary air adjustment

SAFETY REQUIREMENTS
Wear safety glasses
Always familiarize yourself with the equipment and operating manuals prior to starting up any system.

PROCEDURE
1. Identify the type of burners in the furnaces assigned by the instructor. They could be
   - atmospheric—slotted port
   - atmospheric—drilled port
   - atmospheric—ribbon
   - atmospheric—in shot
   - power burner
2. Adjust the primary air on the burners assigned by an instructor.
3. Close off the primary air shutter on one burner. What happens? Why?
4. Open the primary air shutters all the way open on all the burners. What happens? Why?
5. Turn off the call for heat. Call for heat and watch the lighting of the burners. DO NOT GET YOUR FACE RIGHT IN FRONT OF THE FURNACE!!! What happens? Why?
6. Readjust the burners to the proper primary air setting. Now, turn the furnace off and then on again. How do the burners light now?

LAB 55.1 GAS FURNACE CHARACTERISTICS

LABORATORY OBJECTIVE
You will identify gas furnaces by their characteristics; including blower configuration, fuel type, efficiency, and type of draft.

FUNDAMENTALS OF HVACR TEXT REFERENCE
Unit 55 Gas Heat

LABORATORY NOTES
You will examine the gas furnaces assigned by the instructor, identify the different furnace characteristics, and record them in the data table below.

REQUIRED TOOLS, EQUIPMENT, AND MATERIALS
Safety glasses
Gas furnaces

SAFETY REQUIREMENTS
Wear safety glasses

PROCEDURE
Examine each gas furnace for the blower configuration, type of fuel, efficiency, and type of draft.

Fuel
The fuel for a gas furnace will be either natural gas or propane, indicated as LP. The furnace data-plate will indicate the type of fuel. Most furnace data-plates are on the inside of the furnace.

Blower Configuration
Locate the blower. The air goes across the blower first, so the blower compartment will be opposite where the air leaves.

Upflow
A vertically oriented furnace with the blower on the bottom is an upflow because it is blowing the air up.

Downflow or Counterflow
A vertically oriented furnace with the blower on top is a downflow because it is blowing the air down.

Horizontal
Air moves horizontally, or sideways, through a horizontal furnace.

Multipoise

A multiposition or multipoise furnace can be installed in more than one position. Multipoise furnaces typically indicate their ability to be installed in multiple positions somewhere on the furnace itself.

Draft

Natural

Natural draft furnaces do not have a draft blower. Instead, they have either a draft hood or a draft diverter. They use metal vent pipe.

Induced

Induced draft furnaces have a draft blower that sucks air through the combustion chamber.

Forced

Forced draft furnaces have a draft blower that blow, or forces air through the combustion chamber.

Efficiency

Gas furnaces fall into three large efficiency categories: 60%, 80%, and 90%.

60% Efficiency

Natural draft, standing pilot furnaces are no longer manufactured, but many still exist. They typically operate at 50%–70% efficiency.

80% Efficiency

Furnaces that do not have a standing-pilot, have an induced draft blower, but still use metal venting are 80% efficient.

90% Efficiency

Furnaces that have plastic PVC vents are 90%–98% efficient.

	Furnace 1	Furnace 2	Furnace 3	Furnace 4
Fuel *Natural* *LP*				
Configuration *Upflow* *Horizontal* *Downflow* *Multipoise*				
Draft *Natural* *Induced* *Forced*				
Efficiency *60%* *80%* *90%*				

LAB 55.2 NATURAL DRAFT FURNACE COMPONENTS

LABORATORY OBJECTIVE
You will identify the components on a natural draft furnace.

FUNDAMENTALS OF HVACR TEXT REFERENCE
Unit 55 Gas Heat

LABORATORY NOTES
You will inspect a natural draft furnace and identify the listed components.

REQUIRED TOOLS, EQUIPMENT, AND MATERIALS
Safety glasses
Natural draft gas furnace

SAFETY REQUIREMENTS
Wear safety glasses

PROCEDURE
You will examine the natural draft gas furnace assigned by the instructor and identify the listed components.

- Gas line
- Gas valve
- Gas manifold
- Pilot light
- Burners
- Heat exchanger
- Indoor blower
- Draft diverter
- Vent

LAB 55.3 NATURAL DRAFT FURNACE SEQUENCE OF OPERATION

LABORATORY OBJECTIVE
You will describe the sequence of operation for a standing pilot, natural draft furnace.

FUNDAMENTALS OF HVACR TEXT REFERENCE
Unit 55 Gas Heat

LABORATORY NOTES
You will inspect a standing pilot, natural draft furnace and describe the sequence of operation to the instructor.

REQUIRED TOOLS, EQUIPMENT, AND MATERIALS
Safety glasses
Natural draft gas furnace

SAFETY REQUIREMENTS
Wear safety glasses
Always familiarize yourself with the equipment and operating manuals prior to starting up any system.

PROCEDURE
Set the thermostat to call for heat and observe the furnace operating sequence. List the order that things happen in the data table.

Natural Draft Furnace Operating Sequence	
Order	Action
1	Thermostat set to heat
2	
3	
4	Thermostat set to OFF
5	
6	

Once the indoor fan is running, set the thermostat to "Off." Observe the order that things shut off and record in the table.

QUESTIONS

1. What controls the gas pressure?

2. How are the gas and combustion air mixed?

3. How is the gas-air mixture ignited?

4. How is the heat from combustion transferred into the house?

5. What happens to the combustion gasses?

LAB 55.4 INDUCED DRAFT FURNACE COMPONENTS

LABORATORY OBJECTIVE
You will identify the components on an induced draft natural gas furnace.

FUNDAMENTALS OF HVACR TEXT REFERENCE
Unit 55 Gas Heat

LABORATORY NOTES
You will inspect an induced draft furnace and identify the listed components to the instructor.

REQUIRED TOOLS, EQUIPMENT, AND MATERIALS
Safety glasses
Induced draft gas furnace

SAFETY REQUIREMENTS
Wear safety glasses

PROCEDURE
You will examine the natural draft gas furnace assigned by the instructor and identify the listed components.

- Gas line
- Gas valve
- Gas manifold
- Igniter (do not touch the igniter)
- Burners
- Heat exchanger
- Inducer motor
- Indoor blower

LAB 55.5 CONDENSING FURNACE COMPONENTS

LABORATORY OBJECTIVE
You will identify the components on condensing gas furnaces with and without sealed combustion chambers.

FUNDAMENTALS OF HVACR TEXT REFERENCE
Unit 55 Gas Heat

LABORATORY NOTES
You will inspect a condensing furnace and identify the listed components to the instructor.

REQUIRED TOOLS, EQUIPMENT, AND MATERIALS
Safety glasses
Condensing gas furnace without sealed combustion chamber
Condensing gas furnace with sealed combustion chamber

SAFETY REQUIREMENTS
Wear safety glasses

PROCEDURE
Step 1
You will examine the condensing gas furnace without a sealed combustion chambers assigned by the instructor and identify the listed components.

- Gas line
- Gas valve
- Gas manifold
- Igniter (do not touch the igniter)
- Burners
- Primary heat exchanger
- Secondary heat exchanger
- Condensate drain
- Inducer motor
- Indoor blower

Step 2
You will examine the condensing gas furnace with a sealed combustion chamber assigned by the instructor and identify the listed components.

- Gas line
- Gas valve
- Sealed combustion chamber
- Primary heat exchanger
- Secondary heat exchanger
- Condensate drain
- Inducer motor
- Indoor blower

558

LAB 55.6 CONDENSING FURNACE SEQUENCE OF OPERATIONS

LABORATORY OBJECTIVE
You will describe the sequence of operation for a condensing gas furnace.

FUNDAMENTALS OF HVACR TEXT REFERENCE
Unit 55 Gas Heat

LABORATORY NOTES
You will inspect a condensing furnace and describe the sequence of operation to the instructor.

REQUIRED TOOLS, EQUIPMENT, AND MATERIALS
Safety glasses
Condensing gas furnace

SAFETY REQUIREMENTS
Wear safety glasses
Always familiarize yourself with the equipment and operating manuals prior to starting up any system.

PROCEDURE
Set the thermostat to call for heat and observe the furnace operating sequence. List the order that things happen in the data table. Once the indoor fan is running, set the thermostat to "Off." Observe the order that things shut off and record in the table.

Condensing Gas Furnace Operating Sequence	
Order	Action
1	Thermostat set to heat
2	
3	
4	
5	Thermostat set to OFF
6	
7	
8	

LAB 56.1 NATURAL DRAFT FURNACE ELECTRICAL COMPONENTS

LABORATORY OBJECTIVE

You will identify common electrical components on a standing pilot, natural draft furnace and explain their function.

FUNDAMENTALS OF HVACR TEXT REFERENCE

Unit 56 Gas Furnace Controls

LABORATORY NOTES

You will inspect a standing pilot, natural draft furnace and identify the listed electrical components.

REQUIRED TOOLS, EQUIPMENT, AND MATERIALS

Safety glasses
Operating natural draft gas furnace

SAFETY REQUIREMENTS

Wear safety glasses

PROCEDURE

You will examine the natural draft gas furnace assigned by the instructor and identify the listed components.

- Gas valve
- Thermocouple
- Limit switch
- Auxiliary limit switch
- Fan switch
- Indoor blower
- Door switch
- Transformer

LAB 56.2 INDUCED DRAFT ELECTRICAL COMPONENTS

LABORATORY OBJECTIVE
You will identify common electrical components on an induced draft furnace and explain their function.

***FUNDAMENTALS OF HVACR* TEXT REFERENCE**
Unit 56 Gas Furnace Controls

LABORATORY NOTES
You will inspect an induced draft furnace and identify the listed electrical components.

REQUIRED TOOLS, EQUIPMENT, AND MATERIALS
Safety glasses
Operating induced draft gas furnace

SAFETY REQUIREMENTS
Wear safety glasses

PROCEDURE
You will examine the natural draft gas furnace assigned by the instructor and identify the listed components.

- Ignition control
- Transformer
- Inducer motor
- Draft pressure switch
- Gas valve
- Igniter
- Flame rod
- Limit switch
- Roll out switch
- Door switch
- Indoor blower

LAB 56.3 INDUCED DRAFT FURNACE SEQUENCE OF OPERATION

LABORATORY OBJECTIVE
You will describe the sequence of operation for an induced draft furnace.

FUNDAMENTALS OF HVACR TEXT REFERENCE
Unit 56 Gas Furnace Controls

LABORATORY NOTES
You will inspect an induced draft furnace and describe the sequence of operation to the instructor.

REQUIRED TOOLS, EQUIPMENT, AND MATERIALS
Safety glasses
Induced draft gas furnace

SAFETY REQUIREMENTS
A. Wear safety glasses
B. Check all circuits for voltage before doing any service work.
C. Never allow gas flow without a flame.

PROCEDURE
1. Set the thermostat to call for heat and observe the furnace operating sequence.
2. List the order that things happen in the data table.

Induced Draft Furnace Operating Sequence	
Order	Action
1	Thermostat set to heat
2	
3	
4	
5	
6	Thermostat set to OFF

7	
8	
9	

3. Once the indoor fan is running, set the thermostat to "Off." Observe the order that things shut off and record in the table.
4. Pull a wire off one side of the draft switch and set the thermostat to call for heat again. Observe the furnace operation. What happens?
5. Turn the thermostat to the off position, replace the draft switch wire and remove the wire from the flame sensor.
6. Set the thermostat to call for heat. Observe the operation. What happens?
7. Show the instructor how to adjust the length of the fan running cycle. Pull one wire off the high limit and set the thermostat to call for heat. Observe the operation. What happens?

QUESTIONS

1. What comes on first?

2. What controls the gas pressure?

3. How are the gas and combustion air mixed?

4. How is the gas-air mixture ignited?

5. How is the heat from combustion transferred into the house?

6. What happens to the combustion gases?

LAB 56.4 DIRECT SPARK IGNITION CONTROLS

LABORATORY OBJECTIVE
You will observe the operating sequence of a direct spark ignition system and simulate common failure symptoms.

FUNDAMENTALS OF HVACR TEXT REFERENCE
Unit 56 Gas Furnace Controls

LABORATORY NOTES
You will inspect a direct spark ignition furnace, observe its operation, and simulate failure conditions.

REQUIRED TOOLS, EQUIPMENT, AND MATERIALS
Safety glasses
Electrical tools
Operating gas furnace with direct spark ignition system

SAFETY REQUIREMENTS
A. Wear safety glasses
B. Check all circuits for voltage before doing any service work.
C. Never allow gas flow without a flame.

PROCEDURE
1. Study the wiring diagram of a furnace using a direct ignition system
2. Operate the DSI control system assigned by the instructor.
3. Does this system use a pre-purge cycle? If so, how long is it?
4. How long did the spark igniter operate?
5. Cycle the system off.
6. Does this system have a post purge cycle? If so, how long is it?
7. Turn the gas valve to the "OFF" position and call for heat once more.
8. How long does an ignition trial take?
9. How many ignition trials does the system have before lockout?
10. Turn the gas valve back to the "ON" position.
11. Reset the control by killing the call for heat and re-establishing it.
12. After flame is proved, turn the gas valve back to the "OFF" position.
13. Does the system attempt to re-establish combustion?
14. How many trial periods does it go through before it locks out?
15. Does it use a purge cycle in between the trials?
16. How long are they?
17. Turn the unit off and return the gas valve to the "ON" position.
18. Give a complete description of the operation cycle of this control.

LAB 56.5 HOT SURFACE IGNITION OPERATION

LABORATORY OBJECTIVE
You will observe the operating sequence of a hot surface ignition system and simulate common failure symptoms.

FUNDAMENTALS OF HVACR TEXT REFERENCE
Unit 56 Gas Furnace Controls

LABORATORY NOTES
You will inspect a hot surface ignition furnace, observe its operation, and simulate failure conditions.

REQUIRED TOOLS, EQUIPMENT, AND MATERIALS
Safety glasses
Electrical tools
Operating gas furnace with a hot surface ignition system

SAFETY REQUIREMENTS
A.   Wear safety glasses
B.   Never allow gas flow without a flame.

PROCEDURE
1.   Study the wiring diagram of a furnace using a hot surface ignition system
2.   Operate the furnace assigned by the instructor.
3.   Does this system use a pre-purge cycle? If so, how long is it?
4.   How long did the igniter operate?
5.   Cycle the system off.
6.   Does this system have a post purge cycle? If so, how long is it?
7.   Turn the gas valve to the "OFF" position and call for heat once more.
8.   How long does an ignition trial take?
9.   How many ignition trials does the system have before lockout?
10.   Turn the gas valve back to the "ON" position.
11.   Reset the control by killing the call for heat and re-establishing it.
12.   After flame is proved, turn the gas valve back to the "OFF" position.
13.   Does the system attempt to re-establish combustion?
14.   How many trial periods does it go through before it locks out?
15.   Does it use a purge cycle in between the trials?
16.   How long are they?
17.   Turn the unit off and return the gas valve to the "ON" position.
18.   Give a complete description of the operation cycle of this control.

LAB 56.6 PILOT TURNDOWN TEST—THERMOCOUPLE TYPE

LABORATORY OBJECTIVE
The purpose of this lab is to test the proper operation of a thermocouple type pilot for a gas furnace.

FUNDAMENTALS OF HVACR TEXT REFERENCE
Unit 56 Gas Furnace Controls

LABORATORY NOTES
A pilot turndown test is a test to determine the smallest possible pilot capable of proving flame presence to the control circuit and lighting the main burner safely. It is performed by lighting the pilot, measuring the pilot flame signal, and observing a series of main burner ignitions using various pilot sizes. Remember a cold burner is more difficult to ignite and in the performance of this test the burner becomes warmed up; you must allow sufficient cool-down time to perform the final test. A pilot turndown test can also point to other pilot problems.

REQUIRED TOOLS, EQUIPMENT, AND MATERIALS
Operating gas furnace
Tool kit
Millivoltmeter

SAFETY REQUIREMENTS
A. Wear safety glasses

B. Never allow gas flow without a flame. If gas is allowed to build up and then suddenly ignite, this can create a serious hazard.

C. If the furnace contains a pilot, always ensure it is properly lit prior to starting the unit.

D. Check all circuits for voltage before doing any service work.

PROCEDURE
Step 1

Obtain any standard thermocouple installed or not installed.

1. Complete the Thermocouple Open Circuit Test (30 mV) Check List and make sure to check each step off in the appropriate box as you finish it. This will help you to keep track of your progress.

Thermocouple Open Circuit Test (30 mV) Check List

STEP	PROCEDURE	CHECK
1	Remove the threaded connection from the gas valve or pilot safety switch if necessary.	
2	Connect a millivolt meter from the outer copper line to the inner core lead at the end of the thermocouple.	
3	Heat the enclosed end of the thermocouple with a torch or a normal pilot assembly if installed.	
4	A voltage of up to 30 millivolts will be read on the meter. If the meter goes down, then reverse the leads.	
5	Record the highest millivoltage. Voltage =_____mV	

Step 2

Obtain a thermocouple installed on a standard combination gas valve or pilot safety switch.

1. Complete the Thermocouple Closed Circuit Test (30 mV) Check List.

Thermocouple Closed Circuit Test (30 mV) Check List

STEP	PROCEDURE	CHECK
1	Obtain the thermocouple adaptor for the valve end.	
2	Light the pilot for a normal pilot flame.	
3	Read the millivolts at the adaptor. Voltage =_____mV	
4	Blow out the pilot light and then relight it within 30 seconds of going out.	
5	Why does gas still come out of the pilot burner after the pilot is out?	
6	Blow out the pilot again and observe the voltage output.	
7	Record the voltage in millivolts when the pilot valve closes stopping the pilot gas so that the pilot will not relight. This value should be between 9 and 12 mV. Voltage =_____mV	
8	Adjust the size of the pilot flame for the smallest pilot flame capable of proving the pilot and lighting the main burner smoothly and safely.	

LAB 56.7 PILOT TURNDOWN TEST—FLAME ROD TYPE

LABORATORY OBJECTIVE
The purpose of this lab is to test the proper operation of a flame rod type pilot for a gas furnace. Perform this test on any furnace equipped with a pilot, a separate spark igniter, and a flame rod.

FUNDAMENTALS OF HVACR TEXT REFERENCE
Unit 56 Gas Furnace Controls

LABORATORY NOTES
A pilot turndown test is a test to determine the smallest possible pilot capable of proving flame presence to the control circuit and lighting the main burner safely. It is performed by lighting the pilot, measuring the pilot flame signal, and observing a series of main burner ignitions using various pilot sizes. Remember a cold burner is more difficult to ignite and in the performance of this test the burner becomes warmed up; you must allow sufficient cool-down time to perform the final test. A pilot turndown test can also point to other pilot problems.

REQUIRED TOOLS, EQUIPMENT, AND MATERIALS
Operating gas furnace
Tool kit
Microammeter

SAFETY REQUIREMENTS
A. Wear safety glasses
B. Never allow gas flow without a flame. If gas is allowed to build up and then suddenly ignite, this can create a serious hazard.
C. If the furnace contains a pilot, always ensure it is properly lit prior to starting the unit.
D. Check all circuits for voltage before doing any service work.

PROCEDURE
Step 1
Perform this test on any furnace equipped with a pilot, a separate spark igniter, and a flame rod.

1. Complete the Flame Rod Pilot System Test Check List and make sure to check each step off in the appropriate box as you finish it. This will help you to keep track of your progress.

Flame Rod Pilot System Test Check List

STEP	PROCEDURE	CHECK
1	Locate the pilot adjustment screw. This is usually a needle valve screw located under a screw cap labeled *pilot adj*.	
2	Turn off main power.	
3	Install the microammeter in series with the flame rod sensor wire to sensor wire terminal of the control box.	
4	Turn on power, and obtain a pilot only flame by turning off manual main burner valve or pull wire labeled main at redundant gas valve. Consult wiring diagram as required to locate main gas wire.	
5	Temporarily wire a small toggle switch in series with the main burner wire and the main gas valve. Turn switch off.	
6	Obtain operation of pilot only flame.	
7	Turn on main burner toggle and observe main burner on.	
8	Read flame signal of pilot only in micro-amps. Current =_____ micro Amps	
9	Turn system off.	

10	Remove and clean flame rod with steel wool.	
11	Read flame signal of pilot only in micro-amps. Current =_____ micro Amps	
12	Has the signal changed from step #8?	
13	Reposition the flame rod to obtain a better contact with the clean blue flame to gain a stronger signal if possible.	
14	Turn the pilot adjustment screw and observe the pilot flame and corresponding micro-amp signal get smaller.	
15	Obtain the smallest flame possible that is capable of proving the pilot flame. Current =_____ micro Amps	
16	Turn on the main burner toggle and observe the main burner light.	
17	Cycle the burner on and off several times and observe the ignition	

18	Enlarge the pilot as required to provide for smooth main burner ignition. Current =_____ micro Amps	

LAB 56.8 PILOT TURNDOWN TEST—FLAME IGNITER TYPE

LABORATORY OBJECTIVE
The purpose of this lab is to test the proper operation of a flame igniter type pilot for a gas furnace. Perform this test on any furnace equipped with a pilot and a combination pilot igniter/pilot proving device.

FUNDAMENTALS OF HVACR TEXT REFERENCE
Unit 56 Gas Furnace Controls

LABORATORY NOTES
A pilot turndown test is a test to determine the smallest possible pilot capable of proving flame presence to the control circuit and lighting the main burner safely. It is performed by lighting the pilot, measuring the pilot flame signal, and observing a series of main burner ignitions using various pilot sizes. Remember a cold burner is more difficult to ignite and in the performance of this test the burner becomes warmed up; you must allow sufficient cool-down time to perform the final test. A pilot turndown test can also point to other pilot problems.

REQUIRED TOOLS, EQUIPMENT, AND MATERIALS
Operating gas furnace
Tool kit

SAFETY REQUIREMENTS
A. Wear safety glasses
B. Never allow gas flow without a flame. If gas is allowed to build up and then suddenly ignite, this can create a serious hazard.
C. If the furnace contains a pilot, always ensure it is properly lit prior to starting the unit.
D. Check all circuits for voltage before doing any service work.

PROCEDURE
Step 1
Perform this test on any furnace equipped with a pilot and a combination pilot igniter/pilot proving device.

1. Complete the Flame Igniter Pilot System Test Check List and make sure to check each step off in the appropriate box as you finish it. This will help you to keep track of your progress.

Flame Igniter Pilot System Test Check List

STEP	PROCEDURE	CHECK
1	Locate the pilot adjustment screw. This is usually a needle valve screw located under a screw cap labeled *pilot adj*.	
2	Turn off main power.	
3	Install an appropriate toggle switch in series with main gas terminal of redundant gas valve.	
4	Turn on power and obtain a pilot only flame.	
5	Turn pilot adjustment valve in (clockwise, cw) and observe the pilot flame get smaller.	
6	Continue turning in until the pilot goes out and pilot igniter goes out.	
7	Turn pilot adjustment out (counter-clockwise, ccw) and observe the pilot relight.	
8	With the pilot on, turn on main gas toggle and observe main gas burner turn on.	
9	With the main burner on, adjust the pilot flame smaller.	
10	Observe the main burner go off when the pilot flame goes out.	
11	Adjust the pilot flame to a size that will light easily and ignite the main burner assembly.	

LAB 56.9 SEPARATE PILOT GAS PRESSURE REGULATOR

LABORATORY OBJECTIVE

The purpose of this lab is to test the proper operation of a separate pilot gas pressure regulator for a gas furnace. This Laboratory Worksheet deals with any commercial burner controller utilizing a separate pilot gas pressure regulator, such as a Honeywell commercial RA89F series controller.

This procedure will use a plug in flame monitor jack or wire meter in series with the flame rod.

FUNDAMENTALS OF HVACR TEXT REFERENCE

Unit 56 Gas Furnace Controls

LABORATORY NOTES

A pilot turndown test is a test to determine the smallest possible pilot capable of proving flame presence to the control circuit and lighting the main burner safely. It is performed by lighting the pilot, measuring the pilot flame signal, and observing a series of main burner ignitions using various pilot sizes.

REQUIRED TOOLS, EQUIPMENT, AND MATERIALS

Operating gas furnace
Tool kit
Microammeter
Plug-in flame monitor

SAFETY REQUIREMENTS

A. Wear safety glasses

B. Never allow gas flow without a flame. If gas is allowed to build up and then suddenly ignite, this can create a serious hazard.

C. If the furnace contains a pilot, always ensure it is properly lit prior to starting the unit.

D. Check all circuits for voltage before doing any service work.

PROCEDURE
Step 1

Perform this test on commercial gas furnace power burners.

1. Complete the Separate Pilot Gas Pressure Regulator Test Check List and make sure to check each step off in the appropriate box as you finish it. This will help you to keep track of your progress.

Separate Pilot Gas Pressure Regulator Test Check List

STEP	PROCEDURE	CHECK
1	Turn off power, main gas valve, and pilot gas valve.	
2	Obtain a microammeter and a plug-in flame monitor jack or wire the microammeter in series with the flame rod.	
3	Remove the primary control cover.	
4	Open the pilot gas valve and turn on the power.	
5	Observe the pilot flame only.	
6	Record the pilot micro-amp reading. Current =_____ micro Amps	
7	Decrease he gas pressure by turning the gas pressure regulator out (counter-clockwise, ccw). Observe the pilot flame get smaller and the flame signal reduce in current (micro-amps)	

8	Turn the pilot down until the controller relay opens. Current =_____micro Amps	
9	Increase the pilot flame size until the controller relay opens. Current =_____micro Amps	
10	Observe the main burner go off when the pilot flame goes out.	
11	Turn on the main burner gas and power.	
12	Observe the main burner ignition.	
13	Adjust the pilot flame as required to obtain smooth burner ignition during cold start. Remember a cold burner is more difficult to ignite and in the performance of this test the burner becomes warmed up; you must allow sufficient cool-down time to perform the final test.	

LAB 56.10 GAS VALVE INSPECTION

LABORATORY OBJECTIVE
You will inspect a selection of gas valves and describe how they work.

FUNDAMENTALS OF HVACR TEXT REFERENCE
Unit 56 Gas Furnace Controls

LABORATORY NOTES
You will inspect a selection of gas valves, disassemble them, and describe their operation to the instructor.

REQUIRED TOOLS, EQUIPMENT, AND MATERIALS
Safety glasses
Allen wrench set
Screw drivers
Gas valve for dismantling

SAFETY REQUIREMENTS
Wear safety glasses

PROCEDURE
1. Disassemble and examine the gas valves assigned by the instructor.
2. Be prepared to identify each valve type and describe its operation. Reassemble the valves after explaining them to the instructor. Clean up your work area.
3. Examine the gas valves on equipment assigned by your instructor. Be prepared to identify the types of the valves and their operation. After discussing the valves with the instructor, put all the covers back on the units and clean up your work area.

LAB 56.11 INSPECTING AND ADJUSTING GAS REGULATORS

LABORATORY OBJECTIVE
You will dismantle and inspect a gas regular to learn how it works and then adjust the gas valve regulator to maintain the manufacturer's specified manifold pressure on a gas furnace.

FUNDAMENTALS OF HVACR TEXT REFERENCE
Unit 56 Gas Furnace Controls

LABORATORY NOTES
You will inspect a gas regulator and describe its operation. You will then adjust a gas regulator to manufacturer's specifications.

REQUIRED TOOLS, EQUIPMENT, AND MATERIALS
Safety glasses
Allen wrench set
Screwdrivers
Gas regulator for dismantling
Manometer
Operating gas furnace

SAFETY REQUIREMENTS
A.   Wear safety glasses
B.   Never allow gas flow without a flame.

PROCEDURE
Step 1 (Inspecting a gas regulator)
1.   Remove the plug on the bottom of the regulator.
2.   Manually push the regulator plunger up until it is firmly seated.
3.   Listen as you do this.
4.   You should hear the sound of air leaving the regulator vent through the orifice. Now release it and listen.
5.   You should hear the sound of air entering the orifice in the regulator vent. Disassemble the gas regulator.
6.   Identify the diaphragm, seat, and plunger.
7.   Note where the gas enters the regulator and where it leaves the regulator.

Step 2 (Adjusting a gas regulator)
1.   Find the specified manifold pressure on the unit data plate.
2.   With the unit off, connect your manometer to the manifold pressure tap on the gas valve.
3.   Operate the furnace and check the operating manifold pressure.
4.   Turn the adjustment in clockwise until you see a change in the manometer reading.
5.   Turn the adjustment out counterclockwise back to the starting point.
6.   Adjust the manifold pressure to meet the pressure specified on the data plate.
7.   Show your results to the instructor and cleanup your work area.

LAB 56.12 WIRE A NATURAL DRAFT FURNACE

LABORATORY OBJECTIVE
You will completely wire a natural draft furnace, including all internal wiring.

FUNDAMENTALS OF HVACR **TEXT REFERENCE**
Unit 56 Gas Furnace Controls

LABORATORY NOTES
You will wire a natural draft furnace, have the work checked by an instructor, and operate it

REQUIRED TOOLS, EQUIPMENT, AND MATERIALS
Safety glasses
Non-contact voltage probe
Multimeter
Wire cutter/stripper/crimper
Natural draft gas furnace with all internal wiring removed
Wire
Electrical terminals

SAFETY REQUIREMENTS

A. Wear safety glasses

B. Check all circuits for voltage before doing any service work.

C. Never allow gas flow without a flame

PROCEDURE

1. The instructor will remove all the electrical wiring from a standing pilot, natural draft furnace.

2. Turn the furnace disconnect off.

3. Check to make sure the power is off.

4. You will wire the furnace according to the furnace electrical wiring diagram.

5. Have instructor check your work BEFORE energizing the furnace.

6. Operate the furnace and verify correct operation.

LAB 56.13 WIRE AN INDUCED DRAFT GAS FURNACE

LABORATORY OBJECTIVE
You will completely wire an induced draft furnace, including all internal wiring.

FUNDAMENTALS OF HVACR TEXT REFERENCE
Unit 56 Gas Furnace Controls

LABORATORY NOTES
You will wire an induced draft furnace, have the work checked by an instructor, and operate it

REQUIRED TOOLS, EQUIPMENT, AND MATERIALS
Safety glasses
Non-contact voltage probe
Multimeter
Wire cutter/stripper/crimper
Induced draft gas furnace with all internal wiring removed
Wire
Electrical terminals

SAFETY REQUIREMENTS

A.   Wear safety glasses

B.   Check all circuits for voltage before doing any service work.

C.   Never allow gas flow without a flame.

PROCEDURE

1.   The instructor will remove all the electrical wiring from an induced draft furnace.

2.   Turn the furnace disconnect off.

3.   Check to make sure the power is off.

4.   You will wire the furnace according to the furnace electrical wiring diagram.

5.   Have instructor check your work BEFORE energizing the furnace.

6.   Operate the furnace and verify correct operation.

LAB 57.1 SIZING GAS LINES

(This lab is done at home.)

LABORATORY OBJECTIVE

You will correctly size gas lines to meet manufacturer's specifications.

FUNDAMENTALS OF HVACR TEXT REFERENCE

Unit 57 Gas Furnace Installation

LABORATORY NOTES

You will size the gas lines for three piping systems ranging from simple to complex.

REQUIRED TOOLS, EQUIPMENT, AND MATERIALS

Manufacturer's installation instructions for gas furnace

SAFETY REQUIREMENTS

Wear safety glasses

PROCEDURE

The National Fire Protection Association, otherwise known as NFPA, is the most widely recognized authority on fire prevention. The sizing procedures shown here are not from any specific government code, but from the NFPA Bulletin 54 on the "Installation of Gas Appliances and Gas Piping."

The five factors used to determine gas-piping size are:

1. allowable pressure loss from the gas source to the unit
2. maximum gas consumption to be provided
3. piping length + equivalent length of all fittings and valves
4. specific gravity of the gas
5. diversity factor (more on this later)

The **maximum allowable pressure drop** is the difference in the gas pressure between the source and the appliance. This is limited by the difference between the minimum source pressure available and the minimum pressure needed at the appliance. For instance, if the gas pressure at the source is 4.5" and the unit requires 4.0" W.C. to operate, the maximum allowable pressure drop could be no more than 1/2" W.C. Generally, gas piping is sized for a maximum pressure loss of between 0.35" W.C. and 0.5" W.C.

The **maximum consumption** of the unit is the most cubic feet per hour that the unit would use. This is easily found by dividing the unit's rated input in BTU's per hour by the BTU content of the gas it is using. This is approximately 1000 BTU's per cubic foot for natural gas, 2500 BTU's per cubic foot for propane. These figures are close enough for most calculations; however, the local gas distributor can give you exact figures for the product they sell.

The **piping length** is the actual length in feet between the gas source and the gas appliance plus the equivalent length in feet of all fittings, valves, or obstructions. The tables listed here have allowance for a "normal" amount of fittings so equivalent length determination is needed only for systems with an unusual number of fittings, valves, or accessories. For our purposes, we'll assume that all systems in this section are "normal" systems.

The **diversity factor** is the ratio of maximum PROBABLE demand to maximum POSSIBLE demand. In other words, how much gas is LIKELY to be used versus how much gas COULD POSSIBLY be used. Take the case of a house with a 100,000 BTU input gas furnace and a 50,000 BTU input gas air conditioner. Since both would logically never be running at the same time, the maximum PROBABLE demand would be the higher demand of the two, the gas furnace at 100,000 BTU's. The maximum POSSIBLE demand would of course be the sum of the two, or 150,000 BTU's. In this case, the diversity factor would be 100,000/150,000 or 2/3. This means that instead of sizing the main line supplying the system for 150,000 BTU's it would be sized for 150,000 BTU's x 2/3 or 100,000 BTU's. We will not be considering diversity factor in our examples.

To use the pipe sizing chart:

1. Determine the length of piping to the appliance that is the greatest distance from the gas source.
2. This distance will be used to size ALL piping.
3. Find a row with a length equal to or greater than that length.
4. Use that row to find a cubic foot capacity equal to or greater than that required for each section of gas pipe.
5. The column at the top gives the nominal pipe size.
6. This row is used for sizing ALL appliances and ALL sections of pipe.

Capacity in Cubic Feet of Gas for 0.6 SG with 0.5″ wc Pressure Drop					
Nominal Pipe Size Rigid Schedule 40 Iron Pipe					
Length	1/2	3/4	1	1 1/4	11/2
10	201	403	726	1260	1900
20	138	277	499	865	1310
30	111	222	401	695	1050
40	95	190	343	594	898
50	84	169	304	527	796
60	76	153	276	477	721
70	70	140	254	439	663
80	65	131	236	409	617
90	61	123	221	383	579
100	58	116	209	362	547
125	51	103	185	321	485
150	46	93	168	291	439

Size the following systems using the specifications given:

GAS PIPING PLAN A	
CFH Consumption	
Furthest run	
Pipe Size	

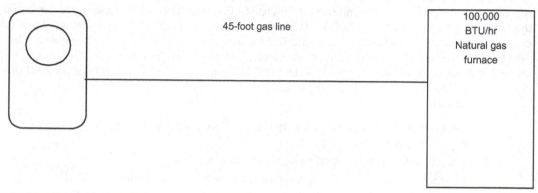

45-foot gas line

100,000 BTU/hr Natural gas furnace

GAS PIPING PLAN B	
Furnace CFH	
Water Heater CFH	
Main CFH	
Longest Run	
Line from Meter to TEE	
Line from TEE to furnace	
Line from TEE to water heater	

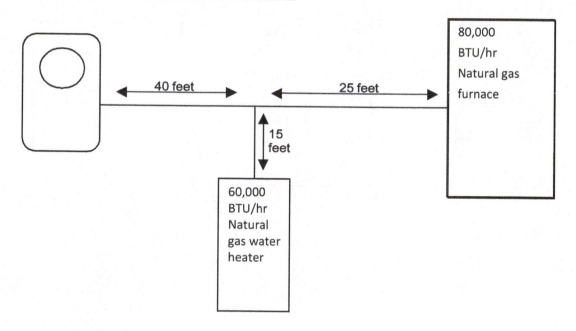

40 feet

25 feet

80,000 BTU/hr Natural gas furnace

15 feet

60,000 BTU/hr Natural gas water heater

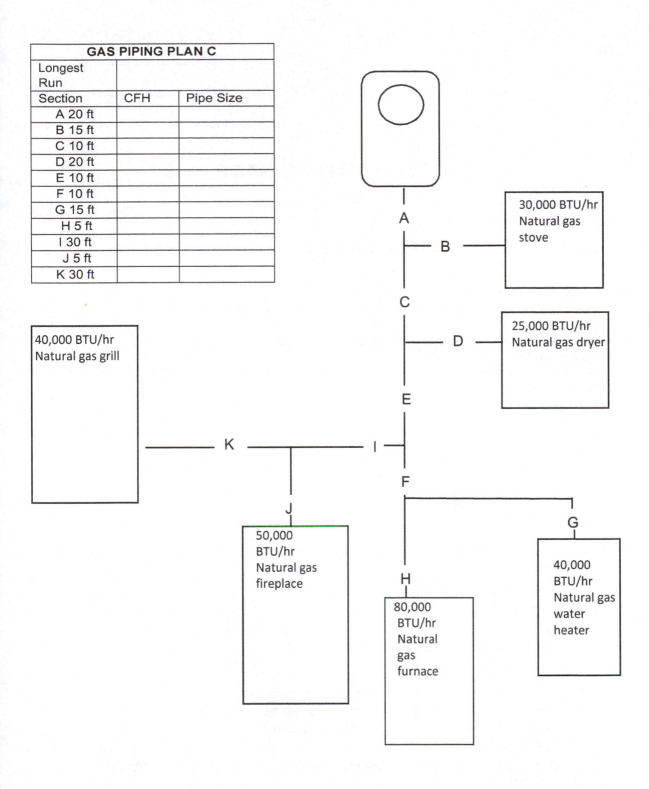

GAS PIPING PLAN C

Longest Run		
Section	CFH	Pipe Size
A 20 ft		
B 15 ft		
C 10 ft		
D 20 ft		
E 10 ft		
F 10 ft		
G 15 ft		
H 5 ft		
I 30 ft		
J 5 ft		
K 30 ft		

40,000 BTU/hr Natural gas grill

30,000 BTU/hr Natural gas stove

25,000 BTU/hr Natural gas dryer

50,000 BTU/hr Natural gas fireplace

80,000 BTU/hr Natural gas furnace

40,000 BTU/hr Natural gas water heater

A
B
C
D
E
F
G
H
I
J
K

LAB 57.2 INSTALLING GAS PIPING

LABORATORY OBJECTIVE
You will size, install, and test gas piping.

FUNDAMENTALS OF HVACR TEXT REFERENCE
Unit 57 Gas Furnace Installation

LABORATORY NOTES
You will size, install, and check the gas line to a gas fired appliance.

REQUIRED TOOLS, EQUIPMENT, AND MATERIALS
Safety glasses
Non-contact voltage probe
Manometer
Allen wrench set
Operational gas furnace
Gas pipe
Pipe fittings
Pipe threading tools
Pipe wrenches
Pipe thread sealing compound
Soap bubbles

SAFETY REQUIREMENTS
A. Wear safety glasses
B. Never allow gas flow without a flame.

PROCEDURE
1. Use the information on the furnace data plate to size the gas line.
2. The gas line should include the following:
    a. Manual shutoff valve
    b. Drip leg
    c. Union near the furnace or flexible gas connector
3. Have the instructor check the piping before turning on the gas.
4. Remove the plug on the gas valve outlet test port.
5. Open the gas line valve.
6. Check all joints for leaks using soap bubbles or an electronic combustible gas leak detector.
7. Check both the gas line inlet pressure and manifold pressure with the furnace off and with the furnace operating.

Gas Pressure Measurements		
	Furnace Off	Furnace Operating
Gas Inlet Pressure		
Manifold Pressure		

Name_____

Date_____

Instructor's OK ☐

LAB 57.3 CHECKING GAS PRESSURE

LABORATORY OBJECTIVE
You will demonstrate your ability to use a manometer to measure gas pressure.

FUNDAMENTALS OF HVACR TEXT REFERENCE
Unit 57 Gas Furnace Installation

LABORATORY NOTES
You will check the gas line pressure and manifold pressure of an operating gas appliance.

REQUIRED TOOLS, EQUIPMENT, AND MATERIALS
Safety glasses
Non-contact voltage probe
Operational gas furnace
Manometer
Allen wrench set

SAFETY REQUIREMENTS
A.   Wear safety glasses
B.   Never allow gas flow without a flame.

PROCEDURE
1.   Turn the power off to the furnace and verify that it is off.
2.   Close the manual gas valve in the gas line.
3.   Remove the plug on the gas valve inlet test port.
4.   Install the manometer in the gas valve inlet test port.
5.   Open the gas line valve.
6.   The manometer should read the incoming gas line pressure.
7.   Turn the power on and set the thermostat to heat.
8.   Read the gas line pressure with the furnace operating.
9.   Turn the thermostat to "OFF."
10.   Turn the power off to the furnace and verify that it is off.
11.   Close the manual gas valve in the gas line.
12.   Replace the plug on the gas valve inlet test port.
13.   Remove the plug on the gas valve outlet test port.
14.   Install the manometer in the gas inlet test port on the gas valve.
15.   Open the gas line valve.
16.   The manometer should read a manifold gas pressure of 0.
17.   Turn the power on and set the thermostat to heat.
18.   Read the manifold gas pressure after the furnace lights.

588

19. Turn the thermostat to OFF.
20. Close the manual gas valve in the gas line.
21. Replace the plug on the gas valve outlet test port.

Gas Pressure Measurements		
	Furnace Off	Furnace Operating
Gas Inlet Pressure		
Manifold Pressure		

LAB 57.4 MEASURING DRAFT

LABORATORY OBJECTIVE

You will measure the draft pressure in a Category I furnace vent.

FUNDAMENTALS OF HVACR TEXT REFERENCE

Unit 57 Gas Furnace Installation

LABORATORY NOTES

You will operate a category I furnace and test the draft pressure

REQUIRED TOOLS, EQUIPMENT, AND MATERIALS

Safety glasses
Hand tools
Operating Category I gas furnace connected vent
Magnehelic, inclined manometer, or draft gauge

SAFETY REQUIREMENTS

A. Wear safety glasses
B. Use safety glasses while drilling hole.
C. Handle carefully as test probe may become hot.

PROCEDURE

1. If the vent stack does not have a hole for testing vent pressure, drill a 5/16" hole in the vent.
2. Operate the furnace assigned by the instructor.
3. Insert the test probe into the hole.
4. The vent should have a negative pressure of at least 0.02" wc.

LAB 57.5 MEASURE GAS USAGE

LABORATORY OBJECTIVE
The purpose of this lab is to demonstrate your ability to determine the gas usage for a furnace.

FUNDAMENTALS OF HVACR TEXT REFERENCE
Unit 57 Gas Furnace Installation

LABORATORY NOTES
You will need to familiarize yourself with gas furnace arrangements and types. Always remember to make the visual check of all system components prior to starting a gas furnace.

REQUIRED TOOLS, EQUIPMENT, AND MATERIALS
Operating gas furnace
Tool kit
Multimeters
Clamp-on ammeter

SAFETY REQUIREMENTS

A. Wear safety glasses
B. Never allow gas flow without a flame. If gas is allowed to build up and then suddenly ignite, this can create a serious hazard.
C. If the furnace contains a pilot, always ensure it is properly lit prior to starting the unit.
D. Check all circuits for voltage before doing any service work.
E. Stand on dry nonconductive surfaces when working on live circuits.
F. Never bypass any electrical protective devices.

PROCEDURE
Step 1
Collect the gas furnace data and fill in the chart.

Gas Furnace And Fuel Meter Data

Furnace Make:		Model Number:	Blower Motor Amperage Rating:
Number of Burners:			
Burner Orifice Size:			
Input Rating in BTUH:			
Burner type (circle one)	Atmospheric	Induced draft	Power
Meter type (circle	1/2 ft³	2 ft³	

Step 2

Complete the Prestart Check List and prepare to start the furnace as outlined in the procedure provided in Lab 51.1.

Step 3

After finishing the Prestart Check List and then starting the furnace to check its operation, you can prepare to clock the main burner. You will follow the same procedure of working through a check list. This will help you to keep track of your progress. Complete the Clock Main Burner Check List and make sure to check each step off in the appropriate box as you finish it.

Clock Main Burner Check List

STEP	PROCEDURE	CHECK
1	Isolate the main burner.	
2	Leave the pilot on for normal operation.	
3	Operate the main burner to be tested.	

4	Observe the rotating dial of the gas meter. This is typically either a 1/2 ft^3 or 2 ft^3 dial	
5	Measure the time in seconds for one revolution of the dial. Time in seconds =	
6	Calculate cubic feet per hour (CFH). CFH = 3600/seconds x 1/2 ft^3 or 2 ft^3 dial) =	
7	Calculate actual burner input. The approximate heating value for natural gas is 1000 BTU per ft^3. You may contact the gas supplier to obtain a more accurate value if desired.	
	Actual burner input = CFH x 1000 BTU/ft^3 = BTUH	
8	Compare actual measured input with the nameplate input rating.	
9	If there is a difference of greater than 5% between the actual measured input as compared to the nameplate input rating, then the gas manifold pressure should be checked.	

LAB 57.6 GAS FURNACE COMBUSTION TESTING

LABORATORY OBJECTIVE
The purpose of this lab is to demonstrate your ability to test a gas furnace to determine proper combustion.

FUNDAMENTALS OF HVACR TEXT REFERENCE
Units 57 Gas Furnace Installation

LABORATORY NOTES
Combustion testing is primarily used when setting power burners on which there is an adjustment that will allow total over-combustion to the burners. The traditional standard setting is at 50% excess air to ensure complete combustion. Consult the manufacturer recommendation for the burner, furnace, or boiler type for more accurate settings. When using one of the newer electronic combustion analyzers, the appearance of CO will show when complete combustion is taking place.

REQUIRED TOOLS, EQUIPMENT, AND MATERIALS
Operating gas furnace
Tool kit
Combustion analyzer
Temperature sensor

SAFETY REQUIREMENTS
A. Wear safety glasses

B. Never allow gas flow without a flame. If gas is allowed to build up and then suddenly ignite, this can create a serious hazard.

C. If the furnace contains a pilot, always ensure it is properly lit prior to starting the unit.

D. Vent pipes will be hot with the furnace operating. Be careful not to touch hot surfaces when inserting, removing, or reading instruments.

PROCEDURE

Step 1
Collect the gas furnace data and fill in the chart.

Gas Furnace Unit Data

Furnace Make:		Furnace model number	Blower Motor Amperage Rating:	
Furnace Configuration (circle one)	Upflow	Counterflow	Lowboy	Horizontal
Blower Type (circle one)	Direct drive	Belt drive		
Blower Motor Type (circle one)	Single speed split phase	Multiple speed PSC	Constant Torque multiple tap	ECM variable speed
System Type (circle one)	Heating only	Heating and humidifying	Heating and cooling	Heating, humidifying, and cooling
Burner Type (circle one)	Atmospheric	Induced draft	Power burner	

Step 2

Complete the Pretart Check List and prepare to start the furnace as outlined in the procedure provided in Lab 51.1. Before starting the furnace complete the Preparation for Combustion Test Check List.

Preparation For Combustion Test Check List

STEP	PROCEDURE	CHECK
1	Prior to starting the furnace locate CO_2 test openings and the temperature probe location for an undiluted vent gas sample.	
2	Insert the thermometer probe and check the CO_2 for position.	
3	Start the furnace and allow it to run until the vent gas temperature is at its normal maximum.	

Step 3

With the furnace now operating, begin taking your initial readings and record them in the Combustion Test Chart.

1. Initial readings are taken prior to making any adjustments.
2. For the yellow flame test, close the primary air shutter until a lazy flame is present.
3. For the excess air test, open the primary air shutter to a maximum amount.
4. For the 8% CO_2 test, close the air shutter to obtain 8%.
5. Calculate the combustion efficiency using a slide rule calculator with the natural gas slide.
6. Which flame is the most efficient?

Combustion Test Chart

Combustion Test Readings	Test 1 (initial)	Test 2 (yellow)	Test 3 (excess)	Test 4 (8% CO_2)
CO_2				
Actual Stack Temperature				
Net Stack (gross–room)				
Combustion Efficiency				

LAB 57.7 MEASURE GAS FURNACE THERMAL EFFICIENCY

LABORATORY OBJECTIVE
The purpose of this lab is to demonstrate your ability to measure thermal efficiency of a gas furnace and to draw a correlation, if there is any, between the airflow through the furnace and the thermal efficiency of the furnace.

FUNDAMENTALS OF HVACR TEXT REFERENCE
Unit 57 Gas Furnace Installation

LABORATORY NOTES
Thermal efficiency is BTU output divided by BTU input. Input is the measure of gas consumed while output is measured by the sensible heat formula which is (temperature in °F) x (1.08) x (airflow in CFM). The thermal efficiency will be calculated at three different airflows.

REQUIRED TOOLS, EQUIPMENT, AND MATERIALS
Operating gas furnace with variable speed blower
Tool kit
Temperature sensor
Airflow hood

SAFETY REQUIREMENTS
A. Wear safety glasses
B. Never allow gas flow without a flame. If gas is allowed to build up and then suddenly ignite, this can create a serious hazard.
C. If the furnace contains a pilot, always ensure it is properly lit prior to starting the unit.
D. Vent pipes will be hot with the furnace operating. Be careful not to touch hot surfaces when inserting, removing, or reading instruments.

PROCEDURE
Step 1
Collect the gas furnace data and fill in the chart.

Gas Furnace Unit Data

Furnace Make:		Furnace model number	Blower Motor Amperage Rating:	
Furnace Configuration (circle one)	Upflow	Counterflow	Lowboy	Horizontal
Blower Type (circle one)	Direct drive	Belt drive		
Blower Motor Type (circle one)	Single speed split phase	Multiple speed PSC	Constant Torque multiple tap	ECM variable speed
System Type (circle one)	Heating only	Heating and humidifying	Heating and cooling	Heating, humidifying, and cooling
Burner Type (circle one)	Atmospheric	Induced draft	Power burner	

Step 2

Complete the Prestart Check List and prepare to start the furnace as outlined in the procedure provided in Lab 51.1. Before starting the furnace complete the Preparation for Efficiency Test Check List.

Preparation For Efficiency Test Check List

STEP	PROCEDURE	CHECK
1	Prepare to measure the airflow on the inlet and/or the outlet of the furnace using an air flow hood. It may be necessary to build a bracket to install the airflow hood.	
2	If the airflow hood is located on the outlet, the airflow measurement must be taken prior to starting the furnace as the airflow hood will be damaged if exposed to temperatures above 140°F (Fan only operation).	
3	Operate the furnace in the normal heating mode. The airflow hood may be left on the return air during this test. Airflow will typically go down as the air is heated due to the air restriction of the furnace and the increased air volume of the heated air.	

Step 3

With the furnace now operating, begin taking your initial readings and record them in the Efficiency Test Chart.

1. Initial readings are taken prior to making any adjustments.
2. For the reduced airflow test, reduce the blower speed or throttle the air damper.
3. For the increased airflow test, increase the blower speed or open the air damper.
4. For the 8% CO_2 test, close the air shutter to obtain 8%.
5. Calculate the combustion efficiency using a slide rule calculator with the natural gas slide.

Efficiency Test Chart

Efficiency Test Readings	Test 1 (initial)	Test 2 (reduced airflow)	Test 3 (increased airflow)
Air Inlet Temperature			
Air outlet Temperature			
Airflow (CFM)			

Step 4

Calculate the BTU output for the three test conditions using the formula:

(outlet temperature – inlet temperature) x (1.08) x (airflow in CFM)

	Test 1 (initial)	Test 2 (reduced airflow)	Test 3 (increased airflow)
BTU Output			

BTU Output

Step 5

Calculate the efficiency for the three test conditions using the formula:

(BTU output from Step 4)/(Burner rated BTU input)

Calculated Efficiency

	Test 1 (initial)	Test 2 (reduced airflow)	Test 3 (increased airflow)
Calculated Efficiency			

1. What correlation can be drawn between the thermal efficiency and the airflow of the furnace?

LAB 57.8 MEASURING TEMPERATURE RISE

LABORATORY OBJECTIVE
You will measure the temperature rise on a gas furnace.

FUNDAMENTALS OF HVACR TEXT REFERENCE
Unit 57 Gas Furnace Installation

REQUIRED TOOLS, EQUIPMENT, AND MATERIALS
Safety glasses
Hand tools
Operating gas furnace

SAFETY REQUIREMENTS
Wear safety glasses

PROCEDURE

1. Measure the return air temperature.
2. Measure the supply air temperature at a point that will not "see" the heat exchanger and pick up radiant heat.
3. Calculate the temperature rise: Temp Rise = Supply air temp - return air temp.
4. Turn the fan switch on the thermostat to "On".
5. This will cause an increase in fan speed on most furnaces.

Furnace Temperature Rise	
Manufacturer's Temperature Rise Specification	
Return Air Temperature	
Supply Air Temperature	
Furnace Operating Temperature rise (Supply Air Temp – Return Air Temp)	

LAB 57.9 GAS FURNACE STARTUP

LABORATORY OBJECTIVE
The purpose of this lab is to demonstrate your ability to go through the necessary sequence for a typical gas furnace startup.

FUNDAMENTALS OF HVACR TEXT REFERENCE
Unit 57 Gas Furnace Installation

LABORATORY NOTES
You will need to familiarize yourself with gas furnace arrangements and types. Always remember to make the visual check of all system components prior to starting a gas furnace.

REQUIRED TOOLS, EQUIPMENT, AND MATERIALS
Operating gas furnace
Tool kit
Multimeters
Clamp-on ammeter

SAFETY REQUIREMENTS

A. Wear safety glasses

B. Never allow gas flow without a flame. If gas is allowed to build up and then suddenly ignite, this can create a serious hazard.

C. If the furnace contains a pilot, always ensure it is properly lit prior to starting the unit.

D. Check all circuits for voltage before doing any service work.

E. Stand on dry nonconductive surfaces when working on live circuits.

F. Never bypass any electrical protective devices.

PROCEDURE

Step 1

Collect the gas furnace data and fill in the chart.

Gas Furnace Unit Data

Furnace Make:		Furnace model number	Blower Motor Amperage Rating:	
Furnace Configuration (circle one)	Upflow	Counterflow	Lowboy	Horizontal
Blower Type (circle one)	Direct drive	Belt drive		
Blower Motor Type (circle one)	Single speed split phase	Multiple speed PSC	Constant Torque multiple tap	ECM variable speed
System Type (circle one)	Heating only	Heating and humidifying	Heating and cooling	Heating, humidifying, and cooling
Burner Type (circle one)	Atmospheric	Induced draft	Power burner	

Step 2

Complete the Prestart Check List and make sure to check each step off in the appropriate box as you finish it. This will help you to keep track of your progress.

Prestart Check List

STEP	PROCEDURE	CHECK
1	Turn thermostat down below the room temperature.	
2	Make sure all power to the furnace is off. Lock and tag the power panel before removing any parts.	
3	Vent connector connected with three screws per joint.	
4	Fuel line installed properly, with no apparent leaks.	
5	Spin all fans to be sure they are loose and turn freely.	
6	Combustion area free from debris.	
7	Electrical connections complete – main power.	
8	Electrical connections complete – thermostat.	

9	All doors and panels available and in place.	
10	Thermostat installed and operating correctly.	

Step 3

After finishing the Prestart Check List you are now ready to start the furnace and check its operation. You will follow the same procedure of working through a check list. This will help you to keep track of your progress. Complete the Start-Up Check List and make sure to check each step off in the appropriate box as you finish it.

Start-Up Check List

STEP	PROCEDURE	CHECK
1	Check the main fuse or breaker and the amperage rating and make sure it is the correct rating for the furnace.	
2	Measure the incoming supply voltage.	
3	If the Steps 1 & 2 have been completed and are satisfactory, then turn on the supply power for the furnace.	
4	Set the thermostat to – Fan On - to obtain the fan only operation and observe that the blower is operating. If the thermostat is not so equipped then consult with your instructor.	

606

5	Light the pilot if so equipped, following the manufacturer's instructions provided with the furnace.	
6	Turn the fan off and the thermostat to heat.	
7	Adjust the thermostat setting to 10°F above the room temperature.	
8	Observe the flame sequence begin and the flame comes on.	
9	Observe that the blower begins running after the heat exchanger comes up to temperature.	
10	Use a clamp-on ammeter to measure the blower motor amperage and compare this value to its rating.	
11	Turn the thermostat setting down below the room temperature and observe the flame go off.	
12	Observe the blower stop running approximately three minutes after the burner turns off.	

LAB 58.1 GAS FURNACE PREVENTATIVE MAINTENANCE (PM)

LABORATORY OBJECTIVE
The purpose of this lab is to demonstrate your ability to conduct the proper preventative maintenance procedure for a gas furnace.

FUNDAMENTALS OF HVACR TEXT REFERENCE
Unit 58 Troubleshooting Gas Furnaces

LABORATORY NOTES
You will need to familiarize yourself with gas furnace arrangements and types. Always remember to make the visual check of all system components prior to starting a gas furnace.

REQUIRED TOOLS, EQUIPMENT, AND MATERIALS
Operating gas furnace
Tool kit
Multimeters
Clamp-on ammeter

SAFETY REQUIREMENTS
A. Wear safety glasses
B. Never allow gas flow without a flame. If gas is allowed to build up and then suddenly ignite, this can create a serious hazard.
C. If the furnace contains a pilot, always ensure it is properly lit prior to starting the unit.
D. Check all circuits for voltage before doing any service work.
E. Stand on dry nonconductive surfaces when working on live circuits.
F. Never bypass any electrical protective devices.

PROCEDURE

Step 1

Collect the gas furnace data and fill in the chart.

Gas Furnace Unit Data

Furnace Make:		Furnace model number	Blower Motor Amperage Rating:	
Furnace Configuration (circle one)	Upflow	Counterflow	Lowboy	Horizontal
Blower Type (circle one)	Direct drive	Belt drive		
Blower Motor Type (circle one)	Single speed split phase	Multiple speed PSC	Constant Torque multiple tap	ECM variable speed
System Type (circle one)	Heating only	Heating and humidifying	Heating and cooling	Heating, humidifying, and cooling
Burner Type (circle one)	Atmospheric	Induced draft	Power burner	

Step 2

Complete the Prestart Check List and prepare to start the furnace as outlined in the procedure provided in Lab 51.1. With the furnace operating complete the Running Check List.

Running Check List

STEP	PROCEDURE	CHECK
1	Observe a normal sequence of operation (burner on, fan on, burner off, fan off). Check for any unusual noise – bearing noise.	
2	Inspect operating pilot and note the color, size, shape, and position.	
3	Inspect burner light off for flame lifting, floating, noise and smooth ignition.	
4	Measure gas manifold pressure.	

Step 3

After finishing the Running Check List, shut down the furnace. Make sure all power to the furnace is off. Lock and tag the power panel before removing any parts. Complete the Blower Maintenance Check List.

Blower Maintenance Check List

STEP	PROCEDURE	CHECK
1	Make sure that the power to the blower motor is secured. Verify this with a multimeter voltage test.	
2	Remove the wires from the blower motor at an accessible location.	
3	Remove screws or bolts securing the blower assembly.	
4	Inspect belt for cracks and signs of wear.	
5	Inspect pulleys for wear, grooving, and alignment.	
6	Spin blower by hand and observe pulleys turn.	
7	Inspect for pulley wobble and alignment.	
8	Listen for bearing noise, drag, or movement.	
9	Inspect blower shaft for signs of wear.	
10	Clean blower and motor with air pressure, brushes, scrapers and cleaning solution as required.	
11	Reassemble blower, taking care to check pulley alignment and correct belt tension.	
12	Check the condition of the heat exchanger before re-installing the blower.	

Step 4

After finishing the Blower Maintenance Check List you can perform the necessary burner maintenance. Again, make sure all power to the furnace is off. Lock and tag the power panel before removing any parts. Complete the Gas Burner Maintenance Check List.

Gas Burner Maintenance Check List

STEP	PROCEDURE	CHECK
1	Remove and clean pilot assembly.	
2	Clean and inspect flame sensor and igniter.	
3	Remove and clean main burners, inspect and mark burners for original location. *Note:* All burners are not interchangeable.	
4	Remove vent connector, draft diverter, and any flue baffles.	
5	Vacuum and brush all soot, rust, and solid particles from the fire side of the heat exchanger.	
6	Insert a light into the combustion area.	
7	If possible turn off the lights in the furnace room.	
8	With the light in each burner, inspect the heat exchanger from both the fan side and the plenum side for holes.	

9	Reinstall the blower assembly after the heat exchanger check.	
10	Reinstall main and pilot burners.	
11	Reconnect vent components.	
12	Prepare to start the furnace for a running check following the start-up procedure from Lab 51.1.	
13	Light pilot and adjust for proper size, configuration, and location.	
14	Turn on main burner and adjust gas pressure as required.	
15	Adjust air shutter to obtain correct flame color and CO_2.	

LAB 58.2 TESTING STANDING PILOT SAFETY DEVICES

LABORATORY OBJECTIVE
You will measure the millivoltage output of a thermocouple.

FUNDAMENTALS OF HVACR TEXT REFERENCE
Unit 58 Troubleshooting Gas Furnaces

REQUIRED TOOLS, EQUIPMENT, AND MATERIALS
Safety glasses
Multimeter that can read DC millivolts
Operating gas furnace with a standing pilot light and thermocouple
Thermocouple millivoltage adapter

SAFETY REQUIREMENTS
A. Wear safety glasses
B. Never allow gas flow without a flame.

PROCEDURE
1. Remove the thermocouple from the gas valve connection.
2. Install the thermocouple millivoltage adapter into the threaded thermocouple connector on the gas valve.
3. Install the thermocouple into the thermocouple adapter.
4. Turn the gas valve know to "Pilot," press in the knob, and light the pilot light. Hold the knob in for at least a minute after lighting the pilot light.
5. Release the knob, the pilot light should remain burning.
6. Measure the millivolt output of the thermocouple on the adapter.

LAB 58.3 CHECKING INTERMITTENT PILOT IGNITION SYSTEMS

LABORATORY OBJECTIVE
You will observe the operating sequence of an intermittent pilot ignition system and simulate common failure symptoms.

FUNDAMENTALS OF HVACR TEXT REFERENCE
Unit 58 Troubleshooting Gas Furnaces

REQUIRED TOOLS, EQUIPMENT, AND MATERIALS
Safety glasses
Electrical tools
Operating gas furnace with intermittent pilot ignition system

SAFETY REQUIREMENTS

A. Wear safety glasses

B. Check all circuits for voltage before doing any service work.

C. Never allow gas flow without a flame

PROCEDURE

1. Study the wiring diagram of a furnace using an intermittent spark pilot system.
2. Explain the sequence of operation to an instructor.
3. Run the furnace and observe the operating sequence.
4. Turn the furnace off and pull a wire loose from the draft switch.
5. Turn the furnace on and set the thermostat to call for heat.
6. What happens?
7. Turn the furnace off, replace the wire on the draft switch, and turn the furnace back on.
8. With the main burners going, turn the gas valve to the OFF position. What happens?
9. Does the system try to relight?
10. Run the furnace. When all burners are lit and the spark is no longer present, pull the wire off of the flame sensor. What happens? Why? Turn the unit off and replace the wire.

LAB 58.4 CHECK/TEST/REPLACE HOT SURFACE IGNITER

LABORATORY OBJECTIVE
The purpose of this lab is to check, test, and replace a hot surface igniter for a gas furnace

FUNDAMENTALS OF HVACR TEXT REFERENCE
Unit 58 Troubleshooting Gas Furnaces

LABORATORY NOTES
Hot surface ignition and flame proving is the most common method of burner control in modern gas furnaces. The major replacement item in this type of system is the hot surface igniter.

Hot surface igniters are sometimes called glow coils because they glow when energized. An igniter that does not glow is the first indication that it has failed. A typical startup sequence will occur as follows: the thermostat calls for heat, the draft fan comes on, the hot surface igniter glows, the gas valve opens (you can hear the click), the gas ignites, the gas flame reaches the flame rod and proves, and the normal heat mode is in progress. When the surface igniter fails to glow, you hear the click of the gas valve opening but no flame will appear and since the flame does not prove, the gas valve will close.

REQUIRED TOOLS, EQUIPMENT, AND MATERIALS
Operating gas furnace
Tool kit
Multimeter

SAFETY REQUIREMENTS
A. Wear safety glasses
B. Check all circuits for voltage before doing any service work.
C. Never allow gas flow without a flame.

PROCEDURE
Step 1
Familiarize yourself with the gas furnace components.
1. Complete Gas Furnace Component Identification Check List and make sure to check each step off in the appropriate box as you finish it. This will help you to keep track of your progress.

Gas Furnace Component Identification Check List

STEP	PROCEDURE	CHECK
1	Locate the inducer fan and induced draft vent system.	
2	Locate the hot surface igniter.	
3	Locate the flame rod for the flame proving system.	
4	Locate the electronic module or the flame system control.	
5	Locate and write down the terminal on the module that the flame rod wire connects to.	
6	Locate the plug or wire connection from the control system to the hot surface igniter.	

616

Step 2

1. Complete the Normal Ignition Sequence Trial Check List and make sure to check each step off in the appropriate box as you finish it.

Normal Ignition Sequence Trial Check List

STEP	PROCEDURE	CHECK
1	Adjust the thermostat to call for heat.	
2	Observe the draft inducer fan come on.	
3	Does the hot surface igniter? (this takes about a minute after the fan comes on) (circle one) YES NO	
4	If the surface igniter glows, the gas valve should open and the burner flame should light and the furnace should operate normally and the trial is complete. If the surface igniter does not glow and the burner fails to light, proceed to the next step.	
5	If the surface igniter does not glow, did you hear the gas valve click as it opened? (circle one) YES NO	

6	If the surface igniter did not glow, the gas valve clicked open, and the burner did not light, then the surface igniter must be checked.	
7	Make sure all power to the furnace is off. Lock and tag the power panel before removing any parts.	
8	Disconnect the wire nuts from the surface igniter leads or unplug it and test the surface igniter for continuity with a multimeter.	
9	If there is continuity, the surface igniter may not be the problem and the voltage to the igniter will need to be verified. This will involve troubleshooting and possibly replacing the module for the flame system control.	
10	If there is no continuity (infinite resistance), then the surface igniter has an open and it will need to be replaced.	
11	Obtain a replacement surface igniter of the same configuration for both the shape of the coil and the connection and replace the defective surface igniter (make sure to supply any shield supplied to protect the coil).	
12	After installing the new surface igniter, turn the power on for the furnace and cycle it through a normal stat mode with the thermostat calling for heat. Observe the main burner ignition and proper operation.	

LAB 58.5 TROUBLESHOOTING HOT SURFACE IGNITION SYSTEMS

LABORATORY OBJECTIVE
Given a hot surface ignition gas furnace with an ignition problem, you will identify the problem, its root cause, and recommend corrective action.

FUNDAMENTALS OF HVACR TEXT REFERENCE
Unit 58 Troubleshooting Gas Furnaces

REQUIRED TOOLS, EQUIPMENT, AND MATERIALS
Safety glasses
Hand tools
Manometer
Multimeter
Hot surface ignition gas furnace with a problem

SAFETY REQUIREMENTS
A. Wear safety glasses

B. Never allow gas flow without a flame.

C. Check all circuits for voltage before doing any service work.

PROCEDURE
Troubleshoot the hot surface ignition gas furnace assigned by the instructor. Be sure to be complete in your description of the problem. You should include:

- What is the furnace is doing wrong?
- What component or condition is causing this?
- What tests did you perform that told you this?
- How would you correct this?

LAB 58.6 TROUBLESHOOTING DSI CONTROL SYSTEMS

LABORATORY OBJECTIVE
Given a DSI gas furnace with an ignition problem, you will identify the problem, its root cause, and recommend corrective action.

FUNDAMENTALS OF HVACR TEXT REFERENCE
Unit 58 Troubleshooting Gas Furnaces

REQUIRED TOOLS, EQUIPMENT, AND MATERIALS
Safety glasses
Hand tools
Manometer
Multimeter
DSI gas furnace with a problem

SAFETY REQUIREMENTS

A. Wear safety glasses

B. Check all circuits for voltage before doing any service work.

C. Never allow gas flow without a flame.

PROCEDURE
Troubleshoot the DSI gas furnace assigned by the instructor. Be sure to be complete in your description of the problem. You should include:

- What is the furnace is doing wrong?
- What component or condition is causing this?
- What tests did you perform that told you this?
- How would you correct this?

LAB 58.7 VENT PROBLEMS

LABORATORY OBJECTIVE
Given a gas furnace with a venting problem, you will identify the problem, its root cause, and recommend corrective action.

***FUNDAMENTALS OF HVACR* TEXT REFERENCE**
Unit 57 Gas Furnace Installation
Unit 58 Troubleshooting Gas Furnaces

REQUIRED TOOLS, EQUIPMENT, AND MATERIALS
Safety glasses
Hand tools
Manometer
Multimeter
DSI gas furnace with a problem

SAFETY REQUIREMENTS
  A.   Wear safety glasses
  B.   Never allow gas flow without a flame.
  C.   Check all circuits for voltage before doing any service work.

PROCEDURE
Troubleshoot the DSI gas furnace assigned by the instructor. Be sure to be complete in your description of the problem. You should include:

- What is the furnace is doing wrong?
- What component or condition is causing this?
- What tests did you perform that told you this?
- How would you correct this?

LAB 58.8 INSPECTING HEAT EXCHANGERS

LABORATORY OBJECTIVE
You will learn to identify defects in furnace heat exchangers.

FUNDAMENTALS OF HVACR TEXT REFERENCE
Unit 58 Troubleshooting Gas Furnaces

REQUIRED TOOLS, EQUIPMENT, AND MATERIALS
Safety glasses
Hand tools
Gas furnace

SAFETY REQUIREMENTS
Wear safety glasses

PROCEDURE
Inspect the heat exchanger(s) assigned by the instructor. Note any cracks, holes, or other aberrations. Also remember to look for heat exchangers, which are sooted up and clean as necessary. Different inspection techniques are listed below.

Flamedance
Normally, the heat exchanger keeps the combustion gas completely separate from the air passing over the heat exchanger; however, a large enough crack can let air blow into the heat exchanger and blow the flames around. If you see a lot of flame movement when the fan comes on, there could be a hole in the heat exchanger. This is especially true if the flames blow in just one of the sections.

Fans Eye View
Probably the best overall view of the heat exchanger is looking at it from the blower's point of view. To do this, you need to remove the blower. Removing the blower is not as difficult as it sounds because most furnace blowers slide in and out and are secured by a handful of screws. Be sure to turn power off and disconnect the wiring to the blower BEFORE removing it. After the blower has been removed, look up into the heat exchanger for cracks and holes. Often, you will need a flashlight and an inspection mirror to really see up inside the heat exchanger. Remember that the unit is still hot. DON'T TOUCH THE SIDES OF THE HEAT EXCHANGER. If you smell burning flesh, remove your hands from the furnace. Pay particular attention to bends and crimps in the heat exchanger because these receive the greatest thermal stress. Also look for rusty areas.

Borescope
If you are still uncertain, the next step is to remove the burners so that you can get a better look inside the heat exchanger. Remove the burners to gain clear access to the heat exchanger. Use a borescope to go up inside each section or tube looking for cracks,

discolored metal, or excessive rust. A large amount of rust on top of the burners and piled up around the bottom of the heat exchanger could mean that it has rusted through. Also, notice any visible differences between the sections. On natural draft furnaces you can also go in through the draft hood or draft diverter.

Smoke Bomb

Finally, there is one sure-fire check: a smoke bomb. With the unit operating, put a smoke bomb in the bottom of the heat exchanger. If there are cracks or holes, smoke will be released into the ductwork and into the house. This can be somewhat unsettling to the homeowners, so it is usually reserved for last. Having smoke pour out of the heating registers produces high levels of anxiety among finicky customers but it does yield a rather positive diagnosis. After you have filled the customer's house with smoke, inform them that that could be carbon monoxide and they could not wake up morning. Then suggest that they contact the sales office because you are disabling their furnaces so that it won't work. Finally, tell them to have a nice day.

90% Furnaces

Suppose when you pulled the blower out all you saw was a coil? You must be checking a high efficiency furnace, which uses a condensing recuperative coil. In this case, the only thing a visual inspection from underneath can possibly show would be possibly a leak in the recuperative coil. If you see water dripping, it could be a leaky recuperative coil. BE SURE that it is not simply a leaky drain. It would be quite embarrassing to condemn a furnace because of a leaky drain.

Draft Pressure Check

On many high efficiency furnaces the heat exchanger is too narrow to see up inside, the underside is hidden by the recuperative coil, and the flames may not be readily visible. What then? Watching for changes in the vent draft pressure of a Category I furnace can also indicate a leak in the heat exchanger. Category I furnaces operate with a negative draft pressure. A leak in the heat exchanger can allow air from the indoor blower to enter the heat exchanger, causing an increase in the pressure inside the heat exchanger and vent. If the pressure in the vent changes from a negative pressure to a positive pressure after the indoor blower starts, there is a leak in the heat exchanger.

LAB 58.9 GAS COMBUSTION EFFICIENCY

LABORATORY OBJECTIVE
You will measure the combustion efficiency of an operating gas furnace.

FUNDAMENTALS OF HVACR TEXT REFERENCE
Unit 58 Troubleshooting Gas Furnaces

REQUIRED TOOLS, EQUIPMENT, AND MATERIALS
Safety glasses
Hand tools
Operating gas furnace
Combustion efficiency analyzer

SAFETY REQUIREMENTS
A. Wear safety glasses
B. While drilling hole.
C. Handle probe carefully as it may become very hot.

PROCEDURE
1. If there is not already a hole in the vent, drill a hole large enough to inset the probe of the combustion analyzer.
2. Turn on the furnace and wait for the fan to come on.
3. *Using Traditional Hourglass Combustion Analyzers !*
   a. Measure the temperature of the air around the furnace. Measure the stack temperature.
   b. Calculate the net stack temperature = stack temperature N ambient temperature Measure the CO_2% or the O_2%.
   c. Line up the O_2% and net stack temperature on the efficiency calculator to get combustion efficiency.

Using Electronic Analyzers

Insert probe into hole and follow manufacturer's instructions.

	Natural Draft	80% Induced Draft	90% Induced Draft
Fuel			
Stack Temperature			
Ambient Temperature			
Net Stack Temperature			
CO2 %			
O2%			
CO ppm free air			
Efficiency			

LAB 58.10 GAS FURNACE SEASONAL START-UP AND SYSTEM CHECK

LABORATORY OBJECTIVE
Given a gas furnace, you will go through the electrical pre-start checks, the gas pre-start checks and the mechanical checks. You will then begin an initial power up and check the voltage and amp draw. You will then perform gas operational checks and airflow operational checks.

FUNDAMENTALS OF HVACR TEXT REFERENCE
Unit 58 Troubleshooting Gas Furnaces

REQUIRED TOOLS, EQUIPMENT, AND MATERIALS
Safety glasses
Hand tools
Manometer
Multimeter

SAFETY REQUIREMENTS
A.   Wear safety glasses
B.   Never allow gas flow without a flame.
C.   Check all circuits for voltage before doing any service work.

PROCEDURE
Complete all pre-start and operational checks and fill in the chart provided below.

Gas Furnace Start and Check Prestart Check List

Electrical Prestart		Gas Prestart Checks		Mechanical Checks	
Minimum Supply		Indoor Temperature		Shipping Materials Removed	
Maximum Supply		Type of Gas		Blower Fan Spins	
Evaporator Fan FLA		Minimum Specified gas line pressure		Air Filter in Place	
MCA (Min Cir Amps)		Maximum Specified Gas Line pressure			
Wire sized to MCA		Actual gas line pressure with unit off			
Max Fuse Size		Specified Temperature Rise			
Actual Fuse Size					

Initial Power-up

Voltage at unit while running			
Amp draw on power wire			
Gas Operational Checks		**Airflow Operational Checks**	
Incoming pressure with unit off		Total External Static Specified	
Incoming pressure with unit operating		Measured total external static measured	
Gas Line pressure drop		Airflow CFM Specified	
Manifold pressure with unit off		Measured CFM	
Manifold pressure with unit operating		Return air temperature	
		Supply air temperature	
		Temp Rise	

LAB 58.11 TROUBLESHOOTING GAS FURNACE SCENARIO

LABORATORY OBJECTIVE
Given a gas furnace with a problem, you will identify the problem, its root cause, and recommend corrective action. Labs 54.7 through 54.12 will all follow the same procedure, however each lab will present a different problem scenario based upon the type of furnace and corrective action required.

FUNDAMENTALS OF HVACR TEXT REFERENCE
Unit 58 Troubleshooting Gas Furnaces

REQUIRED TOOLS, EQUIPMENT, AND MATERIALS
Safety glasses
Hand tools
Manometer
Multimeter
Gas furnace with a problem

SAFETY REQUIREMENTS
A.   Wear safety glasses
B.   Never allow gas flow without a flame.
C.   Check all circuits for voltage before doing any service work.

PROCEDURE
Troubleshoot the gas furnace assigned by the instructor. Be sure to be complete in your description of the problem. You should include:

- What is the furnace is doing wrong?
- What component or condition is causing this?
- What tests did you perform that told you this?
- How would you correct this?

LAB 59.1 WARM-AIR OIL FURNACE COMPONENTS

LABORATORY OBJECTIVE
You will identify the components on a warm-air oil-fired furnace.

FUNDAMENTALS OF HVACR TEXT REFERENCE
Unit 59 Oil-Fired Heating Systems

REQUIRED TOOLS, EQUIPMENT, AND MATERIALS
Safety glasses
Warm-Air
Oil-fired furnace

SAFETY REQUIREMENTS
Wear safety glasses

PROCEDURE
You will examine the warm-air oil-fired furnace assigned by the instructor and identify the listed components.

- Flame retention burner
- Burner blower
- Oil pump
- Ignition transformer
- Electrodes
- Oil burner nozzle
- Heat exchanger
- Refractory
- Oil burner primary control
- Cadmium sulfide cell (Cad Cell)
- Oil furnace fan and limit switch

LAB 60.1 OIL BURNER TUNE-UP

LABORATORY OBJECTIVE

The purpose of this lab is to demonstrate your ability to conduct the proper burner tune-up for an oil-fired furnace.

FUNDAMENTALS OF HVACR TEXT REFERENCE

Unit 60 Residential Oil Heating Service

LABORATORY NOTES

You will need to familiarize yourself with oil furnace arrangements and types. Always remember to make the visual check of all system components prior to starting an oil furnace.

REQUIRED TOOLS, EQUIPMENT, AND MATERIALS

Operating oil furnace
Tool kit
Multimeters
Clamp-on ammeter

SAFETY REQUIREMENTS

A. Wear safety glasses

B. never allow fuel oil to flow into the combustion chamber without a flame. If fuel oil is allowed to build up and then suddenly ignite, this can create a serious hazard.

C. Check all circuits for voltage before doing any service work.

D. Stand on dry nonconductive surfaces when working on live circuits.

E. Never bypass any electrical protective devices.

PROCEDURE

Step 1

Collect the oil-fired furnace data and fill in the chart.

Oil-Fired Furnace Unit Data

Furnace Make:	Furnace Model Number:	Burner make	Burner model number	Nozzle GPH
Nozzle spray pattern (circle one)	Solid		Hollow	
Nozzle spray angle				
Blower Motor Amp Rating				
Furnace Configuration (circle one)	Upflow	Counterflow	Lowboy	Horizontal
Blower Type (circle one)	Direct drive		Belt drive	
Blower Motor Type (circle one)	Single speed split phase	Multiple speed PSC	Constant Torque multiple tap	ECM variable speed

Step 2

Complete the Pre-Start Check List and the Start-Up Check List as outlined in the procedure provided in Lab 55.1. After finishing the Start-Up Check List, shut down the furnace. Make sure all power to the furnace is off. Lock and tag the power panel before removing any parts. Complete the Oil Burner Inspection And Tune-Up Check List.

Oil Burner Inspection And Tune-Up Check List

STEP	PROCEDURE	CHECK
1	Make sure all power to the furnace is off. Lock and tag the power panel before removing any parts.	
2	Open or remove transformer.	

631

3	Remove nozzle assembly.	
4	Remove and clean electrodes.	
5	Remove nozzle and record nozzle data. Make_____ GPH _____ Angle_____ Pattern _____	
6	Replace the nozzle with a new nozzle that matches the data specifications.	
7	Install and adjust electrodes.	
8	Position electrodes to manufacturer recommended setting. (Typically these dimensions are approximately 1/2 inches above, 1/16 inches forward, and 1/8 inches apart.)	
9	Remove and clean cad cell.	
10	Remove and record resistance of cad cell while the face of the cad cell is covered. Resistance =___Ohms	

11	Remove and record resistance of cad cell while the face of the cad cell exposed to room light. Resistance =___Ohms	
12	Install nozzle assembly.	
13	Slide nozzle assembly to midpoint of forward/backward adjustment.	
14	Swing up or reinstall ignition transformer.	
15	Make sure that the transformer contact springs touch electrodes as the transformer is positioned.	
16	Secure transformer in place with at least one screw for testing.	
17	Turn on power to the furnace and burner and observe ignition. DO not get too close as oil burners can puff back.	
18	Observe a normal continuous burn for one minute.	
19	Adjust the air shutter back and forth slowly. Close to see smoke and then open until smoky flame disappears.	
20	Slide nozzle assembly forward and back until the flame is quiet and no longer jagged and smoky.	
21	Refer to Lab 54.2 for final burner adjustments.	

LAB 60.2 FINAL OIL BURNER ADJUSTMENT

LABORATORY OBJECTIVE
The purpose of this lab is to demonstrate your ability to conduct the proper final burner adjustments after a tune-up for an oil-fired furnace.

FUNDAMENTALS OF HVACR TEXT REFERENCE
Unit 60 Residential Oil Heating Service

LABORATORY NOTES
You will need to familiarize yourself with oil furnace arrangements and types. Always remember to make the visual check of all system components prior to starting an oil furnace.

REQUIRED TOOLS, EQUIPMENT, AND MATERIALS
Operating oil furnace
Tool kit
Draft gauge
Smoke gun
CO analyzer

SAFETY REQUIREMENTS
A. Wear safety glasses

B. Never allow fuel oil to flow into the combustion chamber without a flame. If fuel oil is allowed to build up and then suddenly ignite, this can create a serious hazard.

C. Check all circuits for voltage before doing any service work.

D. Stand on dry nonconductive surfaces when working on live circuits.

E. Never bypass any electrical protective devices.

PROCEDURE
Step 1

Complete the Prestart Check List and the Start-Up Check List as outlined in the procedure provided in Lab 57.1. Complete the Oil Burner Inspection and Start-Up Check List as outlined in Lab 56.1.

Obtain a draft gauge, smoke gun, and CO analyzer to perform final settings on the oil burner.

Final Burner Adjustment Check List

STEP	PROCEDURE	CHECK
1	With the burner operating, use the draft gauge and adjust the barometric damper to obtain a -0.1 over fire draft.	
2	Measure and record the draft at breach of the burner. A restriction of greater than -0.4 through the heat exchanger indicates the heat exchanger may be plugged with soot and needs to be cleaned. Final measured draft at breach _____	
3	If necessary inspect and clean the fire side of the heat exchanger. Make sure all power to the furnace is off. Lock and tag the power panel before removing any parts.	
4	Once cleaning is complete, restart the burner and perform a smoke spot test with the smoke gun. The reading should be a #0.	
5	Install the CO meter into the breach of the furnace.	
6	Readjust the air shutter to obtain 10% CO_2.	

7	Turn the burner on and off, observing several ignitions.	
8	Verify a quick clean quiet ignition.	
9	Remove and put away all test equipment and tools Install all panels and doors.	

LAB 60.3 OIL FURNACE PREVENTATIVE MAINTENANCE

LABORATORY OBJECTIVE
The purpose of this lab is to demonstrate your ability to conduct the proper preventative maintenance for an oil-fired furnace.

***FUNDAMENTALS OF HVACR* TEXT REFERENCE**
Unit 60 Residential Oil Heating Service

LABORATORY NOTES
You will need to familiarize yourself with oil furnace arrangements and types. Always remember to make the visual check of all system components prior to starting an oil furnace.

REQUIRED TOOLS, EQUIPMENT, AND MATERIALS
Operating oil furnace
Tool kit
Multimeters
Clamp-on ammeter
Oil pressure gauge
Vacuum

SAFETY REQUIREMENTS
- A. Wear safety glasses
- B. Never allow fuel oil to flow into the combustion chamber without a flame. If fuel oil is allowed to build up and then suddenly ignite, this can create a serious hazard.
- C. Check all circuits for voltage before doing any service work.
- D. Stand on dry nonconductive surfaces when working on live circuits.
- E. Never bypass any electrical protective devices.

PROCEDURE
Step 1
Collect the oil-fired furnace data and fill in the chart.

Oil-Fired Furnace Unit Data

Furnace Make:	Furnace Model Number:	Burner make	Burner model number	Nozzle GPH
Nozzle spray pattern (circle one)	Solid		Hollow	
Nozzle spray angle				
Blower Motor Amp Rating				
Furnace Configuration (circle one)	Upflow	Counterflow	Lowboy	Horizontal
Blower Type (circle one)	Direct drive		Belt drive	
Blower Motor Type (circle one)	Single speed split phase	Multiple speed PSC	Constant Torque multiple tap	ECM variable speed

Step 2

Complete the Prestart Check List and the Start-Up Check List as outlined in the procedure provided in Lab 57.1. After finishing the Start-Up Check List, shut down the furnace. Make sure all power to the furnace is off. Lock and tag the power panel before removing any parts. Complete the Blower Maintenance Check List.

Blower Maintenance Check List

STEP	PROCEDURE	CHECK
1	Make sure that the power to the blower motor is secured. Verify this with a multimeter voltage test.	
2	Remove the wires from the blower motor at an accessible location.	
3	Remove screws or bolts securing the blower assembly.	

4	Inspect belt for cracks and signs of wear.	
5	Inspect pulleys for wear, grooving, and alignment.	
6	Spin blower by hand and observe pulleys turn.	
7	Inspect for pulley wobble and alignment.	
8	Listen for bearing noise, drag, or movement.	
9	Inspect blower shaft for signs of wear.	
10	Clean blower and motor with air pressure, brushes, scrapers and cleaning solution as required.	
11	Oil motor and blower bearings as required.	

12	Reassemble blower, taking care to check pulley alignment and correct belt tension.	
13	Check the condition of the heat exchanger before re-installing the blower.	

Step 3

After finishing the Blower Maintenance Check List you can perform the necessary oil burner maintenance. Again, make sure all power to the furnace is off. Lock and tag the power panel before removing any parts. Complete the Oil Burner Maintenance Check List.

Oil Burner Maintenance Check List

STEP	PROCEDURE	CHECK
1	Make sure all power to the furnace is off. Lock and tag the power panel before removing any parts.	
2	Remove and clean nozzle assembly.	
3	Replace nozzle with a manufacturer recommended nozzle.	
4	Place the old nozzle into a small zipper bag with the new nozzle box. Write your name on the bag, date it, and leave it with the furnace. The bag will keep it from smelling and then on the next service visit, it will be evident what was put in and what was taken out.	

5	Position electrodes to manufacturer recommended setting. (Typically these dimensions are approximately 1/2 inches above, 1/16 inches forward, and 1/8 inches apart.)	
6	Remove the vent connector, barometric damper, and flue baffles or cleanout plugs.	
7	Remove or swing out burner assembly.	
8	Vacuum and brush all soot, rust, and solid particles from the fire side of the heat exchanger. Be careful not to damage the combustion chamber refractory.	
9	Insert a light into the combustion chamber and if possible turn off the lights in the furnace room.	
10	With the light in as far back as possible, inspect the heat exchanger from the fan side and the plenum side for holes, light, or leaky gaskets.	
11	After inspecting the heat exchanger reinstall the blower.	
12	Reinstall the nozzle assembly while the burner is still swung away from the furnace.	

13	After the burner assembly is back in place the fuel oil system can be checked. Disconnect the supply line from the pump to the burner assembly and install an oil pressure gauge on the pump supply line.	
14	Remove the lockout tag and turn on the burner and check the initial fuel oil pressure. Fuel Oil Pressure = _____	
15	Adjust to the manufacturer recommended fuel oil pressure (typically 100 psig). Fuel Oil Pressure = _____	
16	Observe the burner turn off due to flame failure.	
17	Observe oil pressure gauge holding 85 psig quick cut-off pressure.	
18	Reinstall the original oil supply line to the nozzle assembly.	

19	Check the fan control settings. The fan should turn on at a supply air temperature of approximately 90°F and off at 135 °F.	
20	The maximum temperature limit setting can be checked by running the burner with the fan off and at approximately 200 °F, the burner should cycle off.	
21	After all maintenance is complete, turn the burner on and off, observing several ignitions. Verify a quick clean quiet ignition. Remove and put away all test equipment and tools Install all panels and doors.	

LAB 61.1 OIL-FIRED FURNACE STARTUP

LABORATORY OBJECTIVE

The purpose of this lab is to demonstrate your ability to go through the necessary sequence for a typical oil furnace startup.

FUNDAMENTALS OF HVACR TEXT REFERENCE

Unit 61 Residential Oil Heating Installation

LABORATORY NOTES

You will need to familiarize yourself with oil furnace arrangements and types. Always remember to make the visual check of all system components prior to starting an oil furnace.

REQUIRED TOOLS, EQUIPMENT, AND MATERIALS

Operating forced hot air oil furnace
Tool kit
Multimeters
Clamp-on ammeter

SAFETY REQUIREMENTS

A. Wear safety glasses

B. Never allow fuel oil to flow into the combustion chamber without a flame. If fuel oil is allowed to build up and then suddenly ignite, this can create a serious hazard.

C. Check all circuits for voltage before doing any service work.

D. Stand on dry nonconductive surfaces when working on live circuits.

E. Never bypass any electrical protective devices.

PROCEDURE

Step 1

Collect the oil furnace data and fill in the chart.

Oil Furnace Unit Data

Furnace Make:	Furnace Model Number:	Burner make	Burner model number	Nozzle GPH
Nozzle spray pattern (circle one)	Solid		Hollow	
Nozzle spray angle				
Blower Motor Amp Rating				
Furnace Configuration (circle one)	Upflow	Counterflow	Lowboy	Horizontal
Blower Type (circle one)	Direct drive		Belt drive	
Blower Motor Type (circle one)	Single speed split phase	Multiple speed PSC	Constant Torque multiple tap	ECM variable speed

Step 2

Complete the Pre-Start Check List and make sure to check each step off in the appropriate box as you finish it. This will help you to keep track of your progress.

Pre-Start Check List

STEP	PROCEDURE	CHECK
1	Turn thermostat down below the room temperature.	
2	Make sure all power to the furnace is off. Lock and tag the power panel before removing any parts.	
3	Vent connector connected with three screws per joint.	
4	Fuel line installed properly, with no apparent leaks.	
5	Spin all fans to be sure they are loose and turn freely.	
6	Combustion area free from debris.	

7	Electrical connections complete – main power.	
8	Electrical connections complete – thermostat.	
9	All doors and panels available and in place.	
10	Thermostat installed and operating correctly.	
11	Fuel oil present in tank.	
12	Barometric damper installed and swinging freely.	

Step 3

After finishing the Prestart Check List you are now ready to start the furnace and check its operation. You will follow the same procedure of working through a check list. This will help you to keep track of your progress. Complete the Start-Up Check List and make sure to check each step off in the appropriate box as you finish it.

Start-Up Check List

STEP	PROCEDURE	CHECK
1	Check the main fuse or breaker and the amperage rating and make sure it is the correct rating for the furnace.	
2	Measure the incoming supply voltage.	
3	If the Steps 1 & 2 have been completed and are satisfactory, then turn on the supply power for the furnace.	

4	Set the thermostat to – Fan On - to obtain the fan only operation and observe that the blower is operating. If the thermostat is not so equipped then consult with your instructor.	
5	Turn the fan off and the thermostat to heat.	
6	Adjust the thermostat setting to 10 °F above the room temperature.	
7	Observe the burner come on.	
8	Observe the fan come on after a reasonable warm-up time.	
9	Use a clamp-on ammeter to measure the blower motor amperage and compare this value to its rating.	
10	Turn down the thermostat and observe the burner shut off, and then fan shut off, in that order.	
11	Turn on and off for several ignitions.	
12	The flame should be orange/white in color, uniform in shape and within the combustion chamber, quiet, quick in ignition and extinction, and no drip from the burner tip.	

LAB 61.2 OIL FURNACE STORAGE TANK MAINTENANCE

LABORATORY OBJECTIVE
The purpose of this lab is to demonstrate your ability to conduct the proper preventative maintenance for an oil-fired furnace oil storage tank.

FUNDAMENTALS OF HVACR TEXT REFERENCE
Unit 61 Residential Oil Heating Installation

LABORATORY NOTES
You will need to familiarize yourself with oil furnace arrangements and types. Always remember to make the visual check of all system components prior to starting an oil furnace.

REQUIRED TOOLS, EQUIPMENT, AND MATERIALS
Oil furnace tank and operating oil furnace
Tool kit
Oil adsorbent pads
Large empty drip pan

SAFETY REQUIREMENTS
A. Wear safety glasses

B. A full oil storage tank will normally hold 250 gallons of fuel oil or more. Always make sure the shutoff valve is closed before changing the tank filter to prevent the possibility of a large oil spill.

C. It is good practice to have oil absorbent pads and a large drip pan available to collect any minor oil spills that may occur when changing oil tank filters.

PROCEDURE
Step 1
Complete the Oil Furnace Storage Tank Maintenance Check List and make sure to check each step off in the appropriate box as you finish it. This will help you to keep track of your progress.

Oil Furnace Storage Tank Maintenance Check List

STEP	PROCEDURE	CHECK
1	Make sure all power to the furnace is off. Lock and tag the power panel before removing any parts.	
2	Visually check the oil storage tank for leaks. Older tanks will rust from the inside out and leaks may develop.	
3	Close the tank shut off valve located between the tank and the filter assembly.	
4	Place an adsorbent pad beneath the filter assembly and place the large empty drip pan beneath the filter assembly to catch any dripping fuel oil.	
5	Carefully remove the filter assembly and check for water. Water is heavier than the fuel oil and will settle to the bottom of the fuel tank.	
6	If water is present, additional fuel may need to be drained from the tank.	
7	All fuel removed along with the old filter element, adsorbent pads, and any rags, must be removed and disposed of properly. Never leave any of these at the job site.	

8	Always replace the filter assembly gaskets with new ones.	
9	After the new filter assembly is in place and tight, the fuel tank shut off valve can be opened and the air bled through the vent screw located on the top of the filter assembly. Vent air from the assembly until fuel comes from the vent screw. Catch any dripping fuel in the drip pan.	
10	Wipe the filter assembly dry and check for any fuel leaks.	
11	After the new filter has been installed, the fuel line from the tank to the oil pump must be purged of air.	
12	Locate the purge valve on the fuel oil pump located on the furnace burner.	
13	Place the drip pan beneath the oil pump purge valve.	
14	Remove the lock on the furnace power and start the burner. Slowly open the oil pump purge valve and fuel oil will squirt out to collect into the drip pan. Allow the oil to flow from the purge valve until all of the air has been purged from the line, then close the purge valve.	
15	If air remains in the fuel line, the burner will not light normally and go out.	

16	After purging the air from the oil line, turn the burner on and off, observing several ignitions. Verify a quick clean quiet ignition.	
17	All fuel removed along with the old filter element, adsorbent pads, and any rags, must be removed and disposed of properly. Never leave any of these at the job site.	

LAB 61.3 OIL FURNACE TWO PIPE CONVERSION

LABORATORY OBJECTIVE
The purpose of this lab is to demonstrate your ability to convert a one pipe oil system to a two pipe oil system for an oil-fired furnace.

***FUNDAMENTALS OF HVACR* TEXT REFERENCE**
Unit 61 Residential Oil Heating Installation

LABORATORY NOTES
A one pipe oil system refers to the oil line from the tank to the burner being a single pipe, usually 3/8 inch OD copper. This system is recommended for no more than 2 feet of vertical lift from the oil level in the tank to the burner pump. Any time the tank runs out of oil the air must be manually bled from the line at the pump.

A two pipe conversion involves running a second line back to the tank and installing a bypass plug within the fuel oil pump. The second line at the tank needs to go all the way to the bottom of the tank. If oil entered the top of the tank and fell to the bottom, you would hear the oil fall and splash. Frequently a tank duplex fitting is used on two pipe systems. The fitting is installed in the top of the tank and has two fittings in it that will allow a 3/8 inch line to be pushed through to the bottom of the tank and then pulled up 3 to 4 inches. This is the correct position to install both the supply line and the return line. This reduces the chance of pulling sludge or water off the bottom of the tank.

REQUIRED TOOLS, EQUIPMENT, AND MATERIALS
Oil furnace tank and operating oil furnace
Tool kit
Oil adsorbent pads
Large empty drip pan
Tank duplex fitting

SAFETY REQUIREMENTS
A. Wear safety glasses
B. A full oil storage tank will normally hold 250 gallons of fuel oil or more. Always make sure the shutoff valve is closed to prevent the possibility of a large oil spill.
C. It is good practice to have oil absorbent pads and a large drip pan available to collect any minor oil spills that may occur.

PROCEDURE

Step 1

Complete the System Inspection Check List and make sure to check each step off in the appropriate box as you finish it. This will help you to keep track of your progress.

System Inspection Check List

STEP	PROCEDURE	CHECK
1	Identify the current fuel oil piping system. (circle one) One Pipe Two Pipe	
2	Measure the lift from the lowest possible operating oil level to the burner pump fitting	
3	Is there a bypass plug at the pump? Every new oil burner comes with a bypass plug in a cloth bag generally attached to the pump with string. Is it still there?	
4	If yes, you have the plug you need. If no, you will need to get one.	
5	Obtain a bypass plug as required.	
6	Obtain a sufficient length of 3/8 inch OD copper tubing and the required brass fittings to make connection at the pump.	

Step 2

After completing the System Inspection Check List you can prepare to convert to a two pipe system. Complete the Two Pipe System Conversion Check List and make sure to check each step off in the appropriate box as you finish it.

Two Pipe System Conversion Check List

STEP	PROCEDURE	CHECK
1	Make sure all power to the furnace is off. Lock and tag the power panel before removing any parts.	
2	Close the tank shut off valve.	
3	Disconnect the existing one line pipe. Place an adsorbent pad large empty drip pan beneath the connection to catch any dripping fuel oil.	
4	Loosen and remove the bolts holding the fuel oil pump in position.	
5	Inspect the pump fitting opening for the return line to the tank (pumps are generally labeled for openings and plug location).	

6	Inspect the pump for the location of the bypass plug. This is generally a 1/8 inch or 1/16 inch female pipe thread inside the return line opening. Refer to the manufacturer data for the pump as required.	
7	Hold the pump at an upward angle. Use an Allen wrench of sufficient length to install and tighten the bypass plug.	
8	Install the copper tubing to the 3/8 inch flare fitting at the pump.	
9	Mount the pump back onto the burner housing.	
10	Run the 3/8 inch line from the pump outlet into the top of the tank and down into the tank at 3 to 4 inches from the bottom.	
11	Snug the fitting to hold the line in place.	
12	Place supports for the line at appropriate locations from the pump to the tank.	

13	After the line has been installed, the tank shut off valve may be opened, and the power restored to the furnace. The burner should start and the fuel line should self-bleed itself of air with the bypass oil returning to the fuel tank.	
14	All fuel removed along adsorbent pads, and any rags, must be removed and disposed of properly. Never leave any of these at the job site.	

LAB 62.1 CHECK/TEST A CAD CELL OIL BURNER PRIMARY CONTROL

LABORATORY OBJECTIVE
The purpose of this lab is to demonstrate your ability to test for the proper operation of a cadmium sulfide cell (cad cell) for an oil-fired furnace.

FUNDAMENTALS OF HVACR TEXT REFERENCE
Unit 62 Troubleshooting Oil Heating Systems

LABORATORY NOTES
The cadmium sulfide cell (cad cell) of an oil burner primary control system proves the presence of an oil flame by observing the visible light from the flame. The cad cell's electrical resistance is greatly reduced in the presence of light. The resistance must be high to enable the primary to initiate a trial for ignition, also called a dark start function. Once a flame is established, the light from the flame causes the cad cell's resistance to drop and the flame will continue. During the trial ignition, a safety switch heater is energized that will open the safety switch contacts and lock out the burner if flame is not proved within the trial for ignition time, usually 30, 45, or 60 seconds. This heater must cool off before the burner can be rest manually and started again. When flame is established and proved by the cad cell, the safety switch heater is de-energized and the contacts remain closed.

REQUIRED TOOLS, EQUIPMENT, AND MATERIALS
Operating oil furnace
Tool kit
Ohmmeter
Clamp-on ammeter
1200 ohm resistor
Timer for timing seconds

SAFETY REQUIREMENTS
A. Wear safety glasses
B. Never allow fuel oil to flow into the combustion chamber without a flame. If fuel oil is allowed to build up and then suddenly ignite, this can create a serious hazard.
C. Check all circuits for voltage before doing any service work.
D. Stand on dry nonconductive surfaces when working on live circuits.
E. Never bypass any electrical protective devices.

PROCEDURE
Step 1
Collect the oil-fired furnace data and fill in the chart.

Oil-Fired Furnace Unit Data

Furnace Make:	Furnace Model Number:	Burner make	Burner model number	Nozzle GPH
Nozzle spray pattern (circle one)	Solid		Hollow	
Nozzle spray angle				
Blower Motor Amp Rating				
Furnace Configuration (circle one)	Upflow	Counterflow	Lowboy	Horizontal
Blower Type (circle one)	Direct drive		Belt drive	
Blower Motor Type (circle one)	Single speed split phase	Multiple speed PSC	Constant Torque multiple tap	ECM variable speed

Step 2

Complete the Pre-Start Check List and the Start-Up Check List as outlined in the procedure provided in Lab 55.1. After finishing the Start-Up Check List, shut down the furnace. Make sure all power to the furnace is off. Lock and tag the power panel before removing any parts. Complete the CAD Cell Test Check List.

CAD Cell Test Check List

STEP	PROCEDURE	CHECK
1	Make sure all power to the furnace is off. Lock and tag the power panel before removing any parts.	
2	Open or remove transformer.	
3	Locate an unplug cad cell from plug mount.	
4	Inspect and wipe clean the lens cover of the cad cell.	

658

5	Remove and record resistance of cad cell while the face of the cad cell is covered. Resistance =_____Ohms	
6	Remove and record resistance of cad cell while the face of the cad cell exposed to room light. Resistance =_____Ohms	
7	Insert the cad cell into the plug assembly.	
8	Locate yellow wires from the cad cell mount at the primary control terminals F and F.	
9	Remove cad cell wires from F and F and connect to the ohmmeter.	
10	Swing transformer slowly closed with the ohmmeter still connected.	

11	Turn on power to the furnace and burner and observe ignition. DO not get too close as oil burners can puff back.	
12	Read and record the ohms of the cad cell exposed to a normal flame. Resistance =_____ohms	
13	The flame will shut down and the burner will lockout within 30 seconds because the cad cell is not connected to the primary control.	
14	Obtain a 1200 ohm resistor and connect one end of the resistor to one F terminal of the primary control.	
15	Push the reset button. Wait a minimum of 2 minutes cool down time.	
16	Observe the burner start and the flame ignite.	
17	Carefully connect the second wire of the 1200 ohm resistor to the other F terminal of the primary control within the 30 second trial for ignition time.	

18	Observe the flame continue for another five minutes.	
19	You will time how long it takes in seconds for the flame to shutdown and the burner to lockout when you remove one lead from the resistor, which will simulate a flame failure. Running time after removing resistor =_____seconds	
20	After completing this test, make sure all power to the furnace is off. Lock and tag the power panel before removing any parts.	
21	Remove the resistor and ohmmeter and reconnect the cad cell.	
22	After the cad cell has been reconnected turn the burner on and off, observing several ignitions. Verify a quick clean quiet ignition.	

LAB 62.2 INSTALL A REPLACEMENT FUEL OIL PUMP

LABORATORY OBJECTIVE
The purpose of this lab is to demonstrate your ability to install a replacement fuel oil pump for an oil-fired furnace.

FUNDAMENTALS OF HVACR TEXT REFERENCE
Unit 62 Troubleshooting Oil Heating Systems

LABORATORY NOTES
The pump on an oil burner is one of the most important parts of the burner assembly. Its job is to pull the fuel oil from the tank and deliver it to the burner nozzle at the recommended supply pressure (typically 100 or 140 psig). Always check the burner nameplate to verify the manufacturer recommended fuel oil supply pressure. Minor adjustments can be made, however if the ump is worn, it must be replaced.

There are dozens of different pumps available. This is because there is more than one pump manufacturer, several major burner manufacturers, and different types of burner configurations. Not every wholesale house has every pump. To find a replacement pump, begin by contacting the dealer of the furnace type of burner that you are working on. You will need the furnace and burner model numbers, and serial numbers.

REQUIRED TOOLS, EQUIPMENT, AND MATERIALS
Oil furnace tank and operating oil furnace
Tool kit
Oil adsorbent pads
Large empty drip pan

SAFETY REQUIREMENTS
A. Wear safety glasses
B. It is good practice to have oil absorbent pads and a large drip pan available to collect any minor oil spills that may occur.

PROCEDURE
Step 1
Collect the oil-fired furnace data and fill in the chart.

Oil-Fired Furnace Unit Data

Furnace Make:	Model Number:	Serial Number:
Burner Make:	Model Number:	Serial Number:
Pump Make:	Model Number:	Serial Number:

Step 2

Complete the System Inspection Check List and make sure to check each step off in the appropriate box as you finish it. This will help you to keep track of your progress.

System Inspection Check List

STEP	PROCEDURE	CHECK
1	Identify the current fuel oil piping system. (circle one) One Pipe Two Pipe	
2	Measure the lift from the lowest possible operating oil level to the burner pump fitting. Lift =_____feet	
3	If the lift is greater than 2 feet on a one pipe system, we will need to convert to a two pipe system (see Lab 55.3).	

Step 3

After completing the System Inspection Check List you can prepare to replace the fuel oil pump. Complete the Fuel Pump Replacement Check List and make sure to check each step off in the appropriate box as you finish it.

Fuel Pump Replacement Check List

STEP	PROCEDURE	CHECK
1	Make sure all power to the furnace is off. Lock and tag the power panel before removing any parts.	
2	Close the tank shut off valve.	
3	Disconnect the existing oil line or lines. Place an adsorbent pad large empty drip pan beneath the connection to catch any dripping fuel oil.	
4	Loosen and remove the bolts holding the fuel oil pump in position.	
5	Change fittings to the new pump (change to a two pipe system if required – see Lab 55.3).	
6	Bolt new pump into position.	
7	After the replacement fuel oil pump has been installed, the tank shut off valve may be opened, and the power restored to the furnace.	

8	On a two pipe system, the burner should start and the fuel line should self- bleed itself of air with the bypass oil returning to the fuel tank.	
9	On a one pipe system, after the replacement pump has been installed, the fuel line from the tank to the oil pump must be purged of air.	
10	Locate the purge valve on the fuel oil pump located on the furnace burner.	
11	Place the drip pan beneath the oil pump purge valve.	
12	Start the burner and slowly open the oil pump purge valve and fuel oil will squirt out to collect into the drip pan. Allow the oil to flow from the purge valve until all of the air has been purged from the line, then close the purge valve.	
13	If air remains in the fuel line, the burner will not light normally and go out.	
14	After purging the air from the oil line, turn the burner on and off, observing several ignitions. Verify a quick clean quiet ignition.	
15	All fuel removed along with the old filter element, adsorbent pads, and any rags, must be removed and disposed of properly. Never leave any of these at the job site.	

LAB 62.3 INSTALL A REPLACEMENT FUEL OIL BURNER

LABORATORY OBJECTIVE
The purpose of this lab is to demonstrate your ability to install a replacement fuel oil burner for an oil-fired furnace.

FUNDAMENTALS OF HVACR TEXT REFERENCE
Unit 62 Troubleshooting Oil Heating Systems

LABORATORY NOTES
Many oil furnaces are constructed of heavy metal and are quite durable. It is not uncommon for the burner to become worn out while the basic furnace is still in good condition. In such cases the entire burner can be replaced. This has the advantage of a matched nozzle assembly and flame cone along with a new pump, motor, transformer, and primary control. This type of replacement is less expensive, faster, and easier than installing an entirely new furnace.

REQUIRED TOOLS, EQUIPMENT, AND MATERIALS
Oil furnace tank and operating oil furnace
Tool kit
Oil adsorbent pads
Large empty drip pan
Vacuum

SAFETY REQUIREMENTS
A. Wear safety glasses
B. It is good practice to have oil absorbent pads and a large drip pan available to collect any minor oil spills that may occur.

PROCEDURE
Step 1
Collect the oil-fired furnace data and fill in the chart.

Oil-Fired Furnace Unit Data

Furnace Make:	Model Number:	Serial Number:
Existing Burner Make:	Model Number:	Serial Number:
New Burner Make:	Model Number:	Serial Number:

Length of Blast Tube Required:	
Identify Piping System: (circle one) One Pipe Two Pipe	

New Combustion Chamber? (circle one) Yes	No
New Thermostat? (circle one) Yes	No

Step 2

After recording the Oil-Fired Furnace Unit Data List you can prepare to remove the old burner assembly. Complete the Burner Removal Check List and make sure to check each step off in the appropriate box as you finish it.

Burner Removal Check List

STEP	PROCEDURE	CHECK
1	Make sure all power to the furnace is off. Lock and tag the power panel before removing any parts.	
2	Close the fuel tank shut off valve.	
3	Disconnect the existing oil line or lines. Place an adsorbent pad large empty drip pan beneath the connection to catch any dripping fuel oil.	
4	Install 3/8-inch flare plugs in both fuel lines to prevent oil leakage.	
5	Carefully bend the fuel lines out of the way. You will reuse the same lines if possible	

6	Disconnect the main power and thermostat wire from the burner.	
7	Remove the mounting bolts holding the burner assembly in place.	
8	Remove the mount plate from front of the furnace.	
9	Use a vacuum to clean any debris from the combustion chamber area. Do not damage the combustion chamber refractory.	
10	Inspect the combustion chamber for any signs of cracks or deterioration. Replace as required.	

Step 3

After removing the old burner assembly you may prepare to install the new burner assembly. Complete the Burner Installation Check List and make sure to check each step off in the appropriate box as you finish it.

Burner Installation Check List

STEP	PROCEDURE	CHECK
1	Install new combustion chamber and components if required.	
2	Hold the burner mounting plate in position and measure the distance to the combustion chamber. Distance = _____	

3	Measure the length of the blast tube on the new burner. Length = _____	
4	Exchange blast tube on the burner if the length is not a match.	
5	Check/install the nozzle for correct GPH, angle, and pattern.	
6	Check and adjust electrode position (refer to Lab 54.1).	
7	Bolt new mounting plate to furnace.	
8	Bolt new burner with blast tube onto mounting plate.	
9	Connect oil lines to burner. Install bypass plug for two pipe systems.	
10	Replace oil filter in fuel supply line and bleed air from line (refer to Lab 55.2).	
11	Perform final burner adjustments (refer to Lab 54.2)	

LAB 63.1 IDENTIFY RESIDENTIAL HOT-WATER BOILER COMPONENTS AND DESCRIBE ITS OPERATION

LABORATORY OBJECTIVE
You will identify the components on an oil-fired hot-water boiler.

FUNDAMENTALS OF HVACR TEXT REFERENCE
Unit 63 Residential Hot-Water Boilers

REQUIRED TOOLS, EQUIPMENT, AND MATERIALS
Safety glasses
Residential oil fired hot water boiler

SAFETY REQUIREMENTS
A. Wear safety glasses
B. Always familiarize yourself with the equipment and operating manuals prior to starting up any system.

PROCEDURE
Step 1
You will examine the natural draft gas furnace assigned by the instructor and identify the listed components.

- Flame retention burner
- Burner blower
- Oil pump
- Ignition transformer
- Electrodes
- Oil burner nozzle
- Heat exchanger
- Refractory
- Oil burner primary control
- Cadmium sulfide cell (Cad Cell)
- Water circulating pump(s)
- Zone valves
- Expansion tank
- Pressure relief valve
- Low water cut-off
- Water pressure reducing valve
- Aquastat
- High temperature limit

Step 2

Set the thermostat to call for heat and observe the boiler operating sequence. List the order that things happen in the data table.

Hot Water Boiler Operating Sequence	
Order	Action
1	Thermostat set to heat
2	
3	
4	Thermostat set to OFF
5	
6	

Once the indoor fan is running, set the thermostat to "Off." Observe the order that things shut off and record in the table.

LAB 64.1 WATER CIRCULATING PUMP COMPLETE SERVICE

LABORATORY OBJECTIVE
The purpose of this lab is to demonstrate your ability to conduct the proper maintenance service procedure for a water circulating pump.

FUNDAMENTALS OF HVACR TEXT REFERENCE
Unit 64 Hydronic Heating Systems
Unit 63 Residential Hot-Water Boilers

LABORATORY NOTES
Hot water heating boilers are dependent on water pumps to distribute the hot water they produce. The pumps are primarily driven by electric motors. The electric motor requires typical motor service, cleaning, lubrication, proper mounting, insulation resistance and amperage checks. The pump is usually connected to the pump with a spider type coupler or flexible spring connected coupler.

Improper alignment is the main reason for coupler problems. Other pump maintenance issues fall into two major categories: impeller problems and shaft seal problems. On smaller pumps it is typical to exchange the entire pump or bearing assembly for a new or rebuilt replacement. Larger pumps can generally be brought in to be rebuilt by the local factory representative. If the pump is to be rebuilt on site, it is recommended that you have attended factory service demonstrations and have been trained in this area.

REQUIRED TOOLS, EQUIPMENT, AND MATERIALS
Circulating pump
Tool kit
T-handle Allen wrench
Strap wrench
Compressed air
Cleaner—degreaser
Rubber hammer
Thin chisel

SAFETY REQUIREMENTS
A. Wear safety glasses

B. Check all circuits for voltage before doing any service work.

C. Be careful to avoid lubricant contact with exposed skin.

D. Always wear safety glasses whenever using compressed air or CO_2 for blowing out dust and dirt.

PROCEDURE

Step 1

Collect the pump data and fill in the chart.

Blower Unit Data

Circulating Pump Make:	Model Number:	
Motor Data:	Horsepower:	Type and Size of Coupler:

Step 2

Complete an initial inspection. Complete the Initial Inspection & Cleaning Check List and make sure to check each step off in the appropriate box as you finish it. This will help you to keep track of your progress.

Initial Inspection & Cleaning Check List

STEP	PROCEDURE	CHECK
1	Make sure that the power to the circulating pump motor is secured. Verify this with a multi-meter voltage test. Refer to Unit 33 in *Fundamentals of HVACR*.	
2	Use CO_2 or compressed air pressure to blow out and clean all open air passages of the motor and bearing assembly.	
3	Use a non-detergent or recommended lubricant for the motor and bearing assembly.	
4	Using a pump sprayer, apply a cleaner degreaser to the motor and pump assembly.	

5	Use cleaning rags to wipe off all excess oil and any accumulated dirt, dust, grease, etc.	
6	Inspect the bearing assembly for any water dripping, metal shavings, scraping or grinding noise, etc.	
7	Inspect motor mounts. Is the motor centered within the rubber mount or is the motor sagging in the rubber?	
8	If the motor is sagging or out of center in the rubber mount, the motor mounts must be replaced.	

Step 3

After finishing the initial circulating pump inspection and cleaning, make sure all power is off. Lock and tag the power panel before removing any parts. Complete the Motor Mount Replacement Check List.

Motor Mount Replacement Check List

STEP	PROCEDURE	CHECK
1	Make sure that the power to the blower motor is secured. Verify this with a multimeter voltage test. Refer to Unit 33 in *Fundamentals of HVACR*.	
2	Using a thin open end wrench, loosen the two top motor mount bolts from inside the bearing assembly.	
3	Loosen and remove the two bottom motor mount bolts.	

4	Use a long T-handle Allen wrench to remove the motor end of the spring coupler.	
5	Supporting the motor in one hand, remove the previously loosened top two motor mount bolts.	
6	Remove the motor from its cradle and the two machine screws connected to the motor mount straps.	
7	Using an old screwdriver or a thin chisel, pry of both rubber motor mounts from the motor.	
8	Use a small metal rubber or plastic hammer to tap in place two new rubber motor mounts.	

Step 4

If the bearings drag and need to be replaced then complete the Bearing Replacement Check List.

Bearing Replacement Check List

STEP	PROCEDURE	CHECK
1	Make sure that the power to the blower motor is secured. Verify this with a multimeter voltage test. Refer to Unit 33 in *Fundamentals of HVACR*.	
2	Turn off the water isolation valves for the pump and drain off any remaining water.	

3	Remove the motor as in Step # 3 from the Motor Mount Replacement Check List.	
4	Remove the bolts holding the bearing assembly to the pump housing.	
5	Remove the bearing assembly along with the pump impeller.	
6	Inspect and clear any debris from the pump housing.	
7	Hold the impeller with a strap wrench and use a socket wrench to remove the bolt holding the impeller to the pump shaft.	
8	Inspect the impeller and replace it if there are any signs of wear or deterioration.	
9	Install the impeller on the new bearing assembly.	
10	Remove any old gaskets and carefully scrape and clean the surface metal. Always install new gaskets.	
11	After installing new gasket, slip the bearing assembly into the pump housing.	

676

12	Install and tighten the bolts holding the bearing assembly to the pump housing.	
13	Install the coupler to the bearing assembly shaft.	
14	Hold the motor in position and slide the coupler on to the motor shaft. Be sure to tighten the set screws into the hollow spot on both the pump and motor shafts.	
15	While still holding the motor, start the two top motor mount bolts first, then the bottom two bolts.	
16	Tighten the motor mount bolts.	
17	Open the water isolation valves and obtain normal pressure on the system. Purge any air from the system.	
18	Start up the pump and obtain normal operation.	
19	Lubricate the bearing assembly and motor as required.	
20	Inspect the bearing assembly for any water leaks.	

LAB 65.1 IDENTIFY HUMIDIFIER TYPES

LABORATORY OBJECTIVE
You will identify the different types of humidifiers.

FUNDAMENTALS OF HVACR TEXT REFERENCE
Unit 65 Humidifiers

LABORATORY NOTES
You will be assigned different types of humidifiers. You will complete a data sheet for each unit. You will be able to discuss the basic characteristics of each of these humidifiers.

REQUIRED TOOLS, EQUIPMENT, AND MATERIALS
Safety glasses
Six-in-one screwdriver for removing system panels
Atomizing humidifier
Plate evaporative humidifier
Rotating-drum evaporative humidifier
Rotating-plate evaporative humidifier
Fan-powered evaporative humidifier
Vaporizing humidifier

SAFETY REQUIREMENTS
A. Wear safety glasses
B. Be careful of sharp edges when removing sheet metal panels.

PROCEDURE
Examine each unit and complete the appropriate data sheet.

Datasheet Humidifier 1

Type	Power	Heating	Nozzle	Construction		Humidistat
Atomizing	Motorized	Heating Element	Nozzle	Plate	Media Bypass	Nylon Ribbon
Evaporative	Non-Motorized	No Heating Element	No Nozzle	Rotating-Drum	Fan Powered	Hair
Vaporizing				Rotating-Plate	Steam	Electronic

Instructor Check_____

678

Datasheet Humidifier 2

Type	Power	Heating	Nozzle	Construction		Humidistat
Atomizing Evaporative Vaporizing	Motorized Non-Motorized	Heating Element No Heating Element	Nozzle No Nozzle	Plate Rotating-Drum Rotating-Plate	Media Bypass Fan Powered Steam	Nylon Ribbon Hair Electronic

Instructor Check_____

Datasheet Humidifier 3

Type	Power	Heating	Nozzle	Construction		Humidistat
Atomizing Evaporative Vaporizing	Motorized Non-Motorized	Heating Element No Heating Element	Nozzle No Nozzle	Plate Rotating-Drum Rotating-Plate	Media Bypass Fan Powered Steam	Nylon Ribbon Hair Electronic

Instructor Check_____

Datasheet Humidifier 4

Type	Power	Heating	Nozzle	Construction		Humidistat
Atomizing Evaporative Vaporizing	Motorized Non-Motorized	Heating Element No Heating Element	Nozzle No Nozzle	Plate Rotating-Drum Rotating-Plate	Media Bypass Fan Powered Steam	Nylon Ribbon Hair Electronic

Instructor Check_____

Datasheet Humidifier 5

Type	Power	Heating	Nozzle	Construction		Humidistat
Atomizing Evaporative Vaporizing	Motorized Non-Motorized	Heating Element No Heating Element	Nozzle No Nozzle	Plate Rotating-Drum Rotating-Plate	Media Bypass Fan Powered Steam	Nylon Ribbon Hair Electronic

Instructor Check_____

Datasheet Humidifier 6

Type	Power	Heating	Nozzle	Construction		Humidistat
Atomizing	Motorized	Heating Element	Nozzle	Plate	Media Bypass	Nylon Ribbon
Evaporative	Non-Motorized	No Heating Element	No Nozzle	Rotating-Drum	Fan Powered	Hair
Vaporizing				Rotating-Plate	Steam	Electronic

Instructor Check_____

680

LAB 66.1 WIRING SEQUENCERS

LABORATORY OBJECTIVE
The purpose of this lab is to learn how to wire sequencers to control an electric furnace.

FUNDAMENTALS OF HVACR TEXT REFERENCE
Unit 66 Electric Heat
Unit 39 Electrical Diagrams
Unit 40 Control Systems

LABORATORY NOTES
You will wire a sequencer to control an electric furnace with a blower and two sets of strip heat. The sequencer should insure that one strip heater always starts before the other. The blower should operate any time a set of strips is energized.

REQUIRED TOOLS, EQUIPMENT, AND MATERIALS
Low voltage thermostat
Electric furnace that contains:

- Transformer
- 24-volt contactor
- Sequencer(s)
- Electric strip heaters
- Blower
- Multimeter
- Electrical hand tools

SAFETY REQUIREMENTS
A. Check all circuits for voltage before doing any service work.
B. Stand on dry nonconductive surfaces when working on live circuits.
C. Never bypass any electrical protective devices.

PROCEDURE
1. Draw a ladder type schematic diagram of an electric furnace controlled by sequencers.
2. Have the instructor check your design.
3. Wire the control system.
4. Have the instructor check the wiring before energizing the circuit.
5. Operate the control system and troubleshoot any problems.

LAB 67.1 ELECTRIC FURNACE STARTUP

LABORATORY OBJECTIVE

The purpose of this lab is to demonstrate your ability to go through the necessary sequence for a typical electric furnace startup.

***FUNDAMENTALS OF HVACR* TEXT REFERENCE**

Unit 67 Electric Heat Installation

LABORATORY NOTES

A typical electric furnace will have multiple stages of electric heat strips, usually approximately 5 KWH each. A 1 KWH heater would produce 3410 BTUH of heat and pull 4.17 amps at 240 V. A 5KWH heater would use five times that or 20.8 amps. Three 5 KW heaters would pull over 60 amps, too much to energize all the heaters at once. Power companies and some codes require electric furnaces be equipped with a time delay device, such as a sequencer, to stagger when individual strips are energized. This control is part of the furnace and not part of the thermostat. Even with a single-stage heat-only furnace, the electric heat elements come on one at a time, spaced apart by a few seconds at least. Supply air temperatures of lower than 120°F can feel rather cool and airflow should be reduced to keep the air temperature to a comfortable level.

REQUIRED TOOLS, EQUIPMENT, AND MATERIALS

Operating electric furnace
Tool kit
Multimeter
Clamp-on ammeter
Temperature sensor

SAFETY REQUIREMENTS

A. Check all circuits for voltage before doing any service work.

B. Stand on dry nonconductive surfaces when working on live circuits.

C. Never bypass any electrical protective devices.

682

PROCEDURE
Step 1
Collect the electric furnace data and fill in the chart.

Electric Furnace Unit Data

Furnace Make:		Model Number:		
Electrical Data	Voltage:	Phase:	Amperage:	KW:
Blower Type (circle one)	Direct drive	Belt drive		
Blower Speed (circle one)	Single speed	Two speed	Three speed	Four speed
System Type (circle one)	Strip heat only	Heat pump and electric backup		
Electric Heaters	1st KW:	2nd KW:	3rd KW:	Total KW:

Step 2

Complete the Pre-Start Check List and make sure to check each step off in the appropriate box as you finish it. This will help you to keep track of your progress.

Pre-Start Check List

STEP	PROCEDURE	CHECK
1	Make sure all power to the furnace is off. Lock and tag the power panel before removing any parts.	
2	Check all electrical connections for tightness.	
3	Spin all fans to be sure they are loose and turn freely.	
4	Check to make sure all airflow passages are unobstructed.	
5	All doors and panels available and in place.	
6	Thermostat installed and operating correctly.	

Step 3

After finishing the Pre-Start Check List you are now ready to start the furnace and check its operation. You will follow the same procedure of working through a checklist. This will help you to keep track of your progress. Complete the Start-Up Check List and make sure to check each step off in the appropriate box as you finish it.

Start-Up Check List

STEP	PROCEDURE	CHECK
1	Check the main fuse or breaker and the amperage rating and make sure it is the correct rating for the furnace.	
2	Measure the incoming supply voltage—Refer to Unit 35 in *Fundamentals of HVACR* for the proper procedure for testing the circuit.	
3	If the Steps 1 & 2 have been completed and are satisfactory, then turn on the supply power for the furnace.	
4	Set the thermostat to "Fan On" to obtain the fan only operation and observe that the blower is operating. If the thermostat is not so equipped then consult with your instructor.	
5	Turn the fan off and the thermostat to heat.	

6	Adjust the thermostat setting to 10°F above the room temperature.	
7	Observe the fan start and the heater banks come on.	
8	Use the clamp-on ammeter to measure and record the amperage as the sequence brings on electric banks. 1st = _____ 2nd = _____ 3rd = _____	
9	Obtain normal heating operation.	
10	Measure and record temperatures. Discharge air temperature = _____ Room air temperature =_____	

11	Calculate temperature rise. Discharge air temp. – Room air temp. = Temperature rise Temperature rise = _____	
12	Turn the thermostat down and observe the heaters come off.	
13	Observe the blower stop three minutes after the heaters come off.	

LAB 68.1 CALCULATE AIRFLOW BY TEMPERATURE RISE

LABORATORY OBJECTIVE
The purpose of this lab is to demonstrate your ability to calculate airflow by the temperature rise methods through a typical electric furnace.

FUNDAMENTALS OF HVACR TEXT REFERENCE
Unit 68 Troubleshooting Electric Heat

LABORATORY NOTES
Electric heaters are nearly 100% efficient and allow the measurement for the amount of heat produced very accurate. The heaters are located within the duct or furnace and there is no heat lost going up the chimney. We can calculate the airflow by the temperature rise method very accurately. The important thing is to measure the voltage, amperage, and temperatures as accurately as possible. The blower amperage must also be kept separate and not added to the heater amperage.

REQUIRED TOOLS, EQUIPMENT, AND MATERIALS
Operating electric furnace
Tool kit
Multimeter
Clamp-on ammeter
Temperature sensor

SAFETY REQUIREMENTS
A. Check all circuits for voltage before doing any service work.
B. Stand on dry nonconductive surfaces when working on live circuits.
C. Never bypass any electrical protective devices.

PROCEDURE

Step 1

Collect the electric furnace data and fill in the chart.

Electric Furnace Unit Data

Furnace Make:		Model Number:		
Electrical Data	Voltage:	Phase:	Amperage:	KW:
Blower Type **(circle one)**	Direct drive	Belt drive		
Blower Speed **(circle one)**	Single speed	Two speed	Three speed	Four speed
System Type **(circle one)**	Heat only	Heat pump and electric backup		
Electric Heaters	1st KW:	2nd KW:	3rd KW:	Total KW:

Step 2

Complete the Prestart Check List and make sure to check each step off in the appropriate box as you finish it. This will help you to keep track of your progress.

Prestart Check List

STEP	PROCEDURE	CHECK
1	Make sure all power to the furnace is off. Lock and tag the power panel before removing any parts.	
2	Check all electrical connections for tightness.	
3	Spin all fans to be sure they are loose and turn freely.	
4	Check to make sure all airflow passages are unobstructed.	
5	All doors and panels available and in place.	
6	Thermostat installed and operating correctly.	
7	Allocate and count each heater element contactor. How many are there and what is the KW rating? _____ of _____KWH	

Step 3

After finishing the Prestart Check List you are now ready to start the furnace and check its operation. You will follow the same procedure of working through a checklist. This will help you to keep track of your progress. Complete the Start-Up Check List and make sure to check each step off in the appropriate box as you finish it.

Start-Up Check List

STEP	PROCEDURE	CHECK
1	Check the main fuse or breaker and the amperage rating and make sure it is the correct rating for the furnace.	
2	Measure the incoming supply voltage—Refer to Unit 35 in *Fundamentals of HVACR* for the proper procedure for testing the circuit.	
3	If the Steps 1 & 2 have been completed and are satisfactory, then turn on the supply power for the furnace.	
4	Set the thermostat to "Fan On" to obtain the fan only operation and observe that the blower is operating. If the thermostat is not so equipped then consult with your instructor.	
5	Turn the fan off and the thermostat to heat.	

6	Adjust the thermostat setting to 10°F above the room temperature.	
7	Observe the fan start and the heater banks come on.	
8	Use the clamp-on ammeter to measure and record the amperage as the sequence brings on electric banks. 1st = _____ 2nd = _____ 3rd = _____	
9	Obtain normal heating operation.	

10	Measure and record temperatures. Discharge air temperature = _____ Room air temperature =_____	
11	Calculate temperature rise. Discharge air temp. – Room air temp. = Temperature rise Temperature rise = _____	
12	Read total amperage of electric heaters only. Amperage = _____	
13	Calculate BTUH. Voltage x Amperage x 3.414 = BTUH BTUH = _____	

14	Calculate airflow by the temperature rise method. Airflow (CFM) = $\dfrac{\text{BTUH (from step 13)}}{(1.08) \times \text{Temp. rise (from step 11)}}$ CFM = _____	
15	Measure the blower motor amperage. Blower amperage = _____	
16	Rated fan motor amperage from nameplate on motor. Rated blower amperage =_____	
17	The measured amperage should be lower than the rated blower amperage.	
18	Turn the thermostat down and observe the heaters come off.	
19	Observe the blower stop three minutes after the heaters come off.	

LAB 69.1 TYPES OF HEAT PUMPS

LABORATORY OBJECTIVE
You will examine different heat pumps and identify their characteristics; including whether they are packaged or split, air source or water source, and the type of auxiliary heat.

***FUNDAMENTALS OF HVACR* TEXT REFERENCE**
Unit 69 Heat Pump System Fundamentals

LABORATORY NOTES
You will examine several heat pumps and determine if they are packaged or split, air source or water source. You will also identify the type of auxiliary heat.

REQUIRED TOOLS, EQUIPMENT, AND MATERIALS
Safety glasses
Different types of heat pumps

SAFETY REQUIREMENTS
A. Wear safety glasses
B. Turn OFF power to unit before proceeding

PROCEDURE
Unit 1

1. Is this unit a packaged unit or a split system?

2. Is this unit an air source unit or a water source unit?

3. What type of auxiliary heat does this system use?

4. If this is a water source unit, is it an open loop or closed loop system?

Unit 2

1. Is this unit a packaged unit or a split system?

2. Is this unit an air source unit or a water source unit?

3. What type of auxiliary heat does this system use?

4. If this is a water source unit, is it an open loop or closed loop system?

Unit 3

1. Is this unit a packaged unit or a split system?

2. Is this unit an air source unit or a water source unit?

3. What type of auxiliary heat does this system use?

4. If this is a water source unit, is it an open loop or closed loop system?

Unit 4

1. Is this unit a packaged unit or a split system?

2. Is this unit an air source unit or a water source unit?

3. What type of auxiliary heat does this system use?

4. If this is a water source unit, is it an open loop or closed loop system?

LAB 69.2 WINDOW UNIT HEAT PUMP REFRIGERATION CYCLE

LABORATORY OBJECTIVE
You will identify the refrigeration cycle components and describe their function in a window unit heat pump.

FUNDAMENTALS OF HVACR TEXT REFERENCE
Unit 69 Heat Pump System Fundamentals

LABORATORY NOTES
You will identify the refrigeration components on a window unit heat pump system and describe their function. You should be prepared to show each component to the instructor and describe what happens to the refrigerant going through each component.

REQUIRED TOOLS, EQUIPMENT, AND MATERIALS
Safety glasses
Window unit heat pump

SAFETY REQUIREMENTS
A. Wear safety glasses
B. Turn OFF power before proceeding

PROCEDURE
1. Examine the window unit heat pump.

2. Locate the compressor.

3. Locate the reversing valve. What does the reversing valve actually reverse?

4. Locate the outdoor coil. What is the function of this coil in cooling?

5. What is the function of this coil in heating?

6. Locate the metering device. What type of metering device does this unit use?

7. Locate the indoor coil. What is the function of this coil in cooling?

8. What is the function of this coil in heating?

9. Start at the compressor discharge and show the refrigerant path through the system until it returns to the suction of the compressor in cooling. Describe the refrigerant changes through the cycle.

10. Start at the compressor discharge and show the refrigerant path through the system until it returns to the suction of the compressor in heating. Describe the refrigerant changes through the cycle.

LAB 69.3 PACKAGED UNIT HEAT PUMP REFRIGERATION CYCLE

LABORATORY OBJECTIVE
You will identify the refrigeration cycle components and describe their function in a packaged unit heat pump.

FUNDAMENTALS OF HVACR TEXT REFERENCE
Unit 69 Heat Pump System Fundamentals

LABORATORY NOTES
You will identify the refrigeration components on a packaged unit heat pump system and describe their function. You should be prepared to show each component to the instructor and describe what happens to the refrigerant going through each component.

REQUIRED TOOLS, EQUIPMENT, AND MATERIALS
Safety glasses
6- in-1 screwdriver
Packaged unit heat pump

SAFETY REQUIREMENTS
A.   Wear safety glasses
B.   Turn OFF power before proceeding

PROCEDURE
1.   Examine the packaged unit heat pump.

2.   Locate the compressor.

3.   Locate the reversing valve. What does the reversing valve actually reverse?

4.   Locate the outdoor coil. What is the function of this coil in cooling?

5.   What is the function of this coil in heating?

6. Locate the outdoor metering device. What type of metering device does this unit use for the outdoor coil?

7. When does this device meter refrigerant?

8. Locate the indoor coil. What is the function of this coil in cooling?

9. What is the function of this coil in heating?

10. Locate the indoor metering device. What type of metering device does this unit use for the indoor coil?

11. When does this device meter refrigerant?

12. What type of refrigerant storage does this system use?

13. Start at the compressor discharge and show the refrigerant path through the system until it returns to the suction of the compressor in cooling. Describe the refrigerant changes through the cycle.

14. Start at the compressor discharge and show the refrigerant path through the system until it returns to the suction of the compressor in heating. Describe the refrigerant changes through the cycle.

LAB 69.4 SPLIT SYSTEM HEAT PUMP REFRIGERATION CYCLE

LABORATORY OBJECTIVE
You will identify the refrigeration cycle components and describe their function in a split system heat pump.

FUNDAMENTALS OF HVACR TEXT REFERENCE
Unit 69 Heat Pump System Fundamentals

LABORATORY NOTES
You will identify the refrigeration components on a split system heat pump and describe their function. You should be prepared to show each component to the instructor and describe what happens to the refrigerant going through each component.

REQUIRED TOOLS, EQUIPMENT, AND MATERIALS
Safety glasses
6- in-1 screwdriver
Split system heat pump

SAFETY REQUIREMENTS
A. Wear safety glasses
B. Turn OFF power before proceeding

PROCEDURE
1. Examine the split system heat pump.
2. Locate the compressor.
3. Locate the reversing valve. What does the reversing valve actually reverse?

4. Locate the outdoor coil. What is the function of this coil in cooling?

5. What is the function of this coil in heating?

6. Locate the outdoor metering device. What type of metering device does this unit use for the outdoor coil?

7.   When does this device meter refrigerant?

8.   What type of refrigerant storage does this system use?

9.   Locate the indoor coil. Where in the airstream is the coil in relation to the blower: Before or after?

10. What is the function of this coil in cooling?

11. What is the function of this coil in heating?

12. Locate the indoor metering device. What type of metering device does this unit use for the indoor coil?

13. When does this device meter refrigerant?

14. Start at the compressor discharge and show the refrigerant path through the system until it returns to the suction of the compressor in cooling. Describe the refrigerant changes through the cycle.

15. Start at the compressor discharge and show the refrigerant path through the system until it returns to the suction of the compressor in heating. Describe the refrigerant changes through the cycle.

LAB 69.5 REVERSING VALVE FUNDAMENTALS

LABORATORY OBJECTIVE
You will describe the operation of a reversing valve and identify the parts of the valve including the common suction, common discharge, pilot tubes, pilot solenoid, and reversing valve coil.

FUNDAMENTALS OF HVACR TEXT REFERENCE
Unit 69 Heat Pump System Fundamentals

LABORATORY NOTES
You will examine a reversing valve that has been cut open. You will identify the common suction line, common discharge line, pilot solenoid, pilot tubes, and reversing valve coil. Using the reversing valve, you will show the instructor how the reversing valve shifts.

REQUIRED TOOLS, EQUIPMENT, AND MATERIALS
Safety glasses
Reversing valve that is cut open to observe the inside parts

SAFETY REQUIREMENTS
Caution: cut metal edges my be sharp

PROCEDURE
1. Examine the reversing valve.
2. Identify the following parts of the valve:
   - Common suction line
   - Common discharge line
   - Pilot solenoid
   - Pilot tubes
   - Solenoid valve
3. Describe where each of the lines on the reversing valve connects to the unit.

4. Describe the how the valve shifts.

LAB 69.6 REVERSING VALVE OPERATION

LABORATORY OBJECTIVE
You will measure the temperature of the lines entering and leaving an operating reversing valve and observe their change when the valve shifts.

***FUNDAMENTALS OF HVACR* TEXT REFERENCE**
Unit 69 Heat Pump System Fundamentals

LABORATORY NOTES
You will operate the heat pump in cooling and heating, measure the temperature of the refrigerant lines entering and leaving the reversing valve, and complete the data sheet.

REQUIRED TOOLS, EQUIPMENT, AND MATERIALS
Safety glasses
Thermocouple temperature tester
6-in-1 screwdriver
Operational heat pump

SAFETY REQUIREMENTS
A. Wear safety glasses
B. Caution: The discharge line can cause burns
C. Caution: avoid moving parts such as fans

PROCEDURE
1. Locate the reversing valve in the heat pump.
2. Examine the diagram and determine when the reversing valve is energized: heating or cooling?
3. Operate the heat pump in cooling.
4. Measure the temperature of the four lines in this order:
    *   Common discharge line (small line by itself)
    *   Line to outdoor coil (one of the outside lines on the side with three large lines)
    *   Common suction line (line in the middle on the side with three large lines)
    *   Line to indoor coil (should be the only line left) Operate the heat pump in heating.
5. Measure the temperature of the four lines in this order:
    *   Common discharge line (small line by itself)
    *   Line to indoor coil (one of the outside lines on the side with three large lines)
    *   Common suction line (line in the middle on the side with three large lines)
    *   Line to indoor coil (should be the only line left)
6. Operate the heat pump in the mode in which the reversing valve is energized (heating or cooling). While the system is operating, carefully remove one wire from the reversing valve solenoid coil. The valve should shift.
7. Replace the wire and the valve should shift back.

LAB 69.7 HEAT PUMP ELECTRICAL COMPONENTS

LABORATORY OBJECTIVE
You will identify common electrical heat pump components and explain their function.

FUNDAMENTALS OF HVACR TEXT REFERENCE
Unit 69 Heat Pump System Fundamentals

LABORATORY NOTES
You will inspect a heat pump, identify the listed components, and explain their function to the instructor.

REQUIRED TOOLS, EQUIPMENT, AND MATERIALS
Safety glasses
Heat pump

SAFETY REQUIREMENTS
A. Wear safety glasses
B. Turn OFF power before proceeding

PROCEDURE
You will examine the heat pump assigned by the instructor and identify the listed components.

- Contactor
- Defrost board
- High pressure switch
- Reversing valve solenoid
- Dual run capacitor
- Compressor starting components
- Anti-short cycling timer
- Outdoor fan motor
- Compressor
- Crankcase heater

LAB 70.1 IDENTIFYING DEFROST CONTROLS

LABORATORY OBJECTIVE
You will identify defrost controls as either an electronic time–temperature defrost control or a demand defrost control.

FUNDAMENTALS OF HVACR TEXT REFERENCE
Unit 70 Air-Source Heat Pump Applications

LABORATORY NOTES
The instructor will assign a selection of air source heat pumps for you to identify the defrost controls and explain their operation.

REQUIRED TOOLS, EQUIPMENT, AND MATERIALS
Safety glasses
6-in-1 screwdriver
Non-contact voltage probe
Multimeter
Heat pumps with defrost controls (one electronic time–temperature and one demand defrost)

SAFETY REQUIREMENTS
A. Wear safety glasses
B. Check all circuits for voltage before doing any service work
C. Stand on dry non-conductive surfaces when working on live circuits

PROCEDURE
1. Turn the power off to the unit.
2. Check with your non-contact voltage detector or multimeter to insure power is off.
3. Locate the defrost thermostat or coil temperature thermistor.
4. Which type does this unit use, defrost thermostat (switch) or coil temperature sensor (thermistor)?
5. Locate the defrost control board. Does the board have a time jumper or pins?
6. Determine if this is a time–temperature defrost board or a demand defrost board based on the presence or absence of time pins and the type of defrost thermostat (switch or sensor).
7. Time–temperature systems use round, physical defrost thermostat switches and have time period jumpers or switches.
8. Demand defrost boards use two thermistor temperature sensors: one for the coil and one for the outdoor ambient temperature.
9. Check the resistance of the defrost thermostat or the thermistor sensors.
10. The defrost thermostat should be open (OL) until it reaches its closing temperature, usually around 28°F.
11. The thermistor sensors should have a resistance in the thousands of ohms that vary with temperature.

LAB 70.2 WIRING DUAL FUEL HEAT PUMP

LABORATORY OBJECTIVE
You will wire the dual fuel heat pump assigned by the instructor. This includes control wiring.

***FUNDAMENTALS OF HVACR* TEXT REFERENCE**
Unit 70 Air-Source Heat Pump Applications

LABORATORY NOTES
You will wire a dual fuel heat pump using either a dual fuel panel or dual fuel capable thermostat.

REQUIRED TOOLS, EQUIPMENT, AND MATERIALS
Safety glasses
6- in-1 screwdriver
Wire cutters/crimpers/strippers
Non-contact voltage detector
Multimeter
Split system heat pump
Gas furnace
Dual fuel panel or dual fuel capable thermostat
Wire
Wire connectors

SAFETY REQUIREMENTS
  A. Check all circuits for voltage before doing any service work
  B. Stand on dry non-conductive surfaces when working on live circuits
  C. Never bypass any electrical protective devices

PROCEDURE
This system should be a gas furnace with a heat pump coil installed on it connected to a heat pump condensing unit. The coordination of the furnace and heat pump will be provided by either a dual fuel panel or a dual fuel capable thermostat.
  1. Turn the power off to both the furnace and the outdoor unit.
  2. Check the power with your non-contact voltage detector or multimeter to make sure they are off.
  3. Each piece will have its own power wire and disconnect switch.
      a. If using a dual fuel panel, the control wiring should be run from the thermostat to the dual fuel panel, and then from the dual fuel panel to both the furnace and the outdoor unit.
      b. If using a dual fuel capable thermostat, the control wiring should be run from the thermostat to the furnace, and then from the furnace to the outdoor unit.
      c. The thermostat may need to be configured to operate in dual fuel mode.
  4. The thermostat wire does not need to be run in conduit.

5. The thermostat wire should NOT run in the same conduit as the power wire. Mount the thermostat.

6. Wire the control wiring at the thermostat, the dual fuel panel, the furnace, and the outdoor unit.

7. Have the instructor check your work BEFORE energizing.

8. Turn on the power to the unit.

9. Operate the unit in both the cooling, first stage heating, and second stage heating cycles.

LAB 70.3 ELECTRIC STRIP HEAT OPERATION

LABORATORY OBJECTIVE
You will operate electric strip heat and measure the key operating characteristics.

FUNDAMENTALS OF HVACR TEXT REFERENCE
Unit 70 Air-Source Heat Pump Applications
Unit 68 Troubleshooting Electric Heat

LABORATORY NOTES
You will operate an electric furnace or a heat pump in emergency heat mode, measure its operating characteristics, and record them.

REQUIRED TOOLS, EQUIPMENT, AND MATERIALS
Safety glasses
6- in-1 screwdriver
Wire cutters/crimpers/strippers
Non-contact voltage detector
Multimeter
Ammeter
Heat pump with electric strip auxiliary heat or electric furnace

SAFETY REQUIREMENTS
A. Check all circuits for voltage before doing any service work
B. Stand on dry non-conductive surfaces when working on live circuits
C. Never bypass any electrical protective devices

PROCEDURE
1. Turn the power off to the unit.
2. Check the voltage with you non-contact voltage detector or multimeter to insure it is off.
3. Disconnect the wires from one side of each strip heater, paying attention to how they are wired on the unit.
4. Measure the resistance of each strip heater.
5. Measure the resistance of the thermal overload limit switches and the fuse links.
6. Replace the wires, making sure they are properly connected and tight.
7. Turn the power back on.
8. Operate the electric furnace, measure and record the following information:
    - Element voltage
    - Element current (individual)
    - Element current(total)
    - Fan current

    Note: There will normally be a delay before each strip heater is energized.

9. What controls the blower motor?

10. When does the blower motor turn on?

11. When does the blower shut off?

12. How many heat strips are there?

13. What is the kilowatt rating of each heater?

14. What is the rated amp draw of each heater?

15. What is the ACTUAL amp draw and kilowatt usage of each heater?

16. Describe the sequence in which the heat strips come on.

LAB 71.1 MEASURING WATER SOURCE HEAT PUMP SYSTEM CAPACITY

LABORATORY OBJECTIVE
You will demonstrate your ability to determine the system capacity of an operating water source heat pump by measuring its water flow and temperature rise.

FUNDAMENTALS OF HVACR TEXT REFERENCE
Unit 71 Geothermal Heat Pumps

LABORATORY NOTES
In this lab we will calculate the capacity of an operating water source heat pump system by measuring the water flow rate and the water temperature difference. A BTU is defined as a 1 degree temperature rise in 1 pound of water at atmospheric pressure. We can calculate the actual capacity accurately by multiplying the water flow in pounds per hour times the temperature rise in degrees Fahrenheit.

REQUIRED, EQUIPMENT, AND MATERIALS
Temperature sensor with temperature clamp and thermocouple clamp
Water source heat pump
1 gallon bucket
Timer or watch that can measure seconds

SAFETY REQUIREMENTS
Always familiarize yourself with the equipment and operating manuals prior to starting up any system.

PROCEDURE

Step 1
Trace out the system and make sure that you understand the operation of the heat pump.

> Start the water loop through the heat pump before starting the unit.
1. After water flow has been established, you may start the heat pump and run it in the cooling mode for 15 minutes.
2. Measure the flow rate of the water. This can be done by allowing the water leaving the heat pump to flow into the one gallon bucket. Time the duration in seconds from the time the bucket is initially empty until it is full. Repeat this procedure for three different sets of readings.

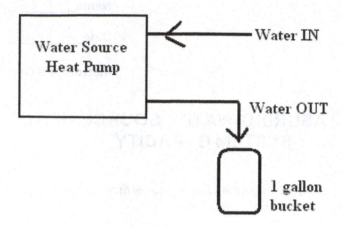

Figure 9-6-1

a)   First measured time in seconds:

b)   Second measured time in seconds:

c)   Third measured time in seconds:

3.   The average time in seconds is equal to the three readings added together and then divided by three.

Average time in seconds = (Time 1 + Time 2 + Time 3) / 3

Average time in seconds =

4.   To determine how many gallons per minute is flowing first convert the seconds to minutes. Take the average time in seconds and divide that value by 60 (this is because there are 60 seconds in one minute).

Gallons per minute = Total seconds / 60

Gallons per minute (GPM) =

5.   Convert gallons per minute (GPM) to pounds per minute by multiplying GPM by 8.34 (this is because there are 8.34 pounds of water in one gallon).

Pounds per minute = GPM x 8.34

Pounds per minute =

712

6. Measure the temperature of the water going in to the heat pump and the temperature of the water leaving the heat pump.

    Temperature of the water IN:

    Temperature of the water OUT:

7. Calculate the temperature difference that is equal to the temperature OUT minus the temperature IN ($°F_{OUT} - °F_{IN}$).

    Temperature Difference (ΔT) =

8. Calculate the system capacity. To do this multiply pounds per minute by the temperature difference. This will then be multiplied by the specific heat of water at atmospheric pressure which is simply 1 (one BTU for every °F for every lb of water). The calculated value will be in BTU per minute.

    (lb/min x ($°F_{OUT} - °F_{IN}$) x 1 BTU/lb/°F) = BTU/min

    System capacity in BTU/min =

9. Calculate the value for tons of refrigeration.

    Remember that one ton of refrigeration is the equivalent of melting one ton (2,000 lbs) of ice in 24 hours. Remember that the latent heat of fusion (ice to water) is 144 BTU/lb. 2,000 lbs of ice multiplied by 144 BTU/lb is equal to 288,000 BTU for one ton in 24 hours. Divide this by 24 hours per day and you have 12,000 BTU/hr. Divide this by 60 minutes per hour and you have 200 BTU/min.

    Therefore a ton of system capacity is equal to:

    1 ton = 288,000 BTU/day

    1 ton = 12,000 BTU/hr

    1 ton = 200 Btu/min

    To find capacity in tons, divide the system capacity in BTU/min by 200 BTU/min-ton.

    System capacity in tons = (BTU/min) / 200 BTU/min-ton

    System capacity in tons =

10. How does the system capacity that you calculated compare to the nameplate rating of the heat pump? Is the heat pump operating at full capacity?

Step 2

Reverse the heat pump so that now it is in the heating mode and run it in this mode for 15 minutes before recording your next measurements.

1. Measure the flow rate of the water. Repeat this procedure for three different sets of readings.

    a) First measured time in seconds:

    b) Second measured time in seconds:

    c) Third measured time in seconds:

2. The average time in seconds is equal to the three readings added together and then divided by three.

    Average time in seconds = (Time 1 + Time 2 + Time 3) / 3

    Average time in seconds =

3. To determine how many gallons per minute is flowing first convert the seconds to minutes. Take the average time in seconds and divide that value by 60 (this is because there are 60 seconds in one minute).

    Gallons per minute = Total seconds / 60

    Gallons per minute (GPM) =

4. Convert gallons per minute (GPM) to pounds per minute by multiplying GPM by 8.34 (this is because there are 8.34 pounds of water in one gallon).

    Pounds per minute = GPM x 8.34

    Pounds per minute =

5. Measure the temperature of the water going in to the heat pump and the temperature of the water leaving the heat pump.

    Temperature of the water IN:

    Temperature of the water OUT:

6. Calculate the temperature difference that is equal to the temperature OUT minus the temperature IN ($^\circ F_{OUT} - ^\circ F_{IN}$).

(*Note:* The temperature OUT minus the temperature IN will have a negative value. This is because in the heating mode the heat pump is absorbing heat from the water rather than rejecting heat. You do not need to use a negative value because you are just calculating the temperature difference so you may ignore the negative sign for this experiment.)

Temperature Difference (ΔT) =

7. Calculate the system capacity. To do this, multiply pounds per minute by the temperature difference. This will then be multiplied by the specific heat of water at atmospheric pressure which is simply 1 (one BTU for every °F for every lb of water). The calculated value will be in BTU per minute.

(lb/min x (°F$_{OUT}$ – °F$_{IN}$) x 1 BTU/lb/°F) = BTU/min

System capacity in BTU/min =

8. Calculate the value for tons of refrigeration.

Remember that one ton of refrigeration is the equivalent of melting one ton (2,000 lbs) of ice in 24 hours. Remember that the latent heat of fusion (ice to water) is 144 BTU/lb. 2,000 lbs of ice multiplied by 144 BTU/lb is equal to 288,000 BTU for one ton in 24 hours. Divide this by 24 hours per day and you have 12,000 BTU/hr. Divide this by 60 minutes per hour and you have 200 BTU/min.

Therefore a ton of system capacity is equal

to: 1 ton = 288,000 BTU/day

1 ton = 12,000 BTU/hr

1 ton = 200 Btu/min

To find capacity in tons, divide the system capacity in BTU/min by 200 BTU/min-ton.

System capacity in tons = (BTU/min) / 200 BTU/min-ton

System capacity in tons =

9. How does the system capacity that you calculated compare to the nameplate rating of the heat pump? Is the heat pump operating at full capacity?

10. How does the system capacity when operating the heat pump in the cooling mode differ from running the heat pump in the heating mode. Explain your answer.

LAB 71.2 GEOTHERMAL HEAT PUMP COMPONENTS

LABORATORY OBJECTIVE
You will identify common water-source heat pump components and explain their function.

***FUNDAMENTALS OF HVACR* TEXT REFERENCE**
Unit 71 Geothermal Heat Pumps

LABORATORY NOTES
You will inspect a geothermal heat pump, identify the listed components, and explain their function to the instructor.

REQUIRED TOOLS, EQUIPMENT, AND MATERIALS
Safety glasses
Water-source heat pump

SAFETY REQUIREMENTS
Wear safety glasses

PROCEDURE
You will examine the water-source heat pump assigned by the instructor and identify the listed components.

- Refrigerant to water coaxial heat exchanger (water coil)
- Water outlet
- Water inlet
- Refrigerant to air heat exchanger (air coil)
- Fan assembly
- Air filter
- Condensate drain
- Condensate overflow sensor
- Bi-flow thermostatic expansion valve
- Bi-flow filter drier
- Reversing valve solenoid
- Control board
- Compressor
- Compressor starting components
- Freeze protection sensor
- Anti-short cycle timer
- High pressure sensor
- Low pressure sensor or temperature sensor for loss of charge

LAB 71.3 GEOTHERMAL HEAT PUMP CYCLE

LABORATORY OBJECTIVE
You will identify the refrigeration cycle components and describe their function in a water-source heat pump.

FUNDAMENTALS OF HVACR TEXT REFERENCE
Unit 71 Geothermal Heat Pumps

LABORATORY NOTES
You will identify the refrigeration components on a water-source heat pump system and describe their function. You should be prepared to show each component to the instructor and describe what happens to the refrigerant going through each component.

REQUIRED TOOLS, EQUIPMENT, AND MATERIALS
Safety glasses
7- in-1 screwdriver
Water-source heat pump

SAFETY REQUIREMENTS
Wear safety glasses

PROCEDURE

1. Examine the water source heat pump.

2. Locate the compressor. What type is it (scroll, reciprocating, rotary)?

3. Locate the reversing valve. What does the reversing valve actually reverse?

4. Locate the refrigerant-to-water coaxial heat exchanger (water coil). What is the function of this coil in the cooling mode?

5. What is the function of this coil in the heating mode?

6. Locate the refrigerant-to-air heat exchanger (air coil). What is the function of this coil in the cooling mode?

7. What is the function of this coil in heating mode?

8. Locate the metering device. Is this a bi-flow metering device? How can you determine this?

9. Locate the filter drier. Is this a bi-flow device? How can you determine this?

10. Locate the fan assembly. Is this a variable speed fan?

11. Locate the water inlet and outlet. What flow rate is the system rated for?

12. Would this heat pump be used for a closed or open loop system?

13. Start at the compressor discharge and show the refrigerant path through the system until it returns to the suction of the compressor in cooling. Describe the refrigerant changes through the cycle.

14. Start at the compressor discharge and show the refrigerant path through the system until it returns to the suction of the compressor in heating. Describe the refrigerant changes through the cycle.

LAB 72.1 WIRING PACKAGED HEAT PUMP

LABORATORY OBJECTIVE
You will wire the packaged heat pump assigned by the instructor. This includes all power and control wiring.

FUNDAMENTALS OF HVACR TEXT REFERENCE
Unit 72 Heat Pump Installation

LABORATORY NOTES
You will wire a packaged heat pump, have your work checked by the instructor, and operate the unit.

REQUIRED TOOLS, EQUIPMENT, AND MATERIALS
Safety glasses
6- in-1 screwdriver
Wire cutters/crimpers/strippers
Non-contact voltage detector
Multimeter
Packaged heat pump
Wire
Wire connectors

SAFETY REQUIREMENTS
 A. Check all circuits for voltage before doing any service work.
 B. Stand on dry non-conductive surfaces when working on live circuits.
 C. Never bypass any electrical protective devices.

PROCEDURE
 1. Turn the power off to the unit.
 2. Check the power with your non-contact voltage detector or multimeter to make sure it is off.
 3. Determine the correct size power wire using the minimum circuit ampacity listed on the unit.
 4. Size the power wire according to this minimum circuit ampacity.
 5. Size the disconnect switch. It should be at least 115% of the minimum circuit ampacity.
 6. Size the circuit breaker or fuse using the maximum over-current protection listed in the unit.
 7. The power wire should be run in liquid tight conduit from the disconnect to the unit.
 8. Make sure the unit is grounded.
 9. The control wiring should be run from the thermostat to the unit.
 10. The thermostat wire does not need to be run in conduit.

11. The thermostat wire should NOT run in the same conduit as the power wire.
12. Mount the thermostat.
13. Wire the control wiring at the thermostat and the unit.
14. Have the instructor check your work BEFORE energizing.
15. Turn on the power to the unit.
16. Operate the unit in both the cooling and heating cycles.

Note: Many units have a built in delay of several minutes the first time the unit is powered up. This will keep it from coming on immediately after the power has been turned off.

Name_____

Date_____

Instructor's OK ☐

LAB 72.2 INSTALL SPLIT SYSTEM HEAT PUMP REFRIGERATION PIPING

LABORATORY OBJECTIVE

You will demonstrate your ability to install refrigeration piping for a split system heat pump and check the piping for leaks.

FUNDAMENTALS OF HVACR TEXT REFERENCE

Unit 72 Heat Pump Installation
Unit 24 Refrigerant System Piping

LABORATORY NOTES

You will run the refrigeration piping for a split system heat pump and pressure test it for leaks using nitrogen and soap bubbles.

REQUIRED TOOLS, EQUIPMENT, AND MATERIALS

Refrigeration gauges
Tubing cutter
Flaring tool
Split system air conditioner
Refrigeration copper
Oxy-acetylene torch
Brazing material
ArmaFlex™ insulation
Nitrogen cylinder, regulator, and pressure relief valve assembly
Soap bubbles or ultrasonic leak detector

SAFETY REQUIREMENTS

A. One hundred percent cotton or leather clothing is the best material to wear while brazing, soldering, or welding.

B. Shirts should have long sleeves and work gloves must be worn (cloth, leather palm, or all leather).

C. When using acetylene, torch pressure should be approximately 5 psig and the cylinder valve should be open no more than 1 1/2 turns.

PROCEDURE

1. Locate the condensing unit and indoor coil.
2. Determine the tubing size needed for the suction and liquid lines. For residential air conditioning, this is normally done by measuring the pipe stub out on the condensing unit.

3. Roll out the tubing by holding the end of the tubing against the ground and straightening to form a long, straight length of tubing long enough to go between the condenser and the evaporator.

4. *Note:* To keep the tubing clean and dry, the caps or plugs on the ends of the tubing should remain in place until just before the tubing is connected.

5. If you are working with a line set, roll the entire length out.

6. If you are working with a roll of refrigeration copper, roll out just the amount you need plus a little extra for security.

7. If you are working with refrigeration copper, slide lengths of ArmaFlex^{TM} insulation over the suction tubing.

8. Pull a length of thermostat wire beside the piping, leaving plenty of wire on each end. Tape the lines and wire together at regular intervals.

9. Pull the line set through the conduit, chase, or wall between the indoor and outdoor units. Connect the lines to the outside pipe stubs.

10. If the lines are brazed, be sure to use nitrogen to minimize internal oxidation inside the tubing. The ArmaFlex^{TM} insulation should be pushed back and taped to keep it clear of the flames during brazing.

11. *Note:* The insulation must be rated for use on heat pump systems. Because the gas line is hot during the heating cycle, not all pipe insulation is rated for heat pump application.

12. Connect the suction line to the inside pipe stub.

13. Brazing should be done after a nitrogen purge to prevent oxidation. Work quickly when brazing the suction line.

14. If the evaporator has an internally installed TEV, overheating the suction line will destroy the TEV.

15. A filter drier should be installed in the liquid line.

16. The filter must be a bi-flow filter rated for heat pump use. Connect the liquid line to the filter drier.

17. Connect a short line from the filter drier to the evaporator coil.

18. Pressurize the lines with nitrogen and check for leaks using soap bubbles or an ultra-sonic leak detector.

LAB 72.3 WIRE THERMOSTAT AND CONTROL WIRING

LABORATORY OBJECTIVE
The purpose of this lab is to learn how to wire a low voltage thermostat to a split system heat pump.

FUNDAMENTALS OF HVACR TEXT REFERENCE
Unit 72 Heat Pump Installation

LABORATORY NOTES
You will wire a two stage heat pump thermostat to control a split system heat pump. You will need to wire the control wiring at the thermostat, blower coil control panel, and condensing unit.

REQUIRED TOOLS, EQUIPMENT, AND MATERIALS
Low voltage heat pump thermostat
Split system heat pump
Multimeter
Electrical hand tools

SAFETY REQUIREMENTS
Check all circuits for voltage before doing any service work.

PROCEDURE
1. Read the installation instructions for the thermostat and heat pump.
2. Turn off the power to the unit.
3. Check the voltage to the unit with a multimeter to be sure the power is off.
4. Mount the thermostat. The thermostat should be level.
5. If the thermostat is mounted on metal, be careful that electrical circuits are not contacting the metal.
6. Wire the thermostat wire to the thermostat sub-base.
7. Wire the thermostat wire to the unit terminal connections or pigtails (depending on the unit).
8. Have an instructor check your wiring.
9. Install thermostat on sub-base.
10. Turn the power on and operate the unit.

LAB 72.4 EVACUATE AND CHARGE A SPLIT SYSTEM HEAT PUMP

LABORATORY OBJECTIVE
You will demonstrate your ability to evacuate and charge a split system heat pump.

FUNDAMENTALS OF HVACR TEXT REFERENCE
Unit 72 Heat Pump Installation
Unit 28 Refrigerant System Evacuation
Unit 29 Refrigerant System Charging

LABORATORY NOTES
You will pull a deep vacuum on a split system air conditioner down to 500 microns and weight in the correct refrigerant charge in liquid form into the high side of the system.

REQUIRED TOOLS, EQUIPMENT, AND MATERIALS
Hand tools
Refrigeration gauges
Temperature tester
Split system heat pump
Vacuum pump
Extension cord
Vacuum gauge
Refrigerant scale

SAFETY REQUIREMENTS
A. Wear safety goggles and gloves when working with refrigerants. Liquid refrigerant can cause frostbite when in contact with eyes or skin.
B. Use low loss hose fittings, or wrap cloth around hose fittings before removing the fittings from a pressurized system or cylinder. Inspect all fittings before attaching hoses.

PROCEDURE
1. Connect your gauges to the system.
2. If the system has a nitrogen holding charge, release the nitrogen to the air.
3. Do NOT release the charge if it is charged with refrigerant.
4. Connect the vacuum gauge and vacuum pump.
5. Evacuate the system down to 500 microns on the vacuum gauge.
6. *Note:* on a new split system you are only evacuating the lines and evaporator coil. With a leak free system, evacuating just the lines and coil should not take very long.
7. While the vacuum is pulling, calculate the total system charge.

8. The total charge should be the condensing unit factory charge + an amount of refrigerant for line length.

9. For most systems, the amount of refrigerant to add for line length is 0.6 ounces per foot over 15 feet of line.

10. Once the vacuum is achieved, close the refrigeration gauges and turn off the vacuum pump.

11. On new split systems, weigh in the amount of refrigerant needed for line length and then open the unit installation valves to let the refrigerant that is in the condenser into the rest of the system.

12. If the system is a dry system or an older system (not newly installed), you need to weigh in the full charge including the factory charge + the line length allowance.

LAB 72.5 HEAT PUMP START AND CHECK

LABORATORY OBJECTIVE
You will demonstrate your ability to perform an initial unit start and check procedure on a heat pump system.

FUNDAMENTALS OF HVACR TEXT REFERENCE
Unit 72 Heat Pump Installation
Unit 89 Installation Techniques

LABORATORY NOTES
You will operate a packaged air conditioning unit, perform an initial unit start and check procedure, and record operational data on the worksheet.

REQUIRED TOOLS, EQUIPMENT, AND MATERIALS
Multimeter
Ammeter
Electrical hand tools
Refrigeration gauges
Temperature tester
Heat pump

SAFETY REQUIREMENTS
Check all circuits for voltage before doing any service work.

PROCEDURE
Pre-Start Inspection—Electrical

Record all electrical data from unit, including:

1. ***Minimum supply voltage, maximum supply voltage, compressor RLA, outdoor fan FLA, indoor blower FLA***

2. Unit MCA (minimum circuit ampacity)—Compare actual wire size in NEC to Unit MCA.

3. Unit MOP (maximum overcurrent protection)—Make sure actual fuse or circuit breaker does not exceed the maximum overcurrent protection.

4. Make sure the unit is properly grounded.

Pre-Start Inspection—Refrigeration
1. Measure and record the outdoor ambient temperature, the indoor temperature, and the indoor wet bulb temperature.
2. Check the type of refrigerant and measure the equalized refrigerant pressures.

726

Pre-Start Inspection—Mechanical

1. Check to see that all packing materials and shipping supports have been removed.
2. Spin the fan blades by hand to verify that they are not hanging or seized.
3. Check to see that an air filter is installed.
4. Pour water in the evaporator condensate drain and verify that the drain work properly.

Initial Power Up

1. Power should be applied to the unit for 24 hours before turning it on. (This requirement is waived for this lab.)
2. Check the incoming voltage to the unit with the thermostat set to off and the disconnect switch on.
3. The voltage should be between the minimum and maximum operating voltages stated on the unit data plate.
4. If the voltage is NOT within the acceptable operating range, turn the disconnect switch off and determine why.
5. If the outdoor temperature is above 60°F, set the thermostat to "Cool" and run the temperature down to call for cooling.
6. If the outdoor temperature is below 60°F, set the thermostat to "Heat" and run the temperature up to call for heat.
7. After the unit starts, recheck the voltage to the unit.
8. It should still be within the acceptable operating voltages listed on the unit data plate.
9. If the voltage is NOT within the acceptable operating range, the voltage drop through the power wire is unacceptably high—turn off the power!
10. Check the amp draw of the power wire feeding the unit and compare it to the unit data plate.
11. The amp draw should be lower than the minimum circuit ampacity stated on the data plate Check the amp draw of the compressor, condenser fan motor, and evaporator fan motor and compare to the RLA and FLA ratings in the data plate.
12. The readings most likely will not be exactly what is on the data plate, but they should be in the same general range.

Operational Checks—General

1. You should check to see that the heat pump operates correctly in all stages, including:
   a. Off—nothing should run. If the unit has been operating, there may be a delay before the unit shuts off.
   b. Fan On—only the indoor blower should operate
   c. Cooling—Compressor and both fans should run. Cool air should be leaving the ducts.
   d. Heating, first stage—Compressor and both fans should run. Warm air leaving the ducts.
   e. Defrost—compressor, indoor blower, and strips should operate. Outdoor fan should not operate. See 73.7 Forcing a Defrost in *Fundamentals of HVACR* for details on how to initiate the defrost cycle.
   f. Heating, Second Stage—Compressor, both fans, and heat strips should all operate. There may be a delay before all strips are energized. Hot air should be leaving the ducts.
   g. Emergency Heat—only the indoor blower and strips should operate. Hot air should be leaving the ducts.

Operational Checks—Airflow

1. If the outdoor temperature is above 60°F, set the thermostat to "Cool" and run the temperature down to call for cooling.
2. If the outdoor temperature is below 60°F, set the thermostat to "Heat" and run the temperature up to call for heat.
3. Read the static pressure difference between the return and supply duct. Compare this total external static reading to the manufacturer's specification.
4. If you do not have a specification, assume 0.5" wc to be "normal." Anything over 0.7" wc is normally too high.
5. Measure the system airflow using a flowhood or other airflow measuring instrument. Compare the actual airflow reading to the manufacturer's specification.
6. If you do not have a manufacturer's specification, the airflow should be approximately 400 CFM per ton.

Operational Checks—Refrigeration

1. Operate the unit in cooling.
2. *Note:* if the outdoor temperature is below 50°F, it may not be possible to check the system refrigerant pressures in cooling.
3. After the unit has operated long enough for the temperatures and pressures to stabilize, record the system suction and discharge pressures.
4. Compare the pressures to the manufacturer's specifications.
5. Measure the suction line temperature and calculate the suction superheat (suction line temperature - evaporator saturation temperature from gauges).
6. Compare the superheat to the manufacturer's specifications.
7. Measure the liquid line temperature and calculate the liquid line subcooling (Condenser saturation temperature from gauges - liquid line temperature)
8. Compare the liquid subcooling to the manufacturer's specifications.
9. Read the temperature of the discharge line leaving the compressor.
10. For most units, any temperature 200°F and higher indicates that the compressor is overheating.
11. Read the temperature of the return air and supply air.
12. Calculate the temperature difference across the coil (return air temperature - supply air temperature)
13. Compare the coil temperature difference to the manufacturer's specifications.
14. This varies depending upon the conditions and airflow.
15. In cooling, normal temperature drops can be anywhere between 10°F and 20°F.
16. In heating, normal temperature rise can be anywhere between 10°F and 30°F.
17. Set the thermostat to "Off" and wait for the system to shut off.
18. Temporarily disable the heat strips by disconnecting the 24-volt wire control wire to the heat strips.
19. Operate the unit in heating and repeat all the measurements.
20. *Note:* If the outdoor temperature is above 75°F, it may not be possible to check the system refrigerant pressures in heating.

HEAT PUMP START AND CHECK		

Pre-Start Check List				
Electrical Pre-Start Checks		**Refrigeration Pre-Start Checks**		**Mechanical Checks**
Minimum Supply Voltage		Outdoor Ambient Temperature		Shipping Materials Removed
Maximum Supply Voltage		Indoor Dry Bulb Temperature		Condenser Fan Spins
Compressor RLA		Indoor Wet Bulb Temperature		Evaporator Fan Spins
Condenser Fan FLA		Type of Refrigerant		Air Filter in Place
Evaporator Fan FLA		Equalized Refrigerant		Evaporator Drain Works
MCA (Min Cir Amps)				
Wire sized to MCA				
Max Fuse Size				
Actual Fuse Size Installed				

Initial Power-up—Cooling		Initial Power-up—Emergency Heat	
Voltage at outdoor unit off		Voltage at indoor unit with unit off	
Voltage at unit while running		Voltage at unit while running	
Amp draw on power wire		Amp draw on power wire	
Compressor amp draw		Amp draw on strips	
Condenser fan motor amp draw		Indoor blower motor amp draw	

Refrigeration Operational Checks	Cool	Heat	Airflow Operational Checks	
Suction Pressure			Total External Static	
Evaporator Saturation			Measured Total External Static	
Condenser Pressure			Airflow CFM Specified	
Condenser Saturation			Measured CFM	
Suction Line Temperature			Return Air Temperature	
Suction Superheat			Supply Air Temperature	
Liquid Line Temperature			Indoor Coil Temperature	
Liquid Subcooling				
Discharge Line Temperature				

LAB 72.6 DETERMINING BALANCE POINT

LABORATORY OBJECTIVE
Given the unit specifications and house heat load, you will determine the system balance point.

FUNDAMENTALS OF HVACR TEXT REFERENCE
Unit 72 Heat Pump Installation
Unit 70 Air-Source Heat Pump Applications

SPECIFICATIONS
House heat load is 45,000 BTUH at 20°F outside. Heat pump capacity is 36,000 BTUH at 47°F and 18,000 BTUH at 17°F.

PROCEDURE
1. Construct a graph with the house load on one axis and the unit capacity on the other axis.
2. Plot the house line using the design load of 45,000 BTUH at 20°F and an assumed load of 0 BTUH 65°F.
3. Plot the unit capacity line using 36,000 BTUH at 47°F and 18,000 BTUH at 17°F.
4. The intersection is the balance point.
5. What is the balance point temperature?

6. What is the house heat load at the balance point?

7. What is the system capacity at the balance point?

8. How much supplemental heat (BTUH) does the system require at the balance point?

9. How many 5 kw electric strip heaters would be required? (Each produces approximately 17,000 BTUH).

LAB 73.1 PACKAGED HEAT PUMP CHARGING–COOLING CYCLE

LABORATORY OBJECTIVE
You will determine if the charge is correct on a packaged heat pump using a manufacturer's charging chart in the cooling cycle.

FUNDAMENTALS OF HVACR TEXT REFERENCE
Unit 73 Troubleshooting Heat Pump Systems
Unit 29 Refrigerant System Charging
Unit 72 Heat Pump Installation

REQUIRED TOOLS, EQUIPMENT, AND MATERIALS
Safety glasses
Gloves
Refrigeration gauges
6-in-1 screwdriver
Temperature tester
Packaged heat pump with a manufacturer's charging chart, which correlates suction pressure, liquid pressure, and outdoor ambient temperature

SAFETY REQUIREMENTS
A. Wear safety goggles and gloves when working with refrigerants. Liquid refrigerant can cause frostbite when in contact with eyes or skin.
B. Use low loss hose fittings, or wrap cloth around hose fittings before removing the fittings from a pressurized system or cylinder. Inspect all fittings before attaching hoses.

PROCEDURE
1. Find the manufacturer's charging chart.
2. Read the instructions on the chart.
3. Pay attention to any specific operating conditions that must be met for the chart to be accurate.
4. Most charts assume that the return air is near comfort conditions.
5. Install your gauges on the assigned packaged heat pump.
6. Start the unit in cooling.
7. Verify that the system has good airflow across both the evaporator and condenser.
8. Measure the outdoor ambient temperature.
9. Allow the unit to run long enough for the pressures to stabilize.
10. Determine the intersection of the suction and liquid pressures on the chart.
11. If the intersection is below the ambient temperature line, the system is undercharged.
12. If the intersection is above the ambient temperature line, the system is overcharged.

Note: Some allowance for error must be made. If the intersection is within 3°F, the system is probably charged correctly.

731

LAB 73.2 HEAT PUMP CHARGING–COOLING CYCLE SUPERHEAT

LABORATORY OBJECTIVE
Using the superheat method, you will determine if the charge is correct on a heat pump with a fixed restriction metering device operating in cooling.

FUNDAMENTALS OF HVACR TEXT REFERENCE
Unit 73 Troubleshooting Heat Pump Systems
Unit 29 Refrigerant System Charging
Unit 72 Heat Pump Installation

REQUIRED TOOLS, EQUIPMENT, AND MATERIALS
Safety glasses
Gloves
Refrigeration gauges
6-in-1 screwdriver
Temperature tester
Heat pump with fixed restriction metering device and a superheat charging chart

SAFETY REQUIREMENTS
A. Wear safety goggles and gloves when working with refrigerants. Liquid refrigerant can cause frostbite when in contact with eyes or skin.
B. Use low loss hose fittings, or wrap cloth around hose fittings before removing the fittings from a pressurized system or cylinder. Inspect all fittings before attaching hoses.

PROCEDURE
1. You will be operating the unit in cooling.
2. Find the manufacturer's charging specifications.
3. Read the instructions on the chart.
4. Pay attention to any specific operating conditions that must be met for the chart to be accurate.
5. Many charts assume that the return air is near comfort conditions.
6. Install your gauges on the assigned heat pump.
7. Start the unit in cooling.
8. Verify that the system has good airflow across both the evaporator and condenser.
9. Measure the outdoor ambient temperature.
10. Read the suction pressure on the compound gauge.
11. Use a PT chart to convert the suction pressure to a saturated temperature.
12. Measure the suction line temperature.
13. The superheat is the temperature difference between the suction line temperature and the evaporator saturation temperature.
14. If the superheat is too low, the unit is overcharged.
15. If the superheat is too high, the unit is undercharged.

Name_____

Date_____

Instructor's OK ☐

LAB 73.3 HEAT PUMP CHARGING–COOLING CYCLE SUBCOOLING

LABORATORY OBJECTIVE
Using the subcooling method, you will determine if the charge is correct on a heat pump with a fixed restriction metering device operating in cooling.

FUNDAMENTALS OF HVACR TEXT REFERENCE
Unit 73 Troubleshooting Heat Pump Systems
Unit 29 Refrigerant System Charging
Unit 72 Heat Pump Installation

REQUIRED TOOLS, EQUIPMENT, AND MATERIALS
Safety glasses
Gloves
Refrigeration gauges
6-in-1 screwdriver
Temperature tester
Heat pump with TEV metering device and subcooling charging chart

SAFETY REQUIREMENTS
A. Wear safety goggles and gloves when working with refrigerants. Liquid refrigerant can cause frostbite when in contact with eyes or skin.
B. Use low loss hose fittings, or wrap cloth around hose fittings before removing the fittings from a pressurized system or cylinder. Inspect all fittings before attaching hoses.

PROCEDURE
1. You will be operating the unit in cooling.
2. Find the manufacturer's charging specifications.
3. Read the instructions on the chart.
4. Pay attention to any specific operating conditions that must be met for the chart to be accurate.
5. Many charts assume that the return air is near comfort conditions.
6. Install your gauges on the assigned heat pump.
7. Start the unit in cooling.
8. Verify that the system has good airflow across both the evaporator and condenser.
9. Read the liquid pressure on the high pressure gauge.
10. Use a PT chart to convert the liquid pressure to a saturated temperature.
11. Measure the liquid line temperature.
12. The subcooling is the temperature difference between the liquid saturation temperature and the liquid line temperature.
13. If the subcooling is too low, the unit is undercharged.
14. If the subcooling is too high, the unit is overcharged.

LAB 73.4 SPLIT SYSTEM HEAT PUMP CHARGING–COOLING CYCLE: COLD AMBIENT TEMPERATURE

LABORATORY OBJECTIVE
You will determine if the charge is correct on a split system heat pump using a liquid line approach temperature method operating the heat pump in a cold ambient temperature.

FUNDAMENTALS OF HVACR TEXT REFERENCE
Unit 73 Troubleshooting Heat Pump Systems
Unit 72 Heat Pump Installation

LABORATORY NOTES
You will operate a heat pump in the cooling cycle in a cold ambient and check its refrigerant charge.

REQUIRED TOOLS, EQUIPMENT, AND MATERIALS
Safety glasses
Gloves
Refrigeration gauges
6-in-1 screwdriver
Temperature tester
Split system heat pump with liquid line ambient temperature approach charging instructions

SAFETY REQUIREMENTS
A. Wear safety goggles and gloves when working with refrigerants. Liquid refrigerant can cause frostbite when in contact with eyes or skin.
B. Use low loss hose fittings, or wrap cloth around hose fittings before removing the fittings from a pressurized system or cylinder. Inspect all fittings before attaching hoses.

PROCEDURE
1. You will be operating the unit in cooling even though the outdoor ambient is cold.
2. Find the manufacturer's specification for the liquid line approach method.
3. Read the instructions on the chart.
4. Pay attention to any specific operating conditions that must be met for the chart to be accurate.
5. Some manufacturers require blocking the condenser to achieve a minimum liquid pressure. Most charts assume that the return air is near comfort conditions.
6. Install your gauges on the assigned heat pump.
7. Start the unit in cooling.
8. Verify that the system has good airflow across both the evaporator and condenser.
9. Measure the outdoor ambient temperature.
10. Measure the liquid line temperature.

11. The approach is the temperature difference between the liquid line temperature and the ambient temperature.
12. If the liquid line is cooler than expected, the unit is overcharged.
13. If the liquid line is warmer than expected, the unit is undercharged.

LAB 73.5 PACKAGED HEAT PUMP CHARGING–HEATING CYCLE

LABORATORY OBJECTIVE
You will determine if the charge is correct on a packaged heat pump using a manufacturer's charging chart in the heating cycle.

FUNDAMENTALS OF HVACR TEXT REFERENCE
Unit 73 Troubleshooting Heat Pump Systems
Unit 72 Heat Pump Installation

LABORATORY NOTES
You will operate a packaged heat pump in the heating cycle and check its refrigerant charge.

REQUIRED TOOLS, EQUIPMENT, AND MATERIALS
Safety glasses
Gloves
Refrigeration gauges
6-in-1 screwdriver
Temperature tester
Packaged heat pump with a manufacturer's charging chart, which correlates suction pressure, liquid pressure, and outdoor ambient temperature.

SAFETY REQUIREMENTS

A. Wear safety goggles and gloves when working with refrigerants. Liquid refrigerant can cause frostbite when in contact with eyes or skin.

B. Use low loss hose fittings, or wrap cloth around hose fittings before removing the fittings from a pressurized system or cylinder. Inspect all fittings before attaching hoses.

PROCEDURE

1. Find the manufacturer's charging chart.
2. Read the instructions on the chart.
3. Pay attention to any specific operating conditions that must be met for the chart to be accurate.
4. Most charts assume that the return air is near comfort conditions.
5. Install your gauges on the assigned packaged heat pump.
6. Start the unit in heating.
7. Verify that the system has good airflow across both the evaporator and condenser.
8. Measure the return air temperature.
9. Allow the unit to run long enough for the pressures to stabilize.
10. Determine the intersection of the suction and liquid pressures on the chart.
11. If the intersection is below the ambient temperature line, the system is undercharged.
12. If the intersection is above the ambient temperature line, the system is overcharged.

    Note: Some allowance for error must be made. If the intersection is within 3°F, the system is probably charged correctly.

LAB 73.6 HEAT PUMP CHARGING–HEATING CYCLE DISCHARGE LINE TEMPERATURE

LABORATORY OBJECTIVE
You will determine if the charge is correct on a heat pump operating in the heating cycle using the discharge line temperature method.

FUNDAMENTALS OF HVACR TEXT REFERENCE
Unit 73 Troubleshooting Heat Pump Systems
Unit 72 Heat Pump Installation

LABORATORY NOTES
You will operate a heat pump in the heating cycle and check its refrigerant charge using the discharge line temperature.

REQUIRED TOOLS, EQUIPMENT, AND MATERIALS
Safety glasses
Gloves
Refrigeration gauges
6-in-1 screwdriver
Temperature tester
Heat pump with discharge line temperature specification

SAFETY REQUIREMENTS
A. Wear safety goggles and gloves when working with refrigerants. Liquid refrigerant can cause frostbite when in contact with eyes or skin.
B. Use low loss hose fittings, or wrap cloth around hose fittings before removing the fittings from a pressurized system or cylinder. Inspect all fittings before attaching hoses.

PROCEDURE
1. You will be operating the unit in heating.
2. Find the manufacturers discharge line temperature specifications.
3. Read the instructions on the chart.
4. Pay attention to any specific operating conditions that must be met for the chart to be accurate.
5. Read the ambient temperature.
6. Start the unit in heating.
7. Verify that the system has good airflow across both the evaporator and condenser.
8. Read the discharge line temperature within 60 of the compressor.
9. It should be 100°F–110°F warmer than the ambient temperature. (Use manufacturers specific specification if available.)

10. If the temperature more than 100°F–110°F warmer than the ambient temperature, the unit is undercharged.

11. If the subcooling is less than 100°F–110°F warmer than the ambient temperature too high, the unit is overcharged.

LAB 73.7 HEAT PUMP CHARGING–HEATING CYCLE TEMPERATURE RISE

LABORATORY OBJECTIVE
You will determine if the charge is correct on a heat pump operating in the heating cycle using the temperature rise method.

FUNDAMENTALS OF HVACR TEXT REFERENCE
Unit 73 Troubleshooting Heat Pump Systems
Unit 72 Heat Pump Installation

LABORATORY NOTES
You will operate a heat pump in the heating cycle and check its refrigerant charge using the air temperature rise.

REQUIRED TOOLS, EQUIPMENT, AND MATERIALS
Safety glasses
Gloves
Refrigeration gauges
6-in-1 screwdriver
Temperature tester
Heat pump with indoor temperature rise specification

SAFETY REQUIREMENTS
A. Wear safety goggles and gloves when working with refrigerants. Liquid refrigerant can cause frostbite when in contact with eyes or skin.

B. Use low loss hose fittings, or wrap cloth around hose fittings before removing the fittings from a pressurized system or cylinder. Inspect all fittings before attaching hoses.

PROCEDURE
1. You will be operating the unit in heating.
2. Find the manufacturer's temperature rise specifications.
3. Read the instructions on the chart.
4. Pay attention to any specific operating conditions that must be met for the chart to be accurate.
5. Read the ambient temperature.
6. Start the unit in heating.
7. Verify that the system has good airflow across both the evaporator and condenser.
8. Read the return air temperature at the return air plenum.
9. Read the supply air temperature at the supply air plenum.
10. Determine the appropriate temperature rise for the outdoor ambient temperature.

11. The actual measured temperature difference should be + or −3°F of the desired temperature rise.

12. If the temperature rise is less than specified, the unit is undercharged or the airflow is above specification.

13. If the temperature rise is more than specified, the airflow is lower than specification.

LAB 73.8 CHECKING DEFROST CONTROLS

LABORATORY OBJECTIVE
You will check an electronic time–temperature defrost control by manually putting it into defrost.

***FUNDAMENTALS OF HVACR* TEXT REFERENCE**
Unit 73 Troubleshooting Heat Pump Systems

LABORATORY NOTES
You will operate a heat pump in the heating cycle, cause frost to form on the outdoor coil, and check the defrost control.

REQUIRED TOOLS, EQUIPMENT, AND MATERIALS
Safety glasses
6- in-1 screwdriver
Non-contact voltage probe
Multimeter
Heat pump with defrost control

SAFETY REQUIREMENTS
Check all circuits for voltage before doing any service work.

PROCEDURE
1. First, you need to get the outdoor coil to freeze.
2. If the weather is cool, you can block the airflow to the outdoor coil using plastic trash bags.
3. If it is hot outside, you may need to disconnect the outdoor fan motor.
   a. Turn the power off to the unit.
   b. Check with your non-contact voltage detector or multimeter to insure power is off.
   c. Remove the fan wire from the defrost control and tape it up.
4. Operate the unit in heating until the coil has a good coating of frost around the defrost thermostat.
5. While the unit is still operating in heating, jump the test pins.
6. The location and labeling of the test pins will vary depending upon the board and manufacturer.
7. The unit may not react immediately, hold on the jumper for up to 2 minutes.
8. The unit should shift into cooling cycle.
9. Remove the jumper and wait for the unit to come out of defrost.
10. Turn the power off after the unit has come back out of defrost.
11. If you removed the outdoor fan wire, replace the fan wire.

LAB 73.9 TROUBLESHOOTING HEAT PUMPS

LABORATORY OBJECTIVE
Given a heat pump with a problem, you will identify the problem, its root cause, and recommend corrective action. Labs 68.9–68.11 will all follow the same procedure, however each lab will present a different problem scenario based upon the unit and corrective action required.

FUNDAMENTALS OF HVACR TEXT REFERENCE
Unit 73 Troubleshooting Heat Pump Systems

LABORATORY NOTES
You will troubleshoot problems in the assigned heat pump. You should determine the root cause and recommend corrective action.

REQUIRED TOOLS, EQUIPMENT, AND MATERIALS
Safety glasses
Hand tools
Manometer
Multimeter
Heat pump with a problem

SAFETY REQUIREMENTS
Check all circuits for voltage before doing any service work.

PROCEDURE
1. Troubleshoot the heat pump assigned by the instructor. Be sure to be complete in your description of the problem. You should include the following:
    a. What is the furnace is doing wrong?

    b. What component or condition is causing this?

    c. What tests did you perform that told you this?

    d. How would you correct this?

LAB 74.1 LEED BUILDING CERTIFICATION

LABORATORY OBJECTIVE
You will research designs, materials, and technology used to achieve recognition as a LEED Building. You will then demonstrate your understanding by completing Data Sheet 74-1-1.

FUNDAMENTALS OF HVACR TEXT REFERENCE
Unit 74 Green Buildings and Systems

LABORATORY NOTES
LEED stands for Leadership in Energy and Environmental Design. LEED is the most widely recognized green building rating system in the world. LEED provides a framework for healthy, highly efficient, and cost-saving green buildings. LEED buildings are recognized as models of sustainability.

REQUIRED TOOLS, EQUIPMENT, AND MATERIALS
Computer with Internet access

PROCEDURE
Refer to Unit 74 Green Buildings and Systems and research on the internet to complete Data Sheet 74-1-1.

Step 1 – Matching projects to LEED rating system
Use the online "Discover LEED" tool to match the listed projects to a LEED Rating system. The tool can be found at https://www.usgbc.org/discoverleed/

Data Sheet 74-1-1 Match LED Rating System	
Project Description	**Appropriate LEED Rating System**
New single family residential	
New retail store	
Planning new neighborhood	
Improving operational efficiency of a school	
Redesign hotel interior	

Step 2 – Levels of LEED Rating

LEED levels are awarded by a point system. Determine how many points are required for each LEED level and complete Data Table 74-1-2 LEED Certification Levels.

Data Table 74-1-2 LEED Certification Levels	
LEED Level	**Points required**
Certified	
Silver	
Gold	
Platinum	

Step 3 – LEED Credit Categories

LEED points are earned in different credit categories. List the different credit categories in Data Sheet 74-1-3 LEED Credit Categories.

Data Table 74-1-3 LEED Credit Categories

LAB 75.1 MANUAL J8 BLOCK LOAD

LABORATORY OBJECTIVE
You will demonstrate how to use the Manual J8ae Speed-Sheet to calculate a block load.

FUNDAMENTALS OF HVACR TEXT REFERENCE
Unit 75 Residential Load Calculations

LABORATORY NOTES
The Manual J Speed-Sheet is an Excel file that automates Manual J calculations. You will use the Manual J8ae Speed-Sheet and tables from Manual J8 abridged edition to calculate the heat loss and gain of a house using the block load procedure.

REQUIRED TOOLS, EQUIPMENT, AND MATERIALS
Manual J8 abridged or tables from Manual J8 abridged
Computer with Microsoft Excel
Manual J8ae Speed-Sheet
Plans for a small one-story house

PROCEDURE
This Excel Sheet has tabs for different parts of the calculation.

Construction Details
You can either use your own plan or the plan in Figure 75-1-1. Table 75-1-1 provides construction details for the plan in Figure 75-1-1.

Design Conditions
1. Start by clicking on the *J1 Form* tab at the bottom.
2. Enter the Project Name and Indoor design conditions in the white spaces at the top of the J1 Form.
3. Select the State by clicking in the white box and then clicking on the arrow to get a drop down list to select the state.
4. After selecting the state, click on the white box to the right of the state to select the city.
5. Most of the J1 Form is completed by the entries made on the Yellow tabs.

Doors
1. Click on the *Doors* tab.
2. Fill in the construction number and details in the left column and the U-value in the right column. You can find these data in manual J Table 4A Construction #11 or in Table 75-1-1.
3. You do not need an entry for every door. Rather, you need an entry for every type of exterior door. If a house has three exterior doors that are all identical in construction, you only need one entry.

Walls

1. Next, click on the Yellow **Walls** tab. (We will do the glass and windows last).
2. Fill in the construction number and details in the left column and the U-value in the middle column, and the Group number in the right column. You can find these data in manual J Table 4A Construction numbers 12–15. Alternatively, you can use the data in Table 75-1-1.
3. You do not need an entry for every wall. Rather, you need an entry for every type of wall. If a house has a basement, you would need up to three entries: the normal frame walls, the basement wall below grade, and the basement wall above grade.

Ceilings

1. Click on the yellow **Ceiling** tab.
2. Fill in the construction number and details in the left column and the U-value in the middle column, and the Cooling Temperature Difference (CLTD) in the right column. You can find these data in manual J, Table 4A Construction numbers 16–18. Or if you like, you can use the data in Table 75-1-1.
3. You do not need an entry for every ceiling. Rather, you need an entry for every type of ceiling. If a house has a cathedral ceiling it would have two entries: one for the regular ceilings and another for the cathedral ceiling.

Floors

1. Click on the yellow **Floors** tab.
2. Fill in the construction number and details in the left column and the U-value in the right column. You can find these data in manual J, Table 4A Construction numbers 19–22. Or if you like, you can use the data in Table 75-1-1.
3. Table 75-1 tells us that this floor is concrete slab on grade with no edge insulation. That is construction number 22 on the Floors tab of the Speedsheet.
4. You do not need an entry for every floor. Rather, you need an entry for every type of floor. Most houses will only have one type of floor.

Windows

1. Click on the yellow **Glass** tab.
2. Fill in the construction number and details in the left column, the U-value in the middle column, and the Cooling Heat Transfer Multiplier (HTM) on the right under the column for the direction the window faces. You can find these data in manual J, Table 3 or use the data in Table 75-1-1.
3. You need an entry for every type of window and every facing. If a house has only one type of window with four facings, it needs four entries.
4. Click on the Green **Glass Schedule** tab.
5. This tab calculates the window shade line from the roof overhang.
6. You need an entry for every facing of each type of window in this table.
7. Click on the first yellow space on the left and choose the window type and direction from the drop down list.
8. Enter the overhang information in the white spaces beside it. This information comes from your plans. If you are using the plans in this lab manual the detail is in Table 75-1-1.
9. Repeat this for every window facing and type.

J1 Form Net Area

1. Click on the **J1 Form** green tab.
2. The white columns labeled **Net Area** are where the area of different materials is entered.
3. The first white column labeled **Net Area** is for the entire house, or block load
4. The subsequent **Net Area** columns are used when calculating a load for each room. We will not be using those columns for this lab because we are doing a block load, calculating the house as one big room.

Line 6A Windows Form J1

1. You should see the different window types and HTMs listed in the blue section in the top left.
2. Find line 6A and the white space under the **Net Area** column next to the block load column.
3. Enter the total square foot area of the first window facing. This is not the area of a single window, but the total of all the similar windows facing that direction.
4. Repeat for each window facing. Most houses will have three window facings since east and west are the same.

Line 7 Doors Form J1

1. Go to line 7 for doors and click on the first yellow line to get a drop down box.
2. Select the type of door.
3. In the white box under **Net Area**, enter the total area of all the doors that are similarly constructed.
4. If the house has more than one type of door, repeat on the second yellow line.

Line 8 & 9 Walls Form J1

1. Click on the first yellow box on line 8 and select the type of wall.
2. If the house has any below grade walls, repeat on line 9 **Below Grade Walls**
3. In the white box under **Net Area** enter the total square foot area of the above grade walls.
4. If the house has a basement, enter the total area of below grade walls in the white box on line 9.

Line 10 Ceiling Form J1

1. Click on the first yellow box on line 10 and select the type of ceiling.
2. In the white box under **Net Area** enter the total square foot area of the ceiling.

Line 11a Passive Floors Form J1

Note: This is for wooden floors built over unconditioned spaces.

1. Click on the first yellow box on line 11 and select the type of floor.
2. In the white box under **Net Area** enter the total square foot area of the floor.

Line 11b Exposed Floors Form J1

Note: This is for wooden floors exposed to the outside.

1. Click on the second yellow box on line 11 and select the type of floor.
2. In the white box under **Net Area** enter the total square foot area of the floor.

Line 11c Slab Floors Form J1

Note: This is for concrete slab floors on grade. This is the type of floor for this exercise.
1. Click on the third yellow box on line 11 and select the type of floor.
2. Enter the total perimeter distance, NOT the square feet of floor.

Line 11d Basement Floors Form J1

Note: This is for concrete slab floors on grade.
1. Click on the fourth yellow box on line 11 and select the type of floor.
2. In the white box under **Net Area** enter the total square foot area of the floor.

Line 12 Infiltration Form J1

1. On line 12, click to select the house tightness and the number of fireplaces.
2. In the large white box fill in the total heated and cooled floor area.
3. In the next white box enter the total above grade volume. You find that by multiplying the floor area by the ceiling height. Do not count basement floor area even if it is heated.

Line 13 Internal Gains Form J1

1. Click on the yellow box to select the appliance **Scenario.**
2. For a block load, you MUST enter the Appliance BTUs shown in the Yellow box in the Net Area column.
3. In the small white box under it, enter the number of bedrooms.
4. The program will calculate the number of occupants based on the number of bedrooms and display the number in the blue box next to occupants. You MUST enter that exact number in the white box under **Net Area** or the program will refuse to total the calculations.

Line 15 Duct Loss Form J1

1. In the yellow boxes, select the duct type and location, the R-value of the duct insulation, and the duct leakage.
2. In the white boxes enter the installed square feet of duct surface area for the supply and return. This is easier said than done. For now, use 40 square feet for each.

Line 16 Ventilation Form J1

1. Check the boxes for water heater or furnace if the house has a gas water heater or gas furnace.
2. Select the amount of ventilation in the yellow drop down box.
   This is air brought in on purpose, not counting normal bathroom and kitchen ventilation fans. Most residential homes do not have any designed active ventilation.

Line 19 Blower Heat Gain Form J1

Select whether or not the manufacturer considers blower heat gain in the cooling capacity. If you are not sure, to be safe, select that the blower heat gain is not considered.

Line 20 shows the total sensible loss (heating) and gain (cooling)

Line 21 shows the total latent gain (cooling)

Figure 75-1-1 House Plans for Block Load

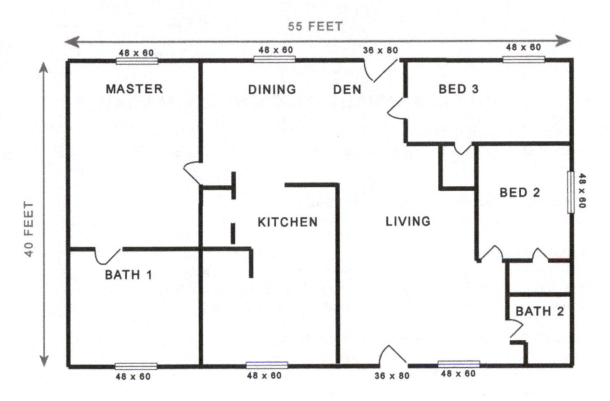

Table 75-1-1 Construction Specifications			
Faces	West		
Component	Description	U Value	Cooling
Doors	Metal with Polystyrene Core	0.29	
Floor	Slab on grade, no edge insulation	1.358 F Value	
Walls	R-11 insulation with OSB Sheathing, no shading	0.097	Group H
Ceiling	8 feet, R30 insulation under ventilated attic with dark roofing	0.032	50° CLTD
Windows	Clear glass, double pane, with wood frame	0.570	East & West 70 HTM South 35 HTM North 21 HTM
Glass Schedule	Overhang = 2 feet Top of Window to Overhang = 1.5 ft		

Name_____

Date_____

Instructor's OK ☐

LAB 75.2 MANUAL J8 ROOM BY ROOM LOAD

LABORATORY OBJECTIVE
You will demonstrate how to use the Manual J8ae speed sheet to calculate a room by room load calculation.

FUNDAMENTALS OF HVACR TEXT REFERENCE
Unit 75 Residential Load Calculations

LABORATORY NOTES
The Manual J Speed-Sheet is an Excel file that automates Manual J calculations. You will use the Manual J8ae Speed-Sheet and tables from Manual J8 Abridged Edition to calculate the heat loss and gain of a house using the room by room procedure.

REQUIRED TOOLS, EQUIPMENT, AND MATERIALS
Manual J8 abridged or tables from Manual J8 abridged
Computer with Microsoft Excel
Manual J8ae Speed-Sheet
Plans for a small one-story house

PROCEDURE
This Excel Sheet has tabs for different parts of the calculation.

The room by room procedure is the same as the block load, but the numbers entered under the **Net Area** column are for each individual room. The **Summary** tab shows the CFM for each room based on the heat loss and gain in each room. The sections labeled **Design Conditions, Doors, Walls, Ceilings, Floors,** and **Windows** are exactly the same for both the block load and the room by room load.

Construction Details
You can either use your own plan or the plan in Figure 75-2-1. Table 75-2-1 provides construction details for the plan in Figure 75-2-1.

Design Conditions
1. Start by clicking on the **J1 Form** tab at the bottom.
2. Enter the Project Name and Indoor design conditions in the white spaces at the top of the J1 Form.
3. Select the State by clicking in the white box and then clicking on the arrow to get a drop down list to select the state.
4. After selecting the state, click on the white box to the right of the state to select the city.
5. Most of the J1 Form is completed by the entries made on the Yellow tabs.

Doors

1. Click on the **Doors** tab.
2. Fill in the construction number and details in the left column and the U-value in the right column. You can find these data in manual J Table 4A Construction #11 or in Table 75-1-1.
3. You do not need an entry for every door. Rather, you need an entry for every type of exterior door. If a house has three exterior doors that are all identical in construction, you only need one entry.

Walls

1. Next, click on the Yellow **Walls** tab. (We will do the glass and windows last)
2. Fill in the construction number and details in the left column and the U-value in the middle column, and the Group number in the right column. You can find these data in manual J Table 4A Construction numbers 12–15. Alternatively, you can use the data in Table 75-1-1.
3. You do not need an entry for every wall. Rather, you need an entry for every type of wall. If a house has a basement, you would need up to three entries: the normal frame walls, the basement wall below grade, and the basement wall above grade.

Ceilings

1. Click on the yellow **Ceiling** tab.
2. Fill in the construction number and details in the left column and the U-value in the middle column, and the Cooling Temperature Difference (CLTD) in the right column. You can find these data in manual J, Table 4A Construction numbers 16–18. Or if you like, you can use the data in Table 75-1-1.
3. You do not need an entry for every ceiling. Rather, you need an entry for every type of ceiling. If a house has a cathedral ceiling it would have two entries: one for the regular ceilings and another for the cathedral ceiling.

Floors

1. Click on the yellow **Floors** tab.
2. Fill in the construction number and details in the left column and the U-value in the right column. You can find these data in manual J, Table 4A Construction numbers 19–22. Or if you like, you can use the data in Table 75-1-1.
3. You do not need an entry for every floor. Rather, you need an entry for every type of floor. Most houses will only have one type of floor.

Windows

1. Click on the yellow **Glass** tab.
2. Fill in the construction number and details in the left column, the U-value in the middle column, and the Cooling Heat Transfer Multiplier (HTM) on the right under the column for the direction the window faces. You can find these data in manual J, Table 3 or use the data in Table 75-2-1.
3. Table 75-2-1 tells us this is a wood floor over an unconditioned crawlspace. That i line 19 on the Floors tab in the Speedsheet.
4. You need an entry for every type of window and every facing. If a house has only one type of window with four facings, it needs four entries.
5. Click on the Green **Glass Schedule** tab.
6. This tab calculates the window shade line from the roof overhang.
7. You need an entry for every facing of each type of window in this table.

8. Click on the first yellow space on the left and choose the window type and direction from the drop down list.
9. Enter the overhang information in the white spaces beside it. This information comes from your plans. If you are using the plans in this lab manual the detail is in Table 75-1-1.
10. Repeat this for every window facing and type.

Room Information
1. Type in the name of each room in the white boxes across the top on the right.
2. *Note:* Only name areas that might reasonably have a supply register. Closets and interior halls normally do not require registers. Their area can be added to adjacent rooms.
3. To see all the room columns, use the right and left arrows at the bottom right of the spread sheet. This scrolls the rooms but leaves the main form in place.

Form J1 Net Area
1. Click on the *J1 Form* green tab.
2. The white columns labeled *Net Area* are where the area of different materials is entered.
3. When doing a Room by Room load, a separate *Net Area* white column is used for each room.

Line 6A Windows
1. You should see the different window types and HTMs listed in the blue section in the top left.
2. For each room: enter the total square foot area of the first window facing in that room. This is not the area of a single window, but the total of all the similar windows facing that direction in each room.
3. Repeat for each room.

Line 7 Doors
1. Go to line 7 for doors and click on the first yellow line to get a drop down box.
2. Select the type of door.
3. For each room: in the white box under *Net Area*, enter the total area of all the exterior doors for that room.
4. All rooms will not have an exterior door. If a room does not have an exterior door, leave that space blank.

Line 8 & 9 Walls
1. Click on the first yellow box on line 8 and select the type of wall.
2. If the house has any below grade walls, repeat on line 9 Below Grade Walls
3. For each room: In the white box under *Net Area* enter the total square foot area of the above grade walls for each room.
4. Repeat for each room in the house.
5. For rooms below ground, enter the total area of the below grade walls in the white box on line 9.

Line 10 Ceiling
1. Click on the first yellow box on line 10 and select the type of ceiling.
2. For each room: In the white box under **Net Area** enter the total square foot area of the ceiling for that room.
3. Repeat for each room.

Line 11a Passive Floors Form J1
Note: This is for wooden floors built over unconditioned spaces. This is the type of floor for the floor plan shown in Figure 75-2-1 of this exercise.
1. Click on the first yellow box on line 11 and select the type of floor.
2. In the white box under **Net Area** enter the total square foot floor area for the room.
3. Repeat for each room.

Line 11b Exposed Floors Form J1
Note: This is for wooden floors exposed to the outside.
1. Click on the second yellow box on line 11 and select the type of floor.
2. In the white box under **Net Area** enter the total square foot floor area for the room.
3. Repeat for each room.

Line 11c Slab Floors Form J1
Note: This is for concrete slab floors on grade.
1. Click on the third yellow box on line 11 and select the type of floor.
2. Enter the total outside perimeter for the room, NOT the square feet of floor.
3. Repeat for each room.

Line 11d Basement Floors Form J1
Note: This is for concrete slab floors on grade.
1. Click on the fourth yellow box on line 11 and select the type of floor.
2. In the white box under **Net Area** enter the total square foot of floor area for the room.
3. Repeat for each room.

Line 12 Infiltration
1. On line 12, click to select the house tightness and the number of fireplaces. In the large white box fill in the total heated and cooled floor area.
2. In the next white box enter the total above grade volume. You find that by multiplying the floor area by the ceiling height. Do not count basement floor area even if it is heated.
3. The Wall Area Ratio (WAR) is automatically calculated based on the ratio of each room's wall area compared to the total wall area. This is used to apportion the infiltration load among the different rooms.

Line 13 Internal Gains
1. Click on the yellow box to select the appliance **Scenario**.
2. You MUST enter enough BTUs in one or more rooms to total exactly the BTUs shown in the Yellow box.
3. In the small white box under it enter the number of bedrooms.

4. The program will calculate the number of occupants based on the number of bedrooms and display the number in the blue box next to occupants.
5. You MUST enter enough people in one or more rooms to total that exact number.

Line 15 Duct Loss

1. In the yellow boxes, select the duct type and location, the R-value of the duct insulation, and the duct leakage.
2. In the white boxes enter the installed square feet of duct surface area for the supply and return.
   This is not easy to determine. For now, use 40 square feet for each.

Line 15 Ventilation

Check the boxes for water heater or furnace if the house has a gas water heater or gas furnace. Select the amount of ventilation in the yellow drop down box.

Line 19 Blower Heat Gain

Select whether or not the manufacturer considers blower heat gain in the cooling capacity. If you are not sure, the safe thing to do is to select that the manufacturer has not considered blower heat gain in the cooling capacity.

Line 20 shows the total sensible loss (heating) and gain (cooling) Line 21 shows the total latent gain (cooling)

SUMMARY

1. Click on the Green *Summary* tab
2. At the top right are two white boxes: Sensible heat ratio and Design CFM
3. The sensible heat ratio is the percentage of equipment cooling capacity that actually cools the air. This is affected by equipment design, airflow, and operating conditions.
4. In humid areas, you should be between 0.6 to 0.8.
5. Lower ratios yield more water removal because more capacity is going into latent cooling. The Design CFM is the airflow the equipment moves.
6. 400 CFM per ton of total cooling capacity (sensible + latent) works on most equipment if you don't have a particular piece of equipment selected.
7. 300 CFM per ton will give better latent cooling performance, but at the cost of operating efficiency.

754

Figure 75-2-1 Room by Room Plan

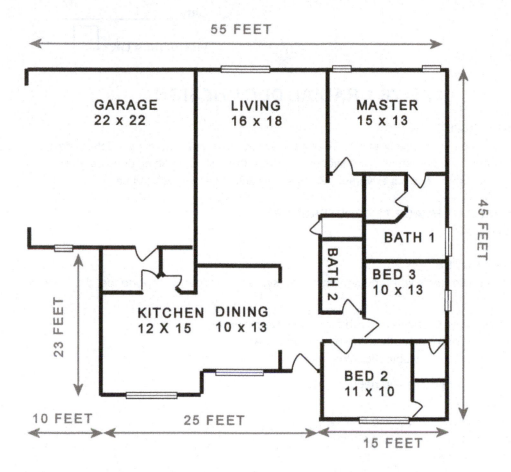

Table 75-2-1 Construction Details			
Faces	West		
Component	Description	U Value	Cooling
Doors	Metal with Polystyrene Core	0.29	
Floor	Wood floor over enclosed crawlspace	0.073	
Walls	R-11 insulation with OSB Sheathing, no shading	0.097	Group H
Ceiling	8 feet, R30 insulation under ventilated attic with dark roofing	0.032	50° CLTD
Windows	Clear glass, double pane, with wood frame	0.570	East & West 70 HTM South 35 HTM North 21 HTM
Glass Schedule	Overhang = 2 feet Top of Window to Overhang = 1.5 ft		

LAB 76.1 RADIAL DUCT DESIGN

LABORATORY OBJECTIVE
You will demonstrate how to use heat loss, heat gain, system capacity, and blower data to design a radial duct system that delivers the conditioned air where the building load is, does not make objectionable noise, and does not adversely affect airflow due to excessive restriction.

FUNDAMENTALS OF HVACR TEXT REFERENCE
Unit 76 Duct Design

LABORATORY NOTES
There are three basic goals when designing a duct system. In order of importance, they are as follows:

1. Allow the system to circulate enough air to operate properly.
2. Distribute the air proportionally to where the heat load is.
3. Keep airflow noise below objectionable levels.

To accomplish all three goals you need to determine
- The system CFM
- The CFM required for each room
- Maximum available system static
- Duct friction rate

REQUIRED TOOLS, EQUIPMENT, AND MATERIALS
A house plan with a completed room by room load study.
You can use Figure 76-1-1 and Table 76-1-1 if you like.
A calculator
A duct friction chart, duct calculator, or duct calculator app
A pencil

PROCEDURE
This lab assumes that the heat load study has been completed and the equipment selected based on the heat load study. You can either provide your own or use the house plan, load study, and equipment data provided in this lab. The overall procedure is
1. determine the duct design static pressure for the supply and return.
2. draw a skeleton drawing of the duct system including the location of all registers and grilles.
3. determine the required CFM for each register based on the load and the system airflow.
4. size the ducts based on required CFM, design static, duct equivalent length, and maximum acceptable velocity.

Determine Total Fixed Pressure Losses

Anything that the air must flow over or through causes a pressure drop. To determine the amount of pressure available to move air through the ductwork we have to first determine how much pressure loss is built into the system already. Common fixed pressure losses include registers and grilles, air filters, and indoor coils. Table 76-1-2 shows typical fixed pressure losses. One question that often comes up is whether or not you need to consider the indoor coil and/or air filter as a fixed pressure loss. The key is remembering that the blower data specifies the total *external* static pressure, meaning you are looking for items outside of the piece of equipment that contains the blower. For a furnace with an air conditioning coil mounted on it, the coil is external to the furnace and you must add its pressure drop to the fixed losses. On the other hand, the coil inside a heat pump indoor unit is inside the box, so it has already been incorporated into the data. It is not considered part of the fixed losses.

Figure 76-1-1 Radial Plan

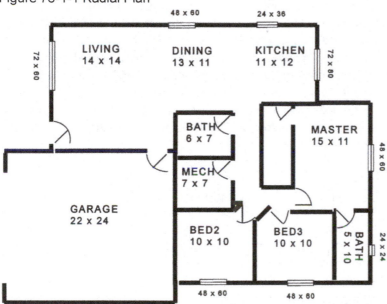

Table 76-1-1 Heat Loss and Gain			
Room	Heat Loss (Heating)	Sensible Heat Gain (Cooling)	Latent Heat Gain (Cooling)
Total House	25594	18898	4549
Living	5507	3830	
Dining	2552	2206	
Kitchen	4911	2977	
Master	3633	2167	
Master Bath	2238	652	
Bedroom 2	3398	2528	
Bedroom 3	2046	2262	
Bath 2	98	112	

Table 76-1-2 Static Pressure Drop of Common HVAC System Components	
Device	Typical Pressure Loss
Standard air filter	0.05 inches wc
High efficiency air filter	0.15 inches wc
Evaporator Coil	0.25 inches wc
Return Air Grille	0.03 inches wc
Supply Air Register	0.03 inches wc
Balancing Damper	0.03 inches wc

Setting Blower Design Point

There are two components of the blower design point: the airflow CFM and the total external static pressure. The system capacity and efficiency are affected by the airflow. As a general rule, more airflow delivers higher efficiency. In cooling, however, the ratio of sensible cooling to latent cooling must also be considered. Too much airflow can reduce the latent cooling to the point that not enough dehumidification is being done. The cooling airflow should ideally be set to deliver the required amount latent cooling according to Manual J. The blower external static is determined by the blower data that matches the design airflow and fan speed. It must be higher than the total fixed pressure losses so that there is pressure available for moving it through the ductwork. You can find your own equipment specifications or use the specifications in Tables 76-1-3 and 76-1-4.

Table 76-1-3 Heat Pump Capacity	
Heating Capacity @ 47°F 24000 BTUH	24000 BTUH
Heating Capacity @ 17°F 11,000 BTUH	14500 BTUH
Sensible Cooling @ AHRI Design and 800 CFM	18720 BTUH
Latent Cooling @ AHRI Design and 800 CFM	5280 BTUH

Table 76-1-4 Heat Pump Blower Data					
Speed Tap	External Static Pressure				
	0.4	0.5	0.6	0.7	0.8
1	545	510	490	410	340
2	700	670	645	595	565
3	760	740	700	670	625
4	885	850	815	800	760
5	1,375	1,335	1,305	1,270	1,240

Duct Design Static

The duct design static is determined by adding up all the fixed air pressure losses and subtracting them from the blower external static. The maximum duct design static pressure is what is left after subtracting fixed pressure losses from the pressure produced by the blower. You do not have to design at the maximum pressure the blower can produce. Designing the duct system at a lower static saves fan energy. Designing at a lower overall external static pressure produces a quieter system with room to adjust the airflow if needed. ECM blowers will automatically adjust to deliver their programmed quantity of air over a wide range of external static pressure, but at the expense of using more energy to overcome extra resistance.

Keep in mind that what you have determined is the total duct system design static, not the pressure to use looking up the duct size. The total includes loss through both the return and supply. The method favored by ACCA's Manual D is to think of the duct system as one long duct from the return grille to the supply register. Basically, adding the total equivalent length of the return and supply together, and then looking up both the supply and return at the same static pressure. Another method is to have separate design pressures for the return and supply while insuring they add up to the total available design static. For systems with more extensive supply ductwork than return ductwork, a common approach is to use two-thirds of the total design static for the supply duct system and one-third of the total design static for the return duct system.

Draw the Duct System

You can draw your duct system directly on Figure 76-1-1, or draw a basic outline of the house plan on a separate sheet of paper and draw the duct system on that. For this lab you are designing a radial system. If you are using the plans in Figure 76-1-1, locate the indoor blower in the Mechanical room. Supply ducts will be in the attic and you will have a central return duct from the mechanical room into the hall. Duct runs will go from the supply plenum to the register without going through main trunk duct. Radial duct systems work well for small houses with just a few runs, all 30 feet or less. Practically speaking, the runs need to be limited to 8 or fewer simply because there is limited area in the supply plenum to cut holes. Also, trying to connect too many runs at the blower can become a tangled mess. One way to reduce the tangle is to use a reducing radial design. Rather than run each run to the plenum, runs can be combined in pairs using wye fittings to reduce the number of ducts and cuts at the plenum.

When using a radial system with the ducts located in the attic, registers are normally placed somewhere near the center of the room. When using a radial system in a crawlspace or basement, the registers are normally located in the floor, under windows on the outside wall because that is where the greatest load is. You should try and avoid register placement that will have registers blowing directly on people.

Determine Register CFM

The CFM for each register is determined by the heat loss and heat gain for the area it serves. To divide the airflow according to load, create airflow divisors by dividing the system capacity by the system CFM for both heating and cooling. These tell you the quantity of BTUH delivered by each CFM of air. You then divide the heating and cooling airflow divisors into the heating and cooling BTUH for each area to get the required heating and cooling CFM. The design CFM will be the higher of the two. As a general rule you normally only need a register in areas that require more than 50 CFM. For example, in Table 76-1-5, Bath 2 has an airflow requirement of only 5 CFM in heating and 7 CFM in cooling. That room does not need a register. That small airflow requirement can be added to a nearby room, such as the Dining Room. Large loads of 400 CFM or more may benefit by splitting the load among multiple registers. Table 76-1-5 shows the CFM airflow divisors and CFM requirements for the plan shown in Figure 76-1-1. You can use Table 76-1-6 if you are using your own house plan and system.

Table 76-1-5 Room Design Airflow

Room	Heat Loss (Heating)	Heating Airflow	Sensible Gain (Cooling)	Cooling Airflow	Design Airflow
Total House	25594	800	18898	800	
Airflow Divisor	32		23.6		
Living	5507	172	3830	162	172
Dining	2552	80	2206	93	93
Kitchen	4911	153	2977	126	153
Master	3633	114	2167	92	114
Master Bath	2238	70	652	28	70
Bedroom 2	3398	106	2528	107	107
Bedroom 3	2046	64	2262	96	96
Bath 2	98	3	112	5	5*

Table 76-1-6 Room Design Airflow

Room	Heat Loss (Heating)	Heating Airflow	Sensible Gain (Cooling)	Cooling Airflow	Design Airflow
Total House					
Airflow Divisor					

Total Equivalent Length

Every time the airflow changes shape or direction there is a pressure drop. The amount of pressure drop is affected by the change being performed, the air velocity, and the static pressure at that point. In short, determining all the different pressure drops could be difficult. We use a system called equivalent length to simplify the process. A table of different duct fittings is used to correlate the pressure lost through each fitting to a length of duct. For example, air moving through a 90° round elbow is equivalent to traveling through 10 feet of duct. To determine the amount of resistance each duct run will impose you need to know the total equivalent length of each run. This is found by adding up the equivalent length of all the fittings in each run and then adding the actual length. Table 76-1-7 gives the equivalent length for a few common fittings used in radial duct systems. You can use columns SR1 through SR8 to add the equivalent length of each supply run. Table 76-1-8 is used for the equivalent length of returns.

Table 76-1-7 Radial Supply Duct System Equivalent Length									
Fitting	EL	SR1	SR2	SR3	SR4	SR5	SR6	SR7	SR8
Plenum Takeoff	35								
Wye	15								
Ells	10								
Boot	5								
Total									

Table 76-1-8 Radial Return Duct System Equivalent Length				
Fitting	EL	R1	R2	R3
Plenum Takeoff	30			
Wye	15			
Ells	10			
Boot	5			
Stackhead	45			
Total				

Adjusted Friction Rate

Duct friction loss charts and duct calculators show the rate of friction loss for a specific type of and size of duct. These charts and calculators are designed for specific duct materials. Sheet metal duct has less resistance than fiberglass, and fiberglass has less resistance than flex duct. It is important that the tool you use is designed for the particular type of duct you plan to use. It is also important to recognize the difference between friction rate and the friction loss for any particular run. The friction rate chart does not tell you what the actual loss through a duct will be because that depends upon the duct length. Rather, it tells you what the loss will be for every 100 feet of duct. A 50-foot section of duct would create only half the static pressure drop shown on the duct design tool while a 200-foot run would create twice the pressure drop. The friction rate you use to look up the duct size must be adjusted for each run based on its total equivalent length using the formula:

Adjusted friction rate = design friction rate x (100/equivalent length)

For a design friction rate of 0.1" wc, the adjusted rate for the 50-foot run would be 0.2" wc and the adjusted rate for the 200-foot run would be 0.05" wc.

Duct Size

The duct size for each register is determined using a friction rate chart, duct calculator, or duct design app. You line up the adjusted friction rate for each run with the design CFM for that run and the tool you are using will give you a round duct size. Frequently the duct size will not be an exact dimension, but may be something like 5.7 inches diameter. Round duct is only available in whole-inch increments. You would choose 6-inch diameter duct. Choosing a slightly larger duct will mean that run may deliver a bit more air, so you add a balancing damper which can be used to reduce the airflow if needed. Choosing a slightly smaller diameter could cause that run to deliver a bit less than design.

Air Velocity

In some cases, a short duct run with a small total equivalent length might call for a small diameter duct based on adjusted pressure drop and desired CFM. Often, this leads to high velocity and noise. As a general rule you want to keep the air velocity at 600 FPM or lower to keep air noise down. The solution to a duct run with a high velocity is to increase the duct size, which decreases both air velocity and noise. Add a balancing damper to balance the run and keep the system balanced.

Name_____

Date_____

Instructor's OK ☐

LAB 76.2 EXTENDED PLENUM DUCT DESIGN

LABORATORY OBJECTIVE
You will demonstrate how to use heat loss, heat gain, system capacity, and blower data to design an extended plenum duct system that delivers the conditioned air where the building load is, does not make objectionable noise, and does not adversely affect airflow due to excessive restriction.

***FUNDAMENTALS OF HVACR* TEXT REFERENCE**
Unit 76 Duct Design

LABORATORY NOTES
There are three basic goals when designing a duct system. In order of importance, they are as follows:

1. Allow the system to circulate enough air to operate properly.
2. Distribute the air proportionally to where the heat load is.
3. Keep airflow noise below objectionable levels.

To accomplish all three goals you need to determine
- The system CFM
- The CFM required for each room
- Maximum available system static
- Duct friction rate

REQUIRED TOOLS, EQUIPMENT, AND MATERIALS
A house plan with a completed room by room load study.
You can use Figure 76-2-1 and Table 76-2-1 if you like.
A calculator
A duct friction chart, duct calculator, or duct calculator app
A pencil

PROCEDURE
This lab assumes that the heat load study has been completed and the equipment selected based on the heat load study. You can either provide your own or use the house plan, load study, and equipment data provided in this lab. The overall procedure is
1. determine the duct design static pressure for the supply and return.
2. draw a skeleton drawing of the duct system including the location of all registers and grilles.
3. determine the required CFM for each register based on the load and the system airflow.
4. size the ducts based on required CFM, design static, duct equivalent length, and maximum acceptable velocity.

763

Determine Total Fixed Pressure Losses

Anything that the air must flow over or through causes a pressure drop. To determine the amount of pressure available to move air through the ductwork we have to first determine how much pressure loss is built into the system already. Common fixed pressure losses include registers and grilles, air filters, and indoor coils. Table 76-2-2 shows typical fixed pressure losses. One question that often comes up is whether or not you need to consider the indoor coil and/or air filter as a fixed pressure loss. The key is remembering that the blower data specifies the total *external* static pressure, meaning you are looking for items outside of the piece of equipment that contains the blower. For a furnace with an air conditioning coil mounted on it, the coil is external to the furnace and you must add its pressure drop to the fixed losses. On the other hand, the coil inside a heat pump indoor unit is inside the box, so it has already been incorporated into the data. It is not considered part of the fixed losses.

Figure 76-2-1 Extended Plenum Plan

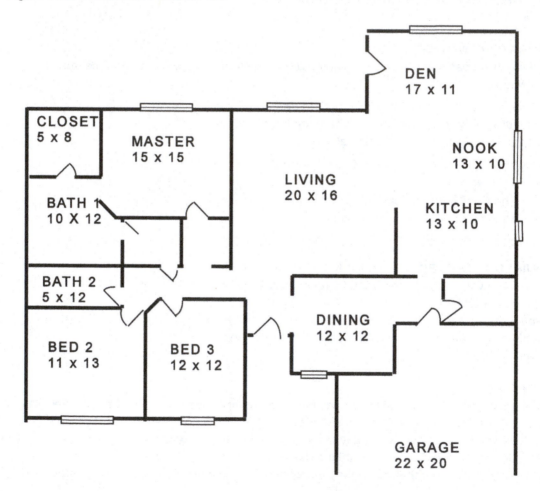

Table 76-2-1 Heat Loss and Gain

Room	Heat Loss (Heating)	Sensible Heat Gain (Cooling)	Latent Heat Gain (Cooling)
Total House	33656	28219	5887
Master	4460	4177	
Master Bath	1835	939	
Bath 2	735	358	
Bedroom 2	3209	3240	
Bedroom 3	2807	2773	
Foyer/Hall	1260	681	
Dining	3706	2427	
Kitchen	2567	4043	
Nook	2095	1492	
Den	4867	3740	
Living	3695	3435	

Table 76-2-2 Static Pressure Drop of Common HVAC System Components

Device	Typical Pressure Loss
Standard air filter	0.05 inches wc
High efficiency air filter	0.15 inches wc
Evaporator Coil	0.25 inches wc
Return Air Grille	0.03 inches wc
Supply Air Register	0.03 inches wc
Balancing Damper	0.03 inches wc

Setting Blower Design Point

There are two components of the blower design point: the airflow CFM and the total external static pressure. The system capacity and efficiency are affected by the airflow. As a general rule, more airflow delivers higher efficiency. In cooling, however, the ratio of sensible cooling to latent cooling must also be considered. Too much airflow can reduce the latent cooling to the point that not enough dehumidification is being done. The cooling airflow should ideally be set to deliver the required amount latent cooling according to Manual J. The blower external static is determined by the blower data that matches the design airflow and fan speed. It must be higher than the total fixed pressure losses so that there is pressure available for moving ir through the ductwork. You can find your own equipment specifications or use the specifications in Tables 76-2-3 and 76-2-4.

Table 76-2-3 Furnace and Air Conditioner Capacity

Furnace Heating Capacity BTUH	38400 BTUH
Sensible Cooling @ AHRI Design and 1200 CFM	26796 BTUH
Latent Cooling @ AHRI Design and 1200 CFM	8000 BTUH

Table 76-2-4 Furnace Blower Data				
External Static Pressure				
0.4	0.5	0.6	0.7	0.8
1117	1078	1035	1002	964

Duct Design Static

The duct design static is determined by adding up all the fixed air pressure losses and subtracting them from the blower external static. The maximum duct design static pressure is what is left after subtracting fixed pressure losses from the pressure produced by the blower. You do not have to design at the maximum pressure the blower can produce. Designing the duct system at a lower static saves fan energy. Designing at a lower overall external static pressure produces a quieter system with room to adjust the airflow if needed. ECM blowers will automatically adjust to deliver their programmed quantity of air over a wide range of external static pressure, but at the expense of using more energy to overcome extra resistance.

Keep in mind that what you have determined is the total duct system design static, not the pressure to use looking up the duct size. The total includes loss through both the return and supply. The method favored by ACCA's Manual D is to think of the duct system as one long duct from the return grille to the supply register. Basically, adding the total equivalent length of the return and supply together, and then looking up both the supply and return at the same static pressure. Another method is to have separate design pressures for the return and supply while insuring they add up to the total available design static. For systems with more extensive supply ductwork than return ductwork, a common approach is to use two-thirds of the total design static for the supply duct system and one-third of the total design static for the return duct system.

Draw the duct system

You can draw your duct system directly on Figure 76-2-1, or draw a basic outline of the house plan on a separate sheet of paper and draw the duct system on that. For this lab you are designing an extended plenum system. If you are using the plan in Figure 76-2-1 the furnace and ductwork will be located in the basement. A trunk duct will run from the supply plenum for most of the length of the house. Branch ducts will run from the trunk duct to the register. Extended plenum duct systems work well for larger homes and buildings that have eight or more supply registers, registers over 30 feet from the blower or both.

When using an extended plenum system with the ducts located in the basement, floor registers are normally placed along the outer wall under windows because that is where the greatest load is. You should try and avoid register placement that will have registers blowing directly on people.

Determine Register CFM

The CFM for each register is determined by the heat loss and heat gain for the area it serves. To divide the airflow according to load, create airflow divisors by dividing the system capacity by the system CFM for both heating and cooling. These tell you the quantity of BTUH delivered by each CFM of air. You then divide the heating and cooling airflow divisors into the heating and cooling BTUH for each area to get the required heating and cooling CFM. The design CFM will be the higher of the two. As a general rule you normally only need a register in areas that require more than 50 CFM. For example, in Table 76-2-5, Bath 2 has an airflow requirement of only 24 CFM in heating and 14 CFM in cooling. That room does not need a register. That small airflow requirement can be added to a nearby room, such as the Foyer and hall. Large loads of 400 CFM or more may benefit by splitting the load among multiple registers. Table 76-2-5 shows the CFM airflow divisors and CFM requirements for the plan shown in Figure 76-2-1. You can use Table 76-2-6 if you are using your own house plan and system.

Table 76-2-5 Room Design Airflow

Room	Heat Loss (Heating)	Heating Airflow	Sensible Gain (Cooling)	Cooling Airflow	Design Airflow
Total House	33656	1117	28219	1117	
Airflow Divisor	30.13		25.26		
Master	4460	148	4177	165	165
Master Bath	1835	30	939	37	37
Bath 2	735	24	358	14	24
Bedroom 2	3209	107	3240	128	128
Bedroom 3	2807	93	2773	110	110
Foyer/Hall	1260	42	681	27	42
Dining	3706	123	2427	96	123
Kitchen	2567	85	4043	160	160
Nook	2095	70	1492	59	70
Den	4867	162	3740	148	162
Living	3695	123	3435	136	136

Table 76-2-6 Room Design Airflow

Room	Heat Loss (Heating)	Heating Airflow	Sensible Gain (Cooling)	Cooling Airflow	Design Airflow
Total House					
Airflow Divisor					

Total Equivalent Length

Every time the airflow changes shape or direction there is a pressure drop. The amount of pressure drop is affected by the change being performed, the air velocity, and the static pressure at that point. In short, determining all the different pressure drops could be difficult. We use a system called equivalent length to simplify the process. A table of different duct fittings is used to correlate the pressure lost through each fitting to a length of duct. For example, air moving through a 90° round elbow is equivalent to traveling through 10 feet of duct. To determine the amount of resistance each duct run will impose you need to know the total equivalent length of each run. This is found by adding up the equivalent length of all the fittings in each run and then adding the actual length. Table 76-2-7 gives the equivalent length for a few common fittings used in extended plenum duct systems. You can use columns SR1 through SR12 to add the equivalent length of each supply run. Table 76-2-8 is used for the equivalent length of returns.

Unique Factors for Extended Plenum Systems

One big difference between a radial system and an extended plenum system is the presence of trunk ducts. The equivalent length of a branch duct includes the length of the branch duct, the length of the trunk duct, and the equivalent length of all the fittings. Included in the fitting equivalent length is the branch duct takeoff from the trunk duct. The air has a tendency to blow right by the first takeoff and slow down by the last takeoff. It is easier for air to make the turn into the branch duct at the end of the trunk duct than at the beginning. This is called the downstream branch effect. To adjust for this, multiply 10 times the number of downstream branches and add that to the equivalent length. The last branch has no downstream branch effect because there are no branches after it. The next to last branch has a downstream branch effect of 10 because there is one branch after it. The third branch from the end has a downstream branch effect of 20 because there are two branches after it. This pattern continues for the number of branches on the trunk duct until you reach the first branch. If a trunk reduces, the new size is considered a new trunk, so the first run on the new size is not considered a downstream branch for any of the previous branches before the trunk reduction. The downstream branch effect starts all over after a trunk size reduction.

Adjusted Friction Rate

Duct friction loss charts and duct calculators show the rate of friction loss for a specific type of and size of duct. These charts and calculators are designed for specific duct materials. Sheet metal duct has less resistance than fiberglass, and fiberglass has less resistance than flex duct. It is important that the tool you use is designed for the particular type of duct you plan to use. It is also important to recognize the difference between friction rate and the friction loss for any particular run. The friction rate chart does not tell you what the actual loss through a duct will be because that depends upon the duct length. Rather, it tells you what the loss will be for every 100 feet of duct. A 50-foot section of duct would create only half the static pressure drop shown on the duct design tool while a 200-foot run would create twice the pressure drop. The friction rate you use to look up the duct size must be adjusted for each run based on its total equivalent length using the formula:

Adjusted friction rate = design friction rate x (100/equivalent length)

For a design friction rate of 0.1" wc, the adjusted rate for the 50-foot run would be 0.2" wc and the adjusted rate for the 200-foot run would be 0.05" wc.

Table 76-2-7 Extended Plenum Supply Duct Equivalent Length

Run	Plenum Takeoff	Trunk Takeoff	Downstream Branch Effect	Ells	Boot	Total
EL	35	30	10 x Down Stream Branches	10	5	
SR1						
SR2						
SR3						
SR4						
SR5						
SR6						
SR7						
SR8						
SR9						
SR10						
SR11						
SR12						

Table 76-2-8 Extended Plenum Return Duct System Equivalent Length

Fitting	EL	R1	R2	R3	R4
Plenum Takeoff	30				
Trunk Takeoff	30				
Ells	10				
Boot	5				
Stackhead	45				
Total					

Duct Size

The duct size for each register is determined using a friction rate chart, duct calculator, or duct design app. You line up the adjusted friction rate for each run with the design CFM for that run and the tool you are using will give you a round duct size. Frequently the duct size will not be an exact dimension, but may be something like 5.7 inches diameter. Round duct is only available in whole-inch increments. You would choose 6-inch diameter duct. Choosing a slightly larger duct will mean that run may deliver a bit more air, so you add a balancing damper which can be used to reduce the airflow if needed. Choosing a slightly smaller diameter could cause that run to deliver a bit less than design.

Air Velocity

In some cases, a short duct run with a small total equivalent length might call for a small diameter duct based on adjusted pressure drop and desired CFM. Often, this leads to high velocity and noise. As a general rule you want to keep the air velocity at 600 FPM or lower to keep air noise down. The solution to a duct run with a high velocity is to increase the duct size, which decreases both air velocity and noise. Add a balancing damper to balance the run and keep the system balanced.

Trunk Duct Size and Velocity

Trunk ducts are sized the same as branch ducts, using a duct calculator, a CFM, and an adjusted static pressure. The trunk CFM is determined by simply adding up the design CFM of all the branch ducts it serves. The trunk design static is just the lowest adjusted static of all the branch ducts it serves. Tables 76-2-9 and 76-2-10 provide an easy way to determine trunk duct CFM and static pressure.

Table 76-2-9 Supply Trunk Sizing								
Branch Ducts	Supply Trunk 1		Supply Trunk 2		Supply Trunk 3		Supply Trunk 4	
	CFM	Static	CFM	Static	CFM	Static	CFM	Static
Total								
Branch Duct 1								
Branch Duct 2								
Branch Duct 3								
Branch Duct 4								
Branch Duct 5								
Branch Duct 6								
Branch Duct 7								
Branch Duct 8								
Branch Duct 9								
Branch Duct 10								
Branch Duct 11								
Branch Duct 12								

Table 76-2-10 Return Trunk Sizing				
Branch Ducts	Return Trunk 1		Return Trunk 2	
	CFM	Static	CFM	Static
Total				
Branch Duct 1				
Branch Duct 2				
Branch Duct 3				
Branch Duct 4				
Branch Duct 5				
Branch Duct 6				

LAB 77.1 WIRING ZONE CONTROL SYSTEM

LABORATORY OBJECTIVE
You will wire a complete zone control system including:

- Zone panel
- Zone thermostats
- Zone dampers
- Furnace control wiring
- Air conditioner control wiring

FUNDAMENTALS OF HVACR TEXT REFERENCE
Unit 77 Zone Control Systems

LABORATORY NOTES
You will wire a complete zone control system. You will wire the transformer, zone thermostats, and zone dampers to the zone control panel. The zone control panel will wire to the furnace and air conditioner. The zone panel requires its own transformer.

REQUIRED TOOLS, EQUIPMENT, AND MATERIALS
Safety glasses and gloves
Electrical hand tools
Multimeter and ammeter

SAFETY REQUIREMENTS
    A. Turn off and lock out power
    B. Check all circuits for voltage before beginning wiring exercise.

PROCEDURE
1. Figure 77-1-1 shows wiring for a typical zone system which uses standard 24 volt thermostats.
2. The zone panel requires its own transformer. The transformer primary can be wired to the line voltage for the furnace. The transformer secondary wires to the zone panel. The zone panel needs to be powered 24-7.
3. The zone thermostats each wire to the zone panel at the terminals designated for the thermostats.
4. The zone dampers wire to the panel at the terminals designated for them.
5. The system control wires that normally go to the thermostat are wired to the zone panel instead. As far as the system is concerned, the zone panel is the thermostat.
6. Some zone panels use an outdoor temperature sensor and a supply air plenum sensor. They are wired to the terminals on the zone panel designated for them.

Figure 77-1-1 Zone Panel Wiring

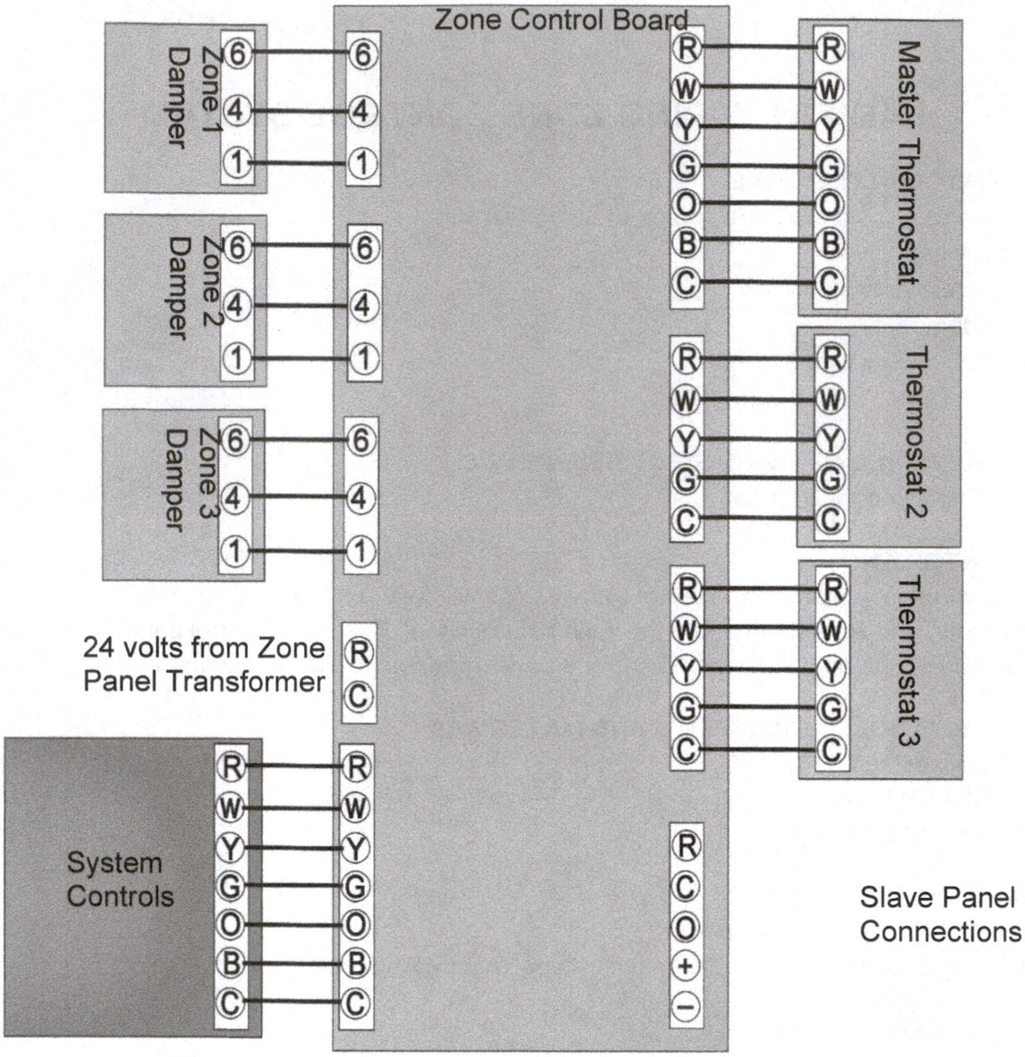

LAB 77.2 WIRING COMMUNICATING ZONE CONTROL SYSTEM

LABORATORY OBJECTIVE

You will wire a complete communicating zone control system including:

- Communicating zone panel
- Communicating thermostat
- Zone temperature sensors
- Zone dampers
- System control wiring
- Communicating system

FUNDAMENTALS OF HVACR TEXT REFERENCE

Unit 77 Zone Control Systems

LABORATORY NOTES

You will wire a complete communicating zone control system. The thermostat, zone panel, and system will all wire to the communicating bus. Zone sensors and zone dampers will wire to the zone panel. Figure 77-2-1 shows the connections for a typical communicating zone system.

REQUIRED TOOLS, EQUIPMENT, AND MATERIALS

Safety glasses and gloves
Electrical hand tools
Multimeter and ammeter

SAFETY REQUIREMENTS

A. Turn off and lock out power
B. Check all circuits for voltage before beginning wiring exercise.

PROCEDURE

1. Figure 77-2-1 shows an example of wiring for a communicating zone system. However, you should be sure to read and follow the manufacturer's instructions for the particular system you are installing.

2. Typically communicating zone control systems uses one communicating thermostat to control all zones plus zone temperature sensors to monitor the temperature in each zone other than zone 1.

3. Some communicating zone panels do not require a separate transformer, such as the one shown in Figure 77-2-1. This panel is powered by the 24 volt lines of the C and D terminals going to the indoor and outdoor units. Other systems do require a separate 24 volt transformer for the zone panel.

4. The zone temperature sensors each wire to the zone panel to the designated terminals.

5. The zone dampers wire to the panel at the terminals designated for them.

6. The same four communicating signals go to the thermostat, zone panel, indoor unit, and outdoor unit.

7. Communicating zone systems modulate system capacity to meet the demand rather than simply turning the system on and off.

Figure 77-1-1 Zone Panel Wiring

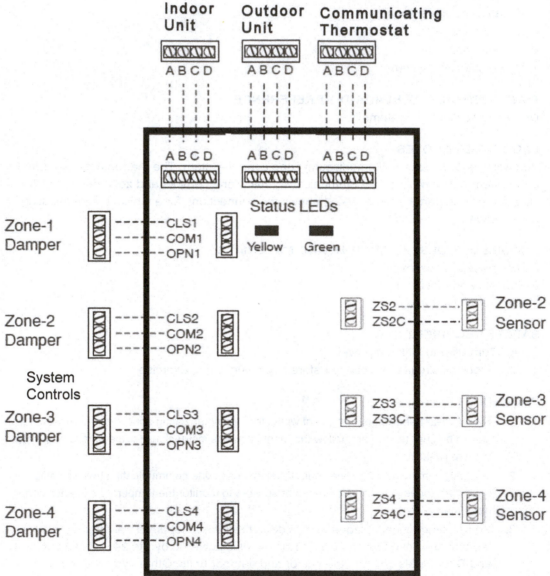

LAB 78.1 ROTATING VANE ANEMOMETER

LABORATORY OBJECTIVE

Demonstrate your ability to use a rotating vane anemometer to measure air flow velocity and air flow volume.

FUNDAMENTALS OF HVACR TEXT REFERENCE

Unit 78 Testing and Balancing Air Systems

LABORATORY NOTES

You will measure the average air velocity in feet per minute (FPM) leaving a duct. You will use that velocity to calculate the volume of air in cubic feet per minute (CFM). You calculate the air volume in CFM by multiplying the velocity in FPM by the duct cross sectional area in square feet.

REQUIRED TOOLS, EQUIPMENT, AND MATERIALS

Ruler and calculator
Rotating vane anemometer register or duct

SAFETY REQUIREMENTS

    A. Safety glasses
    B. The edges of metal duct can be sharp and cause severe cuts.

PROCEDURE

    1. Turn on the blower.

    2. Place the anemometer in the airstream and turn it on.

    3. Set the anemometer to average the reading and continue to hold it in the airstream.

    4. Move the anemometer slowly across the face of the duct or grille.

    5. Make sure to read the whole face of the duct or grille.

    6. Push the hold button and take the average reading in FPM.

    7. Calculate the area of the duct or grille in square feet.

    8. Multiply the FPM by the square feet to arrive at the volume in cubic feet per minute CFM.

LAB 78.2 FLOW HOOD

LABORATORY OBJECTIVE
Demonstrate your ability to use a flow hood to measure air volume.

FUNDAMENTALS OF HVACR TEXT REFERENCE
Unit 78 Testing and Balancing Air Systems

LABORATORY NOTES
You will measure air volume in cubic feet per minute (CPM) at supply registers and return grilles using the flow hood.

REQUIRED TOOLS, EQUIPMENT, AND MATERIALS
Flow hood
Duct system, with supply registers and return air grilles

SAFETY REQUIREMENTS
Wear safety glasses

PROCEDURE

1. Turn on the blower.
2. Assemble the hood on the flow hood.
3. Set the flow hood to read supply air on the highest scale.
4. Place the flow hood over the supply register and look at the reading.
5. If the reading is low, adjust the setting until you have a measureable reading.
6. Record the reading.
7. Repeat on another supply register.
8. Set the flow hood to measure return air on the highest scale.
9. Place the flow hood over the return grille and look at the reading.
10. If the reading is low, adjust the setting until you have a measureable reading.
11. Record the reading.

LAB 78.3 USING MAGNEHELIC GAUGE

LABORATORY OBJECTIVE
Demonstrate your ability to use a magnehelic gauge to measure total pressure, static pressure, and velocity pressure in a duct system.

FUNDAMENTALS OF HVACR TEXT REFERENCE
Unit 45 Fans and Airflow
Unit 78 Testing and Balancing Air Systems

LABORATORY NOTES
You will measure the total pressure, static pressure, and velocity pressure in a duct system. You will calculate the velocity using the velocity pressure, and then you will calculate the air volume in CFM using the calculated velocity and duct cross sectional area.

REQUIRED TOOLS, EQUIPMENT, AND MATERIALS
Ruler calculator
Magnehelic gauge
Duct system
Static pressure pickup tube
Pitot tube
Drill
1/4" drill bit

SAFETY REQUIREMENTS
Make sure to wear safety glasses while drilling hole.

PROCEDURE
1. Drill a hole in the duct large enough to allow the static pressure tube and the pitot tube to pass through, approximately 5/16 inch in diameter.
2. Insert the static pressure tube into the duct with the tip facing the airflow.
3. Position the magnehelic horizontally so that the needle rests at 0.
4. Connect the rubber hose from the high-pressure connection on the magnehelic to the static pressure tube.
5. Turn on the blower and red the static pressure in inches of water column (inches wc).
6. Replace the static pressure tip with a pitot tube.
7. Connect a rubber hose from the end of the pitot tube to the magnehelic high pressure connection.
8. Read the total pressure in inches of water column (inches wc).

9. Connect another rubber hose from the side of the pitot tube to the magnehelic low pressure connection.

10. Read the velocity pressure in inches of water column (inches wc).

11. Convert velocity pressure to velocity using the formula

    Velocity = 4005 x $\sqrt{\text{velocity pressure}}$

12. Calculate air volume by multiplying the air velocity by the cross-sectional area.

    Cross sectional area in square feet = (length inches x width inches)/144

    CFM = FPM velocity x Area Square Feet.

LAB 79.1 COMMERCIAL AIR CONDITIONING SYSTEMS–PACKAGED ROOFTOP UNIT

LABORATORY OBJECTIVE
You will identify the major components and arrangement for a commercial packaged rooftop air conditioning unit.

FUNDAMENTALS OF HVACR TEXT REFERENCE
Unit 79 Commercial Air-Conditioning Systems

LABORATORY NOTES
If the lab does not have a commercial rooftop unit, then an acceptable alternative would be to request permission from a local building owner for access to their rooftop.

REQUIRED TOOLS, EQUIPMENT, AND MATERIALS
Safety glasses
Commercial packaged rooftop unit

SAFETY REQUIREMENTS
A. Turn off and lock out power before proceeding.
B. Proper safety gear and equipment if accessing an actual rooftop unit
C. Take all necessary precautions when inspecting an operating unit

PROCEDURE
1. Examine the rooftop unit and describe how it would be (or is) mounted to the building.

2. Determine how to gain access to the interior compartments.

3. Open the unit and find the nameplate to determine the tonnage rating for the unit.

_____ tons

4. How are the supply and return ducts arranged?

From the bottom of the unit   From the side of the unit

5. Locate the compressor(s) and identify the type.

Single compressor   Dual compressor   Reciprocating   Scroll

Direct drive   Semi-hermetic   Fully hermetic

6. Locate the metering device(s) and identify the type.

Mechanical expansion valve   Electronic expansion valve   Fixed type

7. Locate the heating components and identify the type.

Electric   Gas fired   Hot water   Steam

8. Locate the condensate drain line. Is the line freeze protected?

9. Are crankcase heaters located on the compressor(s)?

10. Is the condenser fan variable speed?

11. Is there any type of outdoor ambient head pressure regulating valve?

LAB 80.1 COMMERCIAL CONTROL SYSTEMS–PACKAGED ROOFTOP UNIT

LABORATORY OBJECTIVE
You will examine the control module for a commercial packaged rooftop unit.

FUNDAMENTALS OF HVACR TEXT REFERENCE
Unit 80 Commercial Control Systems

LABORATORY NOTES
If the lab does not have a commercial rooftop unit, then an acceptable alternative would be to request permission from a local building owner for access to their rooftop.

REQUIRED TOOLS, EQUIPMENT, AND MATERIALS
Safety glasses
Commercial packaged rooftop unit

SAFETY REQUIREMENTS
A. Turn off and lock out power before proceeding.
B. Proper safety gear and equipment if accessing an actual rooftop unit
C. Take all necessary precautions when inspecting an operating unit

PROCEDURE

1. Locate the control module for the unit.

2. What type of user interface does it have?

3. Are any user instructions labeled on or near the control module?

4. Is the control module programmable?

5. List the types of sensor inputs wired to the module (temperature, humidity, etc.)

6. How many output connections does the module have and can you determine what they control?

LAB 81.1 CHILLED WATER SYSTEMS–CONTROL VALVES AND STEAM TRAPS

LABORATORY OBJECTIVE
You will examine and determine the operation of chill water control valves and steam traps.

FUNDAMENTALS OF HVACR TEXT REFERENCE
Unit 81 Chilled-Water Systems

REQUIRED TOOLS, EQUIPMENT, AND MATERIALS
Safety glasses
Chill water control valve
Hot water control valve
Steam control valve
Steam trap

SAFETY REQUIREMENTS
A. Wear Safety Glasses
B. Avoid handling piping on live systems

PROCEDURE
Step 1
1. Inspect a chill water control valve and determine the following:
   a. Is the valve operated manually, pneumatically, or electronically?

   b. Does the valve modulate or is it only a 2-position valve (fully open or fully closed)?

   c. Is the valve direct seating, reverse seating (a reverse seating valve closes as the stem travels upward) or a 3-way type?

Step 2
2. Inspect a hot water control valve and determine the following:
   a. Is the valve operated manually, pneumatically, or electronically?

   b. Does the valve modulate or is it only a 2-position valve (fully open or fully closed)?

   c. Is the valve direct seating, reverse seating (a reverse seating valve closes as the stem travels upward) or a 3-way type?

Step 3

3. Inspect a steam control valve and determine the following:
    a. Is the valve operated manually, pneumatically, or electronically?

    b. Does the valve modulate or is it only a 2-position valve (fully open or fully closed)?

    c. Is the valve direct seating, reverse seating (a reverse seating valve closes as the stem travels upward) or a 3-way type?

Step 4

4. Inspect a steam trap:
    a. Determine the steam trap type (mechanical, thermodynamic, thermostatic).

    b. What is the purpose of the steam trap?

    c. Does the steam trap modulate flow or does it only have two positions (open or closed)?

    d. If the steam trap failed to open what would be the result?

    e. If the steam trap failed to close what would be the result?

LAB 82.1 TESTING COOLING TOWER WATER

LABORATORY OBJECTIVE
You will perform a water test.

FUNDAMENTALS OF HVACR TEXT REFERENCE
Unit 82 Cooling Towers

LABORATORY NOTES
Testing cooling tower water will provide different results as compared to testing standard tap water. This is due to the chemical additives used for water treatment. If there is no cooling tower water available for the test, regular tap water may be used. The method for testing would remain the same whether using a sample of tap water or a sample of cooling tower water, however there will be no traces of any chemical additives when using standard tap water.

REQUIRED TOOLS, EQUIPMENT, AND MATERIALS
Safety glasses
Protective gloves
Cooling water test kit

SAFETY REQUIREMENTS
A. Wear Safety Glasses and Gloves
B. Read Safety Data Sheet for products used.

PROCEDURE

1. Following the instructions supplied with the test kit, test the water sample for Alkalinity.

    Result _____

2. Following the instructions supplied with the test kit, test the water sample for Chlorides.

    Result _____

3. Following the instructions supplied with the test kit, test the water sample for Total Calcium.

    Result _____

4. Following the instructions supplied with the test kit, test the water sample for Nitrites.

    Result _____

5. Following the instructions supplied with the test kit, test the water sample for Phosphonate.

    Result _____

6. Following the instructions supplied with the test kit, test the water sample for pH.

    Result _____

7. Following the instructions supplied with the test kit, test the water sample for total dissolved solids (TDS).

    Result _____

Name_____

Date_____

Instructor's OK ☐

LAB 83.1 SETTING THE LOW-PRESSURE CUT-OUT

LABORATORY OBJECTIVE
The student will demonstrate how to properly set the cut-out and cut-in points on a low-pressure cut-out switch.

FUNDAMENTALS OF HVACR TEXT REFERENCE
Unit 83 Commercial Refrigeration Systems

LABORATORY NOTES
For this lab exercise there needs to be an operating refrigeration system that has a low-pressure switch. The low-pressure cut-out switch, which senses compressor suction pressure, opens on a drop in pressure. It is set to cut out at a protective low limit pressure, but remains closed at normal operating pressures. The low-pressure cut-out switch is also known as a loss of charge switch.

REQUIRED TOOLS, EQUIPMENT, AND MATERIALS
Operating refrigeration unit with low-pressure cut-out switch and king valve

Gauge manifold

SAFETY REQUIREMENTS
A. Always read the equipment manual to become familiarized with the refrigeration system and its accessory components prior to startup.
B. Wear safety goggles and gloves when working with refrigerants. Liquid refrigerant can cause frostbite when in contact with eyes and skin.
C. Use low loss hose fittings, or wrap cloth around hose fittings before removing the fittings from a pressurized system or cylinder. Inspect all fittings before attaching hoses.

PROCEDURE
Step 1
Familiarize yourself with the major components in the refrigeration system including the condenser, compressor, evaporator, and metering device. Determine where the high side and low side connections for the system are located. Figures 83-1-1 and 83-2-1 show the low-pressure switch and its location in a schematic.

786

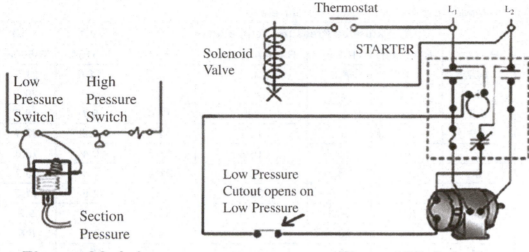

Figure 83-1-1   **Figure 83-2-1**

A. The operating manual with the refrigeration specifications should indicate what the expected suction pressure should be. Many low pressure control settings exist and different systems call for different settings.

B. If you have no information on the system, you may be able to calculate the expected suction pressure based upon the refrigerant type for the system you are working on.

C. The refrigerant temperature in the evaporator should generally be no more than 15°F colder than the medium being cooled.

D. As an example, if a freeze box is kept at 0°F, then the refrigerant temperature should be about minus 15°F.

E. You should be able to determine the lowest evaporator pressure from the P-T chart, Table 83-1-1. If the space to be cooled was for vegetables to be kept at 45°F, then the corresponding refrigerant temperature would be 45°F − 15°F = 30°F which corresponds to an evaporator pressure of 25.6 psig for HFC R-134a.

Table 83-1-1 Vapor Pressure/Temperature of Refrigerants

				Saturated Pressure–Temperature			
Temp °F	R-134a	R-1234yf	R-290	R-404A (vapor)	R-507A	R-513A	R-600a
–40	*14.7*	*11.5*	1.9	4.0	6.2	*11.6*	*21.5*
–35	*12.4*	*8.9*	3.9	6.5	8.8	*9.0*	*20.2*
–30	*9.8*	*6.0*	5.7	9.6	11.0	*6.1*	*18.7*
-25	*6.9*	*2.8*	8.1	12.7	14.1	*2.9*	*17.1*
–20	*3.7*	0.4	10.7	16.8	17.6	0.4	*15.6*
–15	0	2.3	13.6	19.7	21.4	2.4	*13.7*
–10	1.9	4.4	16.7	23.6	25.5	4.5	*11.5*
–5	4.1	6.7	20.1	27.8	30.0	6.9	*9.2*
0	6.5	9.2	23.7	32.6	34.8	9.5	*6.6*
5	9.1	11.9	27.6	37.7	40.0	12.4	*3.8*
10	11.9	14.9	31.8	43.1	45.7	15.5	*0.7*
15	15.0	18.1	36.3	49.0	51.7	18.8	1.3
20	18.4	21.6	41.1	55.3	58.2	22.4	3.1
25	22.1	25.4	46.3	62.1	65.2	26.4	5
30	26.1	29.4	51.8	69.3	72.7	30.6	7.1
35	30.4	33.8	57.7	77.1	80.7	35.2	9.4
40	35.0	38.4	63.9	85.4	89.2	40.1	11.8
45	40.1	43.4	70.6	94.2	98.3	45.4	14.4
50	45.4	48.8	77.6	103.6	107.9	51.0	17.2

Bold italic numbers are inches of mercury vacuum. All other pressures are psig at sea level.

Step 2

Determine the low-pressure cut-out setting as follows:

A   If the low-pressure cut-out is in place to control the space temperature, then first determine the maximum and minimum temperature of the space.

   a)   Example: The space is to be kept at a temperature of between 35°F to 45°F. As explained in the previous section, the refrigerant temperature would be approximately 15°F colder than the space and would therefore vary between 20°F to 30°F. Using HFC R-134a, this would correspond to an evaporator pressure of from 18 to 25.6 psig. The low-pressure cut-out would be set to start the compressor at the warmer temperature and higher pressure of 25.6 psig and then stop the compressor at the lower temperature and lower pressure of 18 psig.

   b)   In this manner, the low-pressure cut-out would cycle the compressor on and off dependent on the cooling load.

B   If the low-pressure cut-out is in place for low charge protection, then first determine the minimum pressure allowed for the system. In this type of system, the space temperature is generally maintained by another control such as a box solenoid valve.

   a) Example: Many refrigeration systems operate at positive pressures to reduce the possibility of air being drawn in. In this case, the low-pressure cut-out is often set at some point slightly above atmospheric pressure.

   b) A common setting for this type of low-pressure cut-out would be to stop the compressor at a low load and low pressure of 2 psig and then restart the compressor as the space warms up and the pressure rises to about 8 psig.

   c) In this manner, the refrigerant pressure in the system is never allowed to fall below atmospheric pressure.

C   If the system loses its refrigerant charge, the space warms up and the compressor continuously runs. Due to a lack of refrigerant circulating through the compressor and its continuous running, the compressor and motor may be damaged.

   a) In this type of undercharged situation, the suction pressure would continue to drop. If the system is equipped with a low-pressure cut-out, the system would automatically shut down at the system cut-out pressure.

D   Depending on the type of refrigeration system, a low-pressure cut-out can operate as both a temperature control and a low charge protection control.

Step 3

Adjust the low-pressure cut-out setting as determined from Step 2 by turning the adjusting screws as follows:

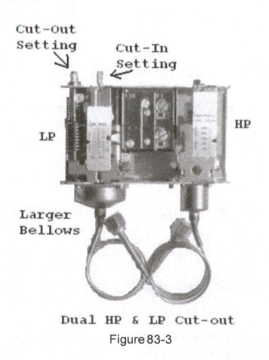

Dual HP & LP Cut-out
Figure 83-3

A. Range adjustment – This is the difference between the minimum and maximum operating pressures within which the control will function. As an example, a cut-out of 2 psig and a cut-in of 8 psig would indicate a typical range setting of a low-pressure cut-out.

B. Differential adjustment – This is the difference between the cut-out and cut-in pressures for the control. The differential of a low-pressure switch set to cut-out at 2 psig and cut-in at 8 psig would be 6 psig.

C. A low-pressure cut-out will have two adjusting screws as shown below. In many cases, one adjusting screw will control the range while the other controls the differential.

Be careful as you make adjustments to the control. In some cases, the set range may change as you change the differential and vice versa.

Step 4

A. If there are no pressure gauges currently on the unit then you must connect a gauge manifold to the high and low side of the system.

B. Start the refrigeration system in the normal cooling mode and allow the system to stabilize.

C. Close the king valve which is located in the liquid line directly following the receiver or condenser. The suction pressure will begin to decrease and eventually the compressor will stop. Record the suction pressure at which the compressor stops:

Suction Pressure at which compressor stops _____

D. After recording the suction pressure at which the compressor stops, slowly open the king valve and observe the suction pressure begin to rise. The compressor should restart. Record the suction pressure at which the compressor restarts:

Suction Pressure at which compressor restarts _____

Step 5

A. Continue making any necessary adjustments until the low-pressure cut-out is correctly set.

B. Allow the system to run and stabilize prior to shutting down and then carefully disconnect the gauge manifold from the unit.

LAB 83.2 SETTING THE HIGH-PRESSURE CUT-OUT

LABORATORY OBJECTIVE
The student will demonstrate how to properly set the cut-out and cut-in points on a high-pressure cut-out switch.

FUNDAMENTALS OF HVACR TEXT REFERENCE
Unit 83 Commercial Refrigeration Systems

LABORATORY NOTES
For this lab exercise there needs to be an operating refrigeration system that has a high-pressure cut-out switch. It can be set to stop the compressor before excessive pressures are reached. Such conditions might occur because of a water supply failure in water cooled condensers or because of a fan motor stoppage on air cooled condensers.

REQUIRED TOOLS, EQUIPMENT, AND MATERIALS
Operating refrigeration unit with high-pressure cut-out switch
Gauge manifold

SAFETY REQUIREMENTS
A. Always read the equipment manual to become familiarized with the refrigeration system and its accessory components prior to startup.
B. Wear safety goggles and gloves when working with refrigerants. Liquid refrigerant can cause frostbite when in contact with eyes and skin.
C. Use low loss hose fittings, or wrap cloth around hose fittings before removing the fittings from a pressurized system or cylinder. Inspect all fittings before attaching hoses.

PROCEDURE
Step 1
Familiarize yourself with the major components in the refrigeration system including the condenser, compressor, evaporator, and metering device. Determine where the high side and low side connections for the system are located. Figures 83-2-1 and 83-2-2 show the high-pressure switch and its location in a schematic.

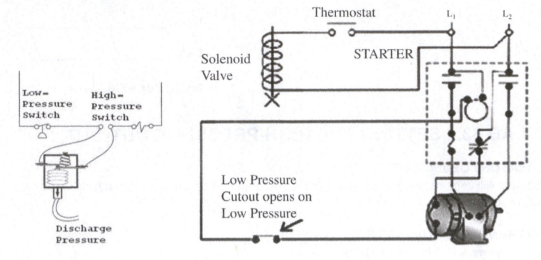

Figure 83-2-1   Figure 83-2-2

A. The operating manual with the refrigeration specifications should indicate what the expected discharge pressure should be. If not, then you must calculate the expected discharge pressure based upon the refrigerant type for the system you are working on.

B. Air cooled condensers – The refrigerant condensing temperature is typically 30°F higher than ambient (room) temperature. If it is 90°F outside, then the refrigerant condensing temperature will equal:

90°F + 30°F = 120°F.

At this temperature, you should be able to determine the condenser pressure (high side gauge) from the P-T chart, Table 83-2-1. For HCFC R-22 it would be expected to be approximately 260 psig. For another type of refrigerant such as R-134a the expected condenser pressure would be somewhat lower, approximately 172 psig.

Table 83-2-1

				Saturated Pressure–Temperature				
Temp °F	R-22	R-134a	R-290	R-404A (liquid)	R-410A (liquid)	R-454B (liquid)	R-507A	R-513A
100	196	124	174	237	319	301	242	132
105	211	135	187	254	342	323	260	143
110	226	146	200	273	366	346	279	155
115	243	158	213	292	392	370	299	167
120	260	171	228	312	419	396	319	180
125	278	185	243	333	448	422	341	193
130	297	199	259	356	478	451	364	208
135	317	223	275	379	509	480	388	223
140	337	229	292	404	543	511	413	238
145	359	246	310	430	577	543	440	255
150	382	263	329	457	614	577	467	272
155	405	281	348	486	652	612	497	290
160	430	300	369	516	693	649	523	309
All pressures are psig at sea level.								

C. Water cooled condensers – The refrigerant condensing temperature is typically 10 degrees higher than the water leaving the condenser. If the water temperature leaving the condenser is 100°F, then the refrigerant condensing temperature will equal:

100°F + 10°F = 110°F.

For R-22 it would be expected to be approximately 226 psig. For another type of refrigerant such as R-134a the expected condenser pressure would be somewhat lower approximately at 147 psig.

Step 2

Determine the high-pressure cut-out setting as follows:

A. The high-pressure cut-out is typically set at 125% of normal discharge pressure.

B. For an air cooled condenser using R-22 the setting would be approximately 1.25 x 260 psig = 325 psig. For R-134a it would be 1.25 x 172 psig = 215 psig.

C. For a water cooled condenser using R-22 the setting would be approximately 1.25 x 226 psig = 283 psig. For R-134a it would be 1.25 x 147 psig = 184 psig.

D. Depending on the type of control switch, the HP cut-out may reset automatically once the pressure reaches the cut-in value (this would be at some point below the normal discharge pressure – in this example about 100 psig).

E. Remember–cycling on the HP cut-out is harmful to the unit!

Step 3

A   Adjust the high-pressure cut-out setting by turning the adjusting screw to the correct setting as determined from step 2.

B   Also adjust the cut-in setting if the high-pressure cut-out is so equipped.

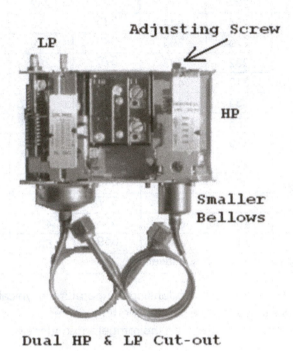

Dual HP & LP Cut-out

Figure 83-2-3

Step 4

A. If there are no pressure gauges currently on the unit then you must connect a gauge manifold to the high and low side of the system.

B. Start the refrigeration system in the normal cooling mode and record the following:

RUNNING

High Side Pressure _____

Low Side Pressure _____

Step 5

A. After the system has stabilized, turn off the cooling air or water and allow the system to cut out on high pressure. Record the cut-out pressure below.

High Side Cut-OutPressure _____

B. Check the cut-in pressure if the switch is so equipped. Turn the cooling air or water back on and the discharge pressure should begin to drop. The unit should restart at the cut-in setting. Record the cut-in pressure below.

High Side Cut-In Pressure _____

Step 6

A. Continue making any necessary adjustments until the high-pressure cut-out is correctly set.

B. Allow the system to run and stabilize prior to shutting down and then carefully disconnect

LAB 83.3 SETTING A BOX THERMOSTAT CONTROL

LABORATORY OBJECTIVE

The student will demonstrate how to properly set a thermostat that is controlling the temperature of a refrigerated space.

FUNDAMENTALS OF HVACR TEXT REFERENCE

Unit 83 Commercial Refrigeration Systems

LABORATORY NOTES

For this lab exercise there needs to be an operating refrigeration system that has a thermostat controlling the temperature of the space being cooled.

REQUIRED TOOLS, EQUIPMENT, AND MATERIALS

Operating refrigeration unit with thermostat control

SAFETY REQUIREMENTS

A. Always read the equipment manual to become familiarized with the refrigeration system and its accessory components prior to startup.

PROCEDURE

Step 1

Familiarize yourself with the major components in the refrigeration system including the condenser, compressor, evaporator, and metering device.

        A. Temperature motor control – This is the simplest type of control system. For this type of system there is only one space being cooled. The thermostat will cycle the compressor motor on and off dependent on the load. This type of system is shown in Figure 83-3-1.

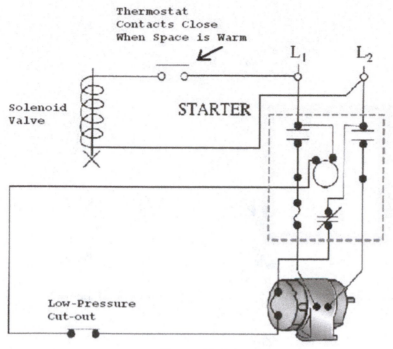

Figure 83-3-1

B. Pressure motor control – This is used on systems that cool multiple spaces at one time. Each individual space will have its own solenoid valve that is controlled by a thermostat. The low-pressure cut-out will cycle the compressor motor on and off dependent on the load.

C. The thermostat will have a sensing bulb that should be located in the space being cooled (Figures 83-3-2 and 83-3-3). The sensing bulb should be in such a location that it is sensing the average temperature of the space. If it is in a walk-in cooler, it should be mounted in a central location. Care should be taken not to damage the thin interconnecting tube located between the sensing bulb and the thermostat contacts.

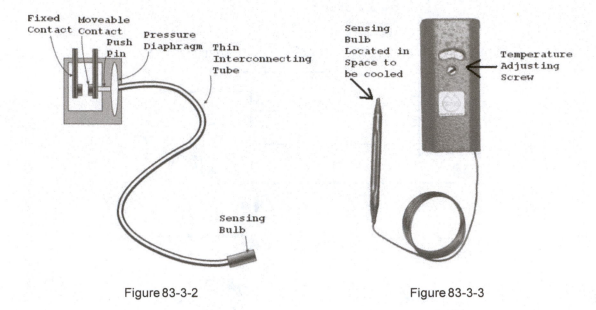

Fixed Contact
Moveable Contact
Push Pin
Pressure Diaphragm
Thin Interconnecting Tube
Sensing Bulb

Sensing Bulb Located in Space to be cooled
Temperature Adjusting Screw

Figure 83-3-2   Figure 83-3-3

Step 2

Set the thermostat as follows:

A. Determine the desired temperature for the space to be cooled and check the thermostat setting.

B. Generally the thermostat can be adjusted with a screwdriver or a small wrench. Some thermostats have a readable scale for temperature. Others are adjusted by counting the number of turns of the screw. As an example, one full turn of the adjusting screw may equal four degrees, etc.

C. Differential adjustment – This is the difference between the cut-out and cut-in pressures for the control. On a temperature-type motor control, this setting will reduce the cycling of the compressor motor on and off. On a pressure-type motor control, this setting will reduce the cycling of the solenoid valve open and close.

   As an Example, assume the space is to be kept at a temperature of 33°F with a 4 degree differential. With a temperature motor control, the compressor will start when the space temperature reaches 35°F and stop again sometime later when the space cools back down to 31°F. This provides for an average temperature of 33°F.

D. Some spaces are allowed higher differentials than others. For example, dairy cooler temperatures should be fairly constant, without large variations in temperatures. Therefore they would be set for a smaller differential.

Step 3

    A. After setting the thermostat, start the refrigeration system in the normal cooling mode and allow the system to stabilize.

    B. Depending on the size of the unit, it may take some time before the setpoint temperature is reached. Record the temperature of the space when the compressor stops or the solenoid valve closes.

_____ °F

    C. Allow the space to warm up and then record the temperature of the space when the compressor starts or the solenoid valve opens.

_____ °F

    D. Continue making any necessary adjustments until the thermostat is correctly set.

LAB 84.1 SUPERMARKET CASES

LABORATORY OBJECTIVE
The student will demonstrate an understanding of a supermarket layout and how the refrigerated cases are arranged.

FUNDAMENTALS OF HVACR TEXT REFERENCE
Unit 84 Supermarket Equipment

LABORATORY NOTES
This lab exercise will take place outside of the lab and require the student to visit a local supermarket.

REQUIRED TOOLS, EQUIPMENT, AND MATERIALS
Notepad and pencil

SAFETY REQUIREMENTS
None

PROCEDURE
Step 1
    A. Walk through a local supermarket taking note of the different display cases and then record the following:

        a. The total number of single deck open display cases.

        b. The total number of multideck open display cases.

        c. The total number of single deck closed display cases.

        d. The total number of glass-doored display cases.

        e. The total number of self-contained refrigeration systems.

Step 2
    A. Draw a sketch of the supermarket layout.

LAB 85.1 DRY ICE

LABORATORY OBJECTIVE
The student will perform a number of experiments using dry ice.

FUNDAMENTALS OF HVACR TEXT REFERENCE
Unit 85 CO_2 Refrigeration Systems

LABORATORY NOTES
Always take care whenever handling dry ice.

REQUIRED TOOLS, EQUIPMENT, AND MATERIALS
Dry ice
Safety goggles
Insulated latex, nitrile, or leather gloves
Non-contact digital laser infrared thermometer
Two standard laboratory beakers
Isopropyl rubbing alcohol
Tongs
Gummi bears
Hammer and cutting board

SAFETY REQUIREMENTS
Dry ice is extremely cold. Prolonged contact with the skin will cause frostbite. Dry ice should be stored in an insulated container that is not air tight. A completely airtight container will expand and possibly explode as the dry ice sublimates. Always provide adequate ventilation and do not directly inhale the CO_2 gas.

PROCEDURE
A. Carefully place some dry ice on the cutting board. Use the infrared thermometer to measure and record the dry ice temperature.

Temperature:_____

B. Fill the first beaker halfway with water. Carefully place some dry ice into the beaker and observe what happens.

C. Fill the second beaker halfway with rubbing alcohol. Carefully place some dry ice into the beaker and observe what happens.

D. Drop a gummi bear into the beaker containing rubbing alcohol and dry ice and allow it to sit for about a minute.

E. Remove the gummi bear using the tongs and place it on the cutting board.

F. Strike the gummi bear with a hammer and observe it shatter into many pieces like glass.

LAB 86.1 HYDROCARBON REFRIGERATION SYSTEM CHECK

LABORATORY OBJECTIVE
The student will perform an operating check on a system containing a hydrocarbon refrigerant.

FUNDAMENTALS OF HVACR TEXT REFERENCE
Unit 86 Hydrocarbon Refrigeration Units

LABORATORY NOTES
Checking system pressures is not a normal maintenance requirement for appliances containing hydrocarbon refrigerants. These systems hold just a little more than 5 ounces of refrigerant. Putting gauges on one of these systems would result in losing enough refrigerant to the hoses and gauges that the system would be undercharged. Instead, correct operation is checked by operating temperatures. Equipment manufacturers can provide detailed data regarding the correct operating temperatures at specific locations on their equipment.

REQUIRED TOOLS, EQUIPMENT, AND MATERIALS
Refrigeration appliance containing a hydrocarbon refrigerant
Refrigeration system operational data
Non-contact digital laser infrared thermometer or similar temperature measuring device
Flammable gas detector
Multimeter

SAFETY REQUIREMENTS
General Precautions Required When Servicing Hydrocarbon Refrigeration Systems
  A. Always read the equipment manual to become familiarized with the refrigeration system and its accessory components prior to startup.
  B. Place a flammable gas detector on the floor near the unit and keep it on during servicing.
  C. Do not turn anything on or off, or plug or unplug anything, until you have verified that there is no flammable gas in the area.
  D. Turn on ventilation.
  E. Post a sign stating that flammable vapors are possible during service and that no open flames or smoking are allowed.
  F. Eliminate any potential ignition sources in the area by turning off all electricity and gas to all appliances in the room.

PROCEDURE

Step 1

A. Review the equipment manual and after all safety precautions have been observed, start the unit.

B. The unit must operate long enough to insure that all the refrigerant dissolved in the compressor oil has vaporized and entered the system.

C. Check the compressor crankcase with an infrared thermometer. If the compressor crankcase is cool, refrigerant is still vaporizing out. The compressor crankcase should be warmer than the ambient temperature.

D. Measure the compressor amperage and then take the temperature readings at the locations as indicated in Figure 86-1-1 and record them in Table 86-1-1.

Figure 86-1-1

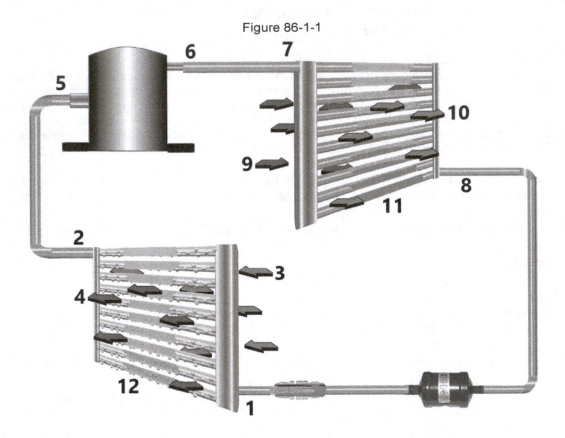

Recorded Temperatures

Table 86-1-1

	Compressor Amperage	Recorded Temperatures
	Ambient	
1	Evaporator In	
2	Evaporator Out	
3	Evaporator Air In	
4	Evaporator Air Out	
5	Compressor Suction	
6	Compressor Discharge	
7	Condenser In	
8	Condenser Out	
9	Condenser Air In	
10	Condenser Air Out	
11	Saturated Suction Temp	
12	Liquid Saturated Temp	

Step 2

Compare recorded temperatures to the manufacturer's refrigeration system operational data and determine if the unit is operating correctly.

LAB 87.1 CLEANING AN ICE MACHINE

LABORATORY OBJECTIVE

The student will clean and sanitize an ice machine.

FUNDAMENTALS OF HVACR TEXT REFERENCE

Unit 87 Ice Machines

LABORATORY NOTES

Some ice machines have built-in features that provide for a regular automatic cleaning or sanitizing cycle. This reduces the need to manually clean the machine as frequently. However it is still important to periodically inspect the machine to determine if manual cleaning is required.

REQUIRED TOOLS, EQUIPMENT, AND MATERIALS

Chemical splash goggles
Chemical-resistant impervious gloves
Cleaning solution
Sanitizing solution
Nylon bristle brush

SAFETY REQUIREMENTS

    A. Always read the SDS for the solutions that you are using.

    B. Wear chemical splash goggles and chemical-resistant impervious gloves when working with cleaning and sanitizing solutions.

    C. Use the solutions in a well-ventilated area.

PROCEDURE

Step 1

    A. Remove all ice from the bin to avoid contamination from the cleaning solution.

    B. The water should be turned off and the sump tank drained.

    C. Mix the cleaning solution in the correct ratio with warm water and pour it into the sump tank.

    D. Turn the unit on to the "cleaning cycle" and the cleaning solution will flow through the system.

    E. Allow the solution to circulate until the unit is clean.

Step 2

  A. After flushing, some built-up scale may need to be manually cleaned.

  B. Use the nylon bristle brush to clean the channels on the evaporator.

  C. Disassemble individual components such as spray tubes and float switches and soak them in the solution.

  D. Clean the inlet water valve and its strainer.

  E. Replace all filters and clean all strainers.

Step 3

  A. After cleaning is completed, circulate the sanitizing solution through the system in the same manner as the cleaning solution.

  B. After cleaning and sanitizing, the machine should be thoroughly rinsed and allowed to air-dry before renewed operation.

LAB 88.1 TESTING POTENTIAL START RELAYS

LABORATORY OBJECTIVE
The student will demonstrate how to properly test a potential starting relay.

***FUNDAMENTALS OF HVACR* TEXT REFERENCE**
Unit 88 Troubleshooting Refrigeration Systems
Unit 34 Electrical Components
Unit 36 Electric Motors

LABORATORY NOTES
A potential starting relay assists in starting the motor by allowing current flow to the starting winding through the start capacitor and the normally closed contacts in the relay.

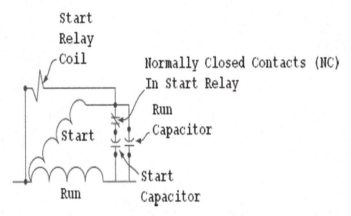

Figure 88-1-1

The potential starting relay coil is energized by the counter electromotive force developed by the motor. The faster the motor turns, the higher the counter electromotive force. When the motor comes up to speed the start relay coil energizes and acting as an electromagnet it will open the normally closed contacts in the start relay. The coil is wired in parallel to the start winding. The contacts are wired in series with the start winding and the start capacitor.

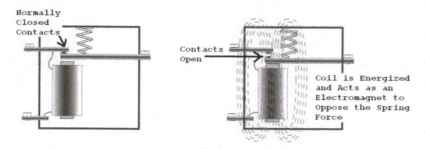

Figure 88-1-2

REQUIRED TOOLS, EQUIPMENT, AND MATERIALS
Motor control circuit
Multimeter

SAFETY REQUIREMENTS

A. Turn the power off if the winding to be tested is installed in an operating system. Lock and tag out the power supply.

B. Confirm the power is secured by testing for 0 voltage with a multimeter.

PROCEDURE

Step 1

Familiarize yourself with electrical meters.

1. Many start relay coils have higher resistance than average control circuit relays.
2. Be sure to test the coil on the R x 100 scale before deciding that the relay is defective.
3. The pull-in/drop-out voltage of the relay is unique. Do not attempt to replace it with an ordinary, similar voltage relay.

Step 2

Test the relay coil as follows:

1. Start with the resistance reading on the highest scale.
2. Record the resistance reading _____ ohms.
3. Compare this reading to the manufacturer's data to see if the coil is satisfactory. A resistance exceeding 5000 ohms is common for most potential start relays.

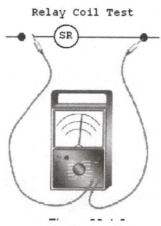

Figure 88-1-3

Test the relay contacts as follows:

1. When testing start relay contacts, the contacts should be closed and the ohmmeter will read zero resistance.

Relay Contact Test

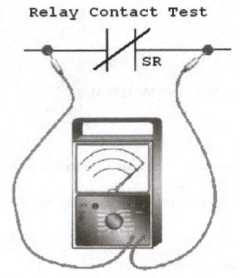

Figure 88-1-4

2. Sometimes while in operation the contacts stick closed and become badly burned. This will make for a poor connection, which will be indicated by a positive resistance reading.

3. If such is the case, replace the relay. Do not attempt to clean the contacts.

4. Always use the identical replacement. An improper substitution can damage the motor.

5. The replacement must be mounted in the same position as the original and connected the same way.

LAB 88.2 OHMING SINGLE PHASE HERMETIC COMPRESSOR MOTORS

LABORATORY OBJECTIVE
The purpose of this lab is to learn how to check a hermetic compressor motors for faults using an ohm meter.

FUNDAMENTALS OF HVACR TEXT REFERENCE
Unit 88 Troubleshooting Refrigeration Systems
Unit 36 Electric Motors

LABORATORY NOTES
We will use an ohm meter to test the windings of hermetic compressors. You will determine if the motor is good, open, shorted, or grounded. If the motor is good, you will use the ohm readings to identify the Common, Start, and Run terminals.

REQUIRED TOOLS, EQUIPMENT, AND MATERIALS
Volt-ohm meter
Electrical hand tools
Hermetic compressors

SAFETY REQUIREMENTS
Motor is disconnected from any power source.

PROCEDURE
1. Take ohm readings between each combination of two terminals (there should be three readings.
2. Write the readings down.
3. If any of the three readings is infinite, the motor winding is open.
4. If any of the three readings is less than 0.3 ohms, the motor is most likely shorted.
   Note: Some very large compressors may have windings with resistances this low.
5. Read the resistance between each terminal and one of the copper lines on the compressor.
6. Any reading other than infinite indicates a grounded motor.
   Note: A reading of 4 million ohms or higher may indicate contamination, not a grounded motor.
7. If the motor is not open, shorted, or grounded you can determine the C, S, and R terminals.
8. The highest reading will be between the Start and Run terminals. Therefore, the terminal that is NOT involved in the highest reading is Common.
9. The lowest reading is between Common and Run. The terminal that is involved in the lowest reading with the Common terminal is Run.

10. The Start terminal will be the terminal left over after identifying both the Common and Run terminals.

11. Complete the table on the next page and be prepared to explain your results to the instructor.

The three most typical terminal arrangements are shown in the boxes below.

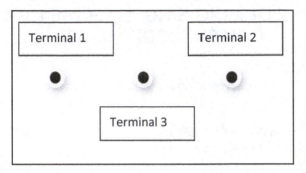

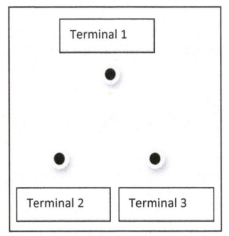

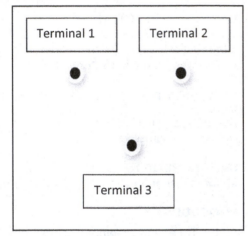

Compressor Motors Resistance Readings							
	TERMINAL READINGS						**CONDITION**
	Term 1 to Term 2	Term 1 to Term 3	Term 2 to Term 3	Terminal 1 to Ground	Terminal 2 to Ground	Terminal 3 to Ground	Good Shorted Open Grounded
Comp 1							
Comp 2							
Comp 3							
Comp 4							

Determining Common, Start, and Run Terminals		
	Terminal 1	Terminal 2	Terminal 3
Comp 1			
Comp 2			
Comp 3			
Comp 4			

LAB 88.3 WIRE AND OPERATE HERMETIC PSC MOTORS

LABORATORY OBJECTIVE
You will learn how to wire and operate hermetic PSC motors used in compressors.

FUNDAMENTALS OF HVACR TEXT REFERENCE
Unit 88 Troubleshooting Refrigeration Systems
Unit 36 Electric Motors

LABORATORY NOTES
You will wire PSC compressors, operate them, and measure their amp draw.

REQUIRED TOOLS, EQUIPMENT, AND MATERIALS
Volt-ohm meter
Clamp-on Ammeter
Electrical hand tools
PSC compressors
Run capacitor

SAFETY REQUIREMENTS
Check all circuits for voltage before beginning exercise.

PROCEDURE

1. Use the wiring diagram on the motor and/or the diagrams in the book to wire the PSC Compressor motors assigned by the instructor.
2. Be sure to check the motor nameplate voltage and compare it to the voltage you are connecting to the motor.
3. Operate the motor.
4. Use a clamp-on ammeter to read the motor current through the common wire.
5. Write down your reading in the chart below.
6. Discuss your results with the instructor.
7. Put up all the motors.

Voltage	Nameplate LRA	Nameplate RLA (if available)	Running Amps

LAB 88.4 TROUBLESHOOTING HERMETIC COMPRESSOR MOTORS

LABORATORY OBJECTIVE
The purpose of this lab is to learn how to troubleshoot PSC hermetic compressor motors for faults using volt, amp, and ohm meters.

FUNDAMENTALS OF HVACR TEXT REFERENCE
Unit 88 Troubleshooting Refrigeration Systems
Unit 36 Electric Motors

LABORATORY NOTES
You will use volt, amp, and ohm meters to identify problems in hermetic PSC compressors. You will determine if the motor is good, open, shorted, grounded, locked, or has a defective run capacitor.

REQUIRED TOOLS, EQUIPMENT, AND MATERIALS
Safety glasses
Gloves
Multimeter with volts, ohms, and microfarads scales
Electric hand tools
Hermetic compressors with PSC motors

SAFETY REQUIREMENTS
Check all circuits for voltage before beginning exercise.

PROCEDURE
The procedure varies depending upon the compressor reaction when you try to energize it

The Compressor does not try to start.
1. You need to determine if the compressor is receiving the correct voltage.
2. Check the voltage on the load side of the compressor contactor.
3. If there is no voltage, the problem is in the circuit to the compressor. Problems could include
   a. Tripped breaker or blown fuse
   b. Defective contactor
   c. Loss of control voltage to contactor due to
      i. open safety—high pressure, low pressure, overload
      ii. incorrect control wiring
      iii. bad control transformer
      iv. bad thermostat
      v. bad control board in unit
4. If there is correct voltage on the load side of the contactor, turn power off.
5. Discharge the run capacitor.

6. Remove the compressor terminal cover and check the wire connections.
7. Check the wires against the manufacturer's diagram.
8. Note where the wires are connected and remove them.
9. Use the procedure from Lab 88-2 to ohm the compressor motor out.
10. If the compressor ohms out, check the capacitor microfarads with the capacitor test function of the multimeter.

The Compressor tries to start but just hums.

Are the pressures equalized?

PSC compressors cannot start against system pressures, the pressures must be equalized.
1. Turn the power off and install gauges to check the system pressures.
2. If the pressures are not equalized, wait for them to equalize and then turn the power back on.
3. Check the amp draw on the common wire while the compressor is trying to start.
   a. If the amp draw spikes and then drops to around the rated load amperage (RLA), the compressor is running. Check the gauges to see if the high side pressure is rising and the low side pressure is dropping.
   b. If the amp draw is considerably lower than the compressor rated load amps (RLA), the compressor may actually be operating. Check the gauges. If it is drawing a low amp draw and not pumping, it has a mechanical failure.
   c. If the amp draw is high, somewhere around locked rotor amps, the compressor is locked—turn off the power.

Is the compressor receiving the correct voltage?

You need to determine if the compressor is receiving the correct voltage. Since it is humming, you know it has voltage, but it may not be correct.
1. You are going to briefly turn the power back on, make four voltage checks on the contactor, and turn the power back off. The voltages to check are:
   a. line side voltage
   b. load side voltage
   c. voltage across each of the sets of contacts (L1 to T1 and L2 to T2)
2. A voltage reading between L1 and T1 or L2 and T2 indicates a defective contactor.
3. If the voltage on the load side of the contactor is correct:
   a. turn off the disconnect.
   b. discharge the run capacitor.
   c. note where the wires are connected to the capacitor and then remove them.
   d. check the capacitor microfarads with the capacitor test function of the multimeter.
4. If the capacitor checks OK, remove the compressor terminal cover and check the wire connections.
   a. Check the wires against the manufacturer's diagram.
   b. Note where the wires are and remove them.
   c. Use the procedure from Lab 88-2 to ohm the compressor motor out.
   d. If the compressor ohms out, the compressor is locked down.
5. If the voltage on the line side of the contactor is lower than the voltage stated on the equipment:
   a. check connections at the contactor.
   b. check the circuit breaker.
   c. check connections at the disconnect.
   d. check the wiring from the disconnect to the unit.
   e. check the wiring from the power panel to the disconnect.

LAB 88.5 IDENTIFYING TYPES OF STARTING RELAYS

LABORATORY OBJECTIVE

You will learn to identify different types of hermetic motor starting relays.

FUNDAMENTALS OF HVACR TEXT REFERENCE

Unit 88 Troubleshooting Refrigeration Systems
Unit 36 Electric Motors

LABORATORY NOTES

You will use an ohm meter to identify and check different types of hermetic motor starting relays.

REQUIRED TOOLS, EQUIPMENT, AND MATERIALS

Volt-ohm meter
Electrical hand tools
Current relays
Potential relays
Solid state relays

SAFETY REQUIREMENTS

None

PROCEDURE

Use the photos and diagrams in the book to identify each relay. Use the ohm meter to check each of the relays.

Current Relays	Coil Resistance	Contact Resistance Upright	Contact Resistance Upside Down
Potential Relays	**Coil Resistance (2 to 5)**	**Contact Resistance (1 to 2)**	

Solid State Relays	Line to Run Resistance	Line to Start Resistance	

LAB 88.6 WIRE AND OPERATE HERMETIC SPLIT PHASE AND CAPACITOR START COMPRESSORS

LABORATORY OBJECTIVE
You will learn how to wire and operate hermetic split phase and capacitor start compressor motors used in HVACR.

FUNDAMENTALS OF HVACR TEXT REFERENCE
Unit 88 Troubleshooting Refrigeration Systems
Unit 36 Electric Motors

LABORATORY NOTES
You will wire split phase compressor motors, starting relay, and overload. You will then operate them, and measure their amp draw.

REQUIRED TOOLS, EQUIPMENT, AND MATERIALS
Volt-ohm meter and ammeter
Electrical hand tools
Split phase or capacitor start hermetic compressors
Current starting relay
Solid state starting relay

SAFETY REQUIREMENTS
Check all circuits for voltage before beginning exercise.

PROCEDURE
1. Use the wiring diagrams in the book to wire the compressor, starting relay, and overload.
2. Be sure to check the compressor nameplate voltage and compare it to the voltage you are connecting to the motor.
3. Operate the motor.
4. Use a clamp-on ammeter to read the motor current.
5. Write down your reading in the chart provided.
6. Discuss your results with the instructor.
7. Put up all the motors.

Voltage	Nameplate LRA	Nameplate RLA (if available)	Running Amps

LAB 88.7 WIRE AND OPERATE CSR MOTORS

LABORATORY OBJECTIVE
You will learn how to wire and operate CSR motors used in compressors.

FUNDAMENTALS OF HVACR TEXT REFERENCE
Unit 88 Troubleshooting Refrigeration Systems
Unit 36 Electric Motors

LABORATORY NOTES
You will wire CSR compressors, operate them, and measure their amp draw.

REQUIRED TOOLS, EQUIPMENT, AND MATERIALS
Volt-ohm meter and ammeter
Electrical hand tools
CSR compressors
Run capacitor
Potential relay
Start capacitor

SAFETY REQUIREMENTS
Check all circuits for voltage before beginning exercise.

PROCEDURE
1. Use the wiring diagram on the motor and/or the diagrams in the book to wire the CSR Compressor motors assigned by the instructor.
2. Be sure to check the motor nameplate voltage and compare it to the voltage you are connecting to the motor.
3. Operate the motor.
4. Use a clamp-on ammeter to read the current through the start capacitor. There should be a measurable current spike when the motor starts, and then the current should drop to 0 after startup.
5. Use a clamp-on ammeter to read the current through the common wire.
6. Write down your reading in the chart below.
7. Discuss your results with the instructor.
8. Put up all the motors.

Voltage	Nameplate LRA	Nameplate RLA (if available)	Running Amps

LAB 88.8 INSTALL HARD START KIT

LABORATORY OBJECTIVE
You will learn how to wire and operate CSR motors used in compressors.

FUNDAMENTALS OF HVACR TEXT REFERENCE
Unit 88 Troubleshooting Refrigeration Systems
Unit 36 Electric Motors

LABORATORY NOTES
You will wire CSR compressors, operate them, and measure their amp draw.

REQUIRED TOOLS, EQUIPMENT, AND MATERIALS
Volt-ohm meter
Clamp-on ammeter
Electrical hand tools
CSR compressors
Run capacitor
Potential relay
Start capacitor

SAFETY REQUIREMENTS
Check all circuits for voltage before beginning exercise.

PROCEDURE
1. Use the wiring diagram on the motor and/or the diagrams in the book to wire a hard start kit to the CSR Compressor motors assigned by the instructor.
2. Be sure to check the motor nameplate voltage and compare it to the voltage you are connecting to the motor.
3. Operate the motor.
4. Use a clamp-on ammeter to read the current through the start capacitor. There should be a measurable current spike when the motor starts, and then the current should drop to 0 after startup.
5. Use a clamp-on ammeter to read the current through the common wire.
6. Write down your reading in the chart below.
7. Discuss your results with the instructor.
8. Put up all the motors.

Voltage	Nameplate LRA	Nameplate RLA (if available)	Running Amps

LAB 89.1 INSTALLING A PACKAGED UNIT

LABORATORY OBJECTIVE
You will install a packaged air conditioning system, including determining location, connecting ductwork, electrical installation, and system startup.

FUNDAMENTALS OF HVACR TEXT REFERENCE
Unit 42 Electrical Installation
Unit 52 Duct Installation
Unit 89 Installation Techniques

LABORATORY NOTES
You will use the unit minimum circuit ampacity to size the conductor and disconnect switch. You will use the maximum overcurrent protection to size the fuse or circuit breaker protection. You will then install the disconnect switch, run conduit between the disconnect and the unit, and run power wire from the disconnect to the unit.

REQUIRED TOOLS, EQUIPMENT, AND MATERIALS
Safety glasses	Wire cutter
Multimeter	Wire crimper/stripper
Ammeter, Wire	6-in-1 screwdriver
Conduit	Disconnect switch
Conduit connections	Air conditioning unit

SAFETY REQUIREMENTS
A. Wear safety glasses
B. Power should be off and locked out
C. Check all circuits for voltage before beginning exercise.
D. Sheet metal duct can cause ever cuts, always wear gloves when handling duct

PROCEDURE

Step 1 – Unit Location
Most packaged units today are convertible between a curb-mount installation on a roof and installation on a pad or frame with ducts connected horizontally. This is determined by the building. Commercial applications on flat roofs are typically curb mounted while residential applications are almost always mounted on a pad or frame.

Curb Mounted Units
Roof curb installation is the most important part of the entire installation. This involves roof framing and penetration. The roofing company should be involved with curb installation. The HVAC mechanic's job is to ensure the curb is installed per manufacturer's instructions, oversee setting the unit on the curb, and make sure the seal is complete between the curb and the unit.

Pad Mounted Units

Most residential installations are pad mounted. These installations are typically used with a crawlspace or basement. The pad should be level. The unit should have at least 3 inches clearance above the ground. Duct clearance is the primary concern for location because the ducts must go through the foundation wall into the crawlspace or basement.

Frame Mounted Units

Packaged heat pumps must be mounted above the normal snow line. In some areas this is difficult to do with a pad and risers. A frame can be built that rises the unit above the snow line. Ducts for frame mounted units generally will not be located under the house, but run up the outside wall and into the attic or into a duct chase.

Step 2 – Duct Connections

On curb mounted units, ducts are connected to the curb. With units on either a pad or a frame, ducts are run from the unit into the house. It is important to completely seal any gap between the ducts and the house for weatherization and to keep out animals and insects. Unsealed wall penetrations are common entry points for unwanted guests. The portion of duct exposed to the outside must be protected from the weather. One common form of protection is to install a metal shroud. A shroud is cover with a top and sides over the ducts between the unit and the house. The top should be cross-broken and higher in the center to shed water.

Step 3 – Electrical Connections

Power Wiring

1. Find the minimum circuit ampacity (MCA) and the maximum overcurrent protection on the unit data plate and record both.

2. Use the National Electrical Code wire sizing charts or the manufacturer's installation literature to look up the correct wire size.

3. Size the disconnect by multiplying 1.15 x MCA.

4. The fuse or circuit breaker should be no larger than the maximum overcurrent protection.

5. Mount the disconnect switch within sight of the unit. It can be mounted on the unit, but it should not be mounted on a service access panel.

6. Run weatherproof conduit between the disconnect switch and the unit.

7. Run the power wire between the disconnect switch and the unit.

8. Single phase systems require 3 wires – 2 power wires and a ground.

9. Three phase systems require 4 wires – 3 power wires and a ground.

10. Have the instructor check your work.

Control Wiring

1. Mount the thermostat 5 feet off the floor on an inside wall out of drafts and sunlight. Run 18-4 solid thermostat wire between the thermostat and the unit.

2. Do not use wire smaller than 18 gauge. Do not use stranded wire.

3. Follow the unit diagram to wire the control wires.

4. Have the instructor check your work.

Step 4 – System Startup

Electrical Check

1. With the disconnect off but the power on, check the voltage into the top of the disconnect.

2. Compare the voltage to the minimum and maximum allowable voltages printed on the nameplate. If the voltage is outside of the allowable voltage range LEAVE THE DISCONNECT SWITCH OFF!

3. If the voltage is within the allowable voltage range, turn on the disconnect switch and check voltage to the unit.

4. If the voltage to the unit is within the allowable voltage range, turn the unit on.

5. Check the voltage to the unit with the unit operating. The voltage should still be within the allowable voltage range. The difference between the voltage at the unit with the unit off and the voltage to the unit with the unit operating is the voltage drop through the circuit. Ideally, it should be kept at 2% or less. For example, if the voltage with the unit off is 240 volts, the voltage with the unit operating should not be less than 235 volts.

Air and Duct Check

Operate the system with the fan on but the system switch set to off so there is no heating or cooling. Measure the temperature of the return air and supply air in the house. Ideally, they should be the same temperature when just circulating air. However, the fact that the air has to go outside and through the unit before coming back inside means that there is usually a little heat loss or gain. Leaky ductwork or poor insulation can cause the supply air to be more than 5°F warmer or cooler the return air. If the loss or gain is 5°F or more you should inspect the ductwork for air leaks or inadequate insulation. Also check the fit of the unit panels. Many packaged units have seals on the panels for the portion of the unit that air passes through. Simply being careless about removing and replacing panels can become a critical issue.

Performance Check

Operate the unit in both heating and cooling Check the airflow, the return air temperature, and the supply air temperature. The airflow and temperature difference in both heating and cooling should fall within the manufacturer's specifications. If the airflow and temperature difference between return and supply fall within the manufacturer's specifications there is no need to check refrigerant pressures. Since packaged units are fully charged at the factory, it is best to leave the refrigerant charge alone unless the system is unable to perform.

LAB 89.2 INSTALLING A SPLIT SYSTEM

LABORATORY OBJECTIVE
You will install a split system air conditioning unit according to the manufacturer's instructions, including all electrical wiring, refrigerant piping, evacuation, refrigerant charging, and system startup.

FUNDAMENTALS OF HVACR TEXT REFERENCE
Unit 24 Refrigerant System Piping

Unit 27 Refrigerant Leak Testing

Unit 28 Refrigeration System Evacuation

Unit 29 Refrigeration System Charging

Unit 42 Electrical Installation

Unit 51 Residential Split-System Air Conditioning Installations

Unit 89 Installation Techniques

LABORATORY NOTES
A split system consists of two parts – an indoor blower coil and an outdoor condensing unit. Each has its own disconnect and power supply. You will need to size the power wire, disconnect, and breaker size for each unit separately. You will run the refrigerant lines between the indoor coil and the outdoor condensing unit, pull a deep vacuum on the lines and coil down to 500 microns and weigh in the correct refrigerant charge. Finally, you will operate the system, perform an initial unit start and check procedure, and record operational data on the worksheet.

REQUIRED TOOLS EQUIPMENT AND MATERIALS
Safety glasses	Split system air conditioning unit
Gloves	Line set
Disconnect switch	Power wire
Control wire	Low-voltage thermostat
Hand tools	Refrigeration gauges
Vacuum pump	Vacuum gauge
Extension cord	Refrigerant scale
Thermometer	Multimeter
Clamp-on ammeter	

SAFETY REQUIREMENTS
A. Wear safety glasses and gloves

B. Turn off and lock out power before opening panel or doing any wiring.

C. Check power supply with non-contact voltage detector or meter BEFORE touching anything or doing any wiring.

D. Disconnect switch and unit should both be grounded.

E. Wear safety glasses and gloves when working with refrigerants. Liquid refrigerant can cause frostbite when in contact with eyes or skin.

PROCEDURE

Step 1 – Refrigerant Lines

1. Locate the condensing unit and indoor coil.
2. Determine the tubing size needed for the suction and liquid lines. For residential air conditioning, this is normally done by measuring the pipe stub outs on the condensing unit.
3. Roll out the tubing by holding the end of the tubing against the ground and straightening to form a long, straight length of tubing long enough to go between the condenser and the evaporator.
4. To keep the tubing clean and dry, the caps or plugs on the ends of the tubing should remain in place until just before the tubing is connected.
5. If you are working with a line set, roll the entire length out.
6. If you are working with a roll of refrigeration copper, roll out just the amount you need plus a little extra for security.
7. If you are working with refrigeration copper, slide lengths of ArmaFlex (trademark) insulation over the suction tubing.
8. Pull a length of thermostat wire beside the piping, leaving plenty of wire on each end. Tape the lines and wire together at regular intervals.
9. Pull the line set through the conduit, chase, or wall between the indoor and outdoor units. Connect the lines to the outside pipe stubs.
10. If the lines are brazed, be sure to use nitrogen to minimize internal oxidation inside the tubing. The Armaflex insulation should be pushed back and taped to keep it clear of the flames during brazing.
11. If the system is a heat pump the insulation must be rated for use on heat pump systems. Because the gas line is hot during the heating cycle, not all pipe insulation is rated for heat pump application.
12. Connect the suction line to the inside pipe stub.
13. Brazing should be done after a nitrogen purge to prevent oxidation. Work quickly when brazing the suction line.
14. If the evaporator has an internally installed TEV, overheating the suction line will destroy the TEV.
15. A filter drier should be installed in the liquid line.
16. The filter must be a bi-flow filter rated for heat pump use. Connect the liquid line to the filter drier.
17. Connect a short line from the filter drier to the evaporator coil.
18. Pressurize the lines with nitrogen and check for leaks using soap bubbles or an ultrasonic leak detector.

Step 2 – Evacuation

1. Use a core tool to connect your gauges to the system with the Schrader cores removed.
2. Connect the vacuum gauge to the core tool side port.
3. Connect your gauge manifold to the vacuum pump using the large vacuum port and a large diameter hose.

4. On new split systems you are only evacuating the lines and evaporator coil. The unit installation valves should remain closed (front-seated).
5. If the system is a dry system or an older system (not newly installed), you need to evacuate the entire system, including the condensing unit. For this the installation valves should be open (back-seated).
6. Evacuate the system down to 500 microns on the vacuum gauge.
   Note: On a new split system you will be evacuating only the lines and indoor coil. This typically should take less than an hour with a clean system that has no leaks.
7. While the vacuum is pulling you can go on to wiring the system.

Step 3 – Electrical

1. Find and record the minimum circuit ampacity (MCA) and the maximum overcurrent protection on the unit data plates of both units (indoor blower and outdoor condensing unit).
2. Use the National Electrical Code wire sizing charts or the manufacturer's installation literature to look up the correct wire size for both units.
3. Size the disconnect switch for each unit by multiplying 1.15 x MCA.
4. The fuse or circuit breaker for each unit should be no larger than the maximum overcurrent protection.
5. Turn off and lock out the power supply before doing ANY wiring.
6. Use a non-contact voltage detector or meter to check the power supply to insure it is off.
7. Mount each disconnect switch (one for each unit) within sight of the unit it controls. The disconnect switches can be mounted on the units, but they should not be mounted on service access panels.
8. Run weatherproof conduit between each disconnect switch and the unit it serves.
9. Run the power wire between each disconnect switch and the unit it serves. Single phase systems require 3 wires – 2 power wires and a ground. Three phase systems require 4 wires – 3 power wires and a ground.
10. Mount the thermostat 5 feet off the floor on an inside wall out of drafts and sunlight.
11. Run 18 gauge solid thermostat wire between the thermostat and the blower coil. The number of conductors will depend upon the equipment. Check the wiring diagram.
12. Run 18 gauge solid thermostat wire between the blower coil and the condensing unit. The number of conductors will depend upon the equipment. Check the wiring diagram.
13. Follow the unit diagram to wire the control wires.
14. Have the instructor check your work.

Step 4 – Charging

1. Calculate any additional required charge. For most systems, the amount of refrigerant shipped with the condensing unit is adequate for 15 feet of line. You will need to add for every to add for every foot over 15 feet of line. Refer to Table 89-2-1 for additional charge per foot of line.
2. Once the vacuum is achieved, close the refrigeration gauges and turn off the vacuum pump.
3. On new split systems, weigh in the amount of refrigerant needed for additional line length and then open the unit installation valves to let the refrigerant that is in the condenser into the rest of the system.

4. If the system is a dry system or an older system (not newly installed), you need to weigh in the full charge, which is the factory charge plus the extra line length allowance.

Table 89-2-1 Line Allowance		
Liquid Line Diameter	R22	R410A
1/4 inch	0.23	0.19
5/16 inch	0.40	0.33
3/8 inch	0.62	0.51
1/2 inch	1.12	1.01
5/8 inch	1.81	1.64

Step 5 – Startup
Pre-start Inspection
1. Measure the outdoor ambient temperature, the indoor temperature, and the indoor wet bulb temperature. Record them in Table 89-2-2.
2. Check to see that all packing materials and shipping supports have been removed.
3. Spin the condenser and evaporator fan blades by hand to verify that they are not hanging or seized.
4. Check to see that an air filter is installed.
5. Pour water in the evaporator condensate drain and verify that the drain work properly
6. Connect refrigeration gauges and verify that the unit has a refrigerant charge.

Initial Power Up
1. Power should be applied to the unit for 24 hours before turning it on. (This requirement is waived for this lab.)
2. Check and record the incoming voltage to the unit with the thermostat set to off and the disconnect switch on. The voltage should be between the minimum and maximum operating voltages stated on the unit data plate.
3. If the voltage is NOT within the acceptable operating range, turn the disconnect switch off and determine why.
4. Set the thermostat to "Cool" and run the temperature down to call for cooling.
5. After the unit starts, recheck and record the voltage to the unit.
6. It should still be within the acceptable operating voltages listed on the unit data plate.
7. If the voltage is NOT within the acceptable operating range, the voltage drop through the power wire is unacceptably high – turn off the power!
8. Check the amp draw of the power wire feeding the unit and compare it to the unit data plate.
9. The amp draw should be lower than the minimum circuit ampacity stated on the data plate.
10. Check the amp draw of the compressor, condenser fan motor, and evaporator fan motor.
11. Compare to the compressor RLA and fan motor FLA ratings on the data plate.

12. The readings most likely will not be exactly what is on the data plate, but they should be in the same range.

Operational Checks

1. If the system has ductwork, read the static pressure difference between the return and supply duct and compare this total external static reading to the manufacturer's specification.
2. If you do not have a specification, assume 0.5" wc to be "normal". Anything over 0.7" wc is usually too high.
3. Measure the system airflow using a flow-hood or other airflow measuring instrument and compare the actual airflow reading to the manufacturer's specification.
4. If you do not have a manufacturer's specification, the airflow should be approximately 400 CFM per ton.
5. After the unit has operated long enough for the temperatures and pressures to stabilize, record the system suction and discharge pressures.
6. Compare the pressures to the manufacturer's specifications.
7. Read the temperature of the return air and supply air.
8. Calculate the temperature drop across the coil (return air temperature – supply air temperature) and compare the coil temperature drop to the manufacturer's specifications. This varies depending upon the conditions and airflow.
9. Normal temperature drops can be anywhere between 10°F to 20°F.
10. Measure and record the system data in Table 89-2-2.

Table 89-2-2 Split System Data					
Pre-Start Check List					
Electrical Pre-Start Checks		**Refrigeration Pre-Start Checks**		**Mechanical Checks**	
Minimum Supply Voltage		Outdoor Ambient Temperature		Remove Shipping Materials	
Maximum Supply Voltage		Indoor Dry Bulb Temperature		Condenser Fan Spins	
Compressor RLA		Indoor Wet Bulb Temperature		Evaporator Fan Spins	
Condenser Fan FLA		Type of Refrigerant		Air Filter in Place	
Evaporator Fan FLA		Equalized Refrigerant Pressure		Evaporator Drain Works	
MCA (Min Cir Amps)					
Wire Sized to MCA					
Max Fuse Size					
Actual Fuse Size Installed					
Initial Power-up					
Voltage at unit with unit off					
Voltage at unit while					

running		
Amp draw on power wire		
Compressor amp draw		
Condenser fan motor amp draw		
Evaporator fan motor amp draw		

Refrigeration Operational Checks		Airflow Operational Checks	
Suction Pressure		Total External Static Specified	
Evaporator Saturation		Measured Total External Static Measured	
Condenser Pressure		Airflow CFM Specified	
Condenser Saturation		Measured CFM	
Suction Line Temperature		Return Air Temperature	
Suction Superheat		Supply Air Temperature	
Liquid Line Temperature		Evaporator Temp Difference	
Liquid Subcooling			
Discharge Line Temperature			

LAB 90.1 BELT DRIVE BLOWER COMPLETE SERVICE

LABORATORY OBJECTIVE
The purpose of this lab is to demonstrate your ability to conduct the proper maintenance service procedure for a belt driven blower.

FUNDAMENTALS OF HVACR TEXT REFERENCE
Unit 38 Motor Application and Troubleshooting
Unit 90 Planned Maintenance

LABORATORY NOTES
Some older residential and most commercial air handlers use belt drive blowers to deliver air through the duct system. The advantage of belt drive blowers over direct drive is the flexibility in choosing blower speed and the ease of manufacturing a blower wheel to withstand the RPM. It is too difficult to match all the factors, RPM, CFM, motor HP, etc. in all commercial applications. Most new residential systems have gone over to direct drive blowers but in commercial systems, belt drives will be around for many years to come. Occasionally a belt drive blower needs a complete overhaul including any or all of the following: motor and bearings, motor pulley, belt, blower pulley, blower bearings, blower shaft, cleaning, balancing, or rebalance.

REQUIRED TOOLS, EQUIPMENT, AND MATERIALS
Belt driven blower assembly	Tool kit
Multimeters	Clamp-on ammeter

SAFETY REQUIREMENTS
A. Check all circuits for voltage before doing any service work.
B. Stand on dry nonconductive surfaces when working on live circuits.
C. Never bypass any electrical protective devices.

PROCEDURE
Step 1 – Blower Data
Examine the blower and fill in Table 90-1-1 with the blower data.

Table 90-1-1 Blower Data			
Blower Motor Data		**Blower Wheel Data**	
Horsepower		Width	
Voltage		Diameter	
Phase		Shaft Size OD	
Frequency		**Pulley and Belt Data**	
LRA		Motor Shaft OD	
FLA		Motor Pulley	
Operating Amps		Blower Pulley	
Initial		Belt Size	
Final		**Lubrication Data**	
		Motor Bearings	Grease / Oil / None
		Blower Bearings	Grease / Oil / None

Step 2 – Perform Initial Inspection

Complete the Initial Inspection Check List. Make sure to check each step off as you finish it to keep track of your progress.

1. Spin the blower wheel slowly by hand.
2. Notice any drag noise or pulley wobble.
3. Install the clamp-on ammeter on one leg of power.
4. Place the blower door in the normal position for normal air flow and load on the motor. Leaving the door off can change the load on the blower, which changes the amp draw.
5. Obtain the fan only operation and record the motor amps.
6. Turn the blower off and remove blower door for further inspection.
7. Observe motor at start-up and watch for belt slipping.
8. Note any running noise, bearing grinding, or any noise other than normal airflow noise.

Step 3 – Blower Removal and Cleaning

1. Turn off and lock out power to the blower motor.
2. Verify this with a multi-meter voltage test.
3. Remove and store all panels and covers.
4. Loosen tension on belt for removal.
5. Remove belt and inspect for any cracks or glazing (caused by slipping).
6. Inspect pulleys for wear and grooving.
7. Inspect blower wheel blades of any accumulated debris.
8. Blow motor air passages with compressed air or CO_2.
9. Wipe down the motor and nameplate with a clean rag.

Step 4 – Blower Disassembly

1. Loosen setscrew from the blower pulley and remove the pulley.
2. Loosen the screws on the set collars.
3. Support the blower wheel to keep it in position and pull the blower shaft out of the blower.
4. Record the length and diameter of the blower shaft.
5. List any special features of blower keyways, flat sides, etc. for replacement.
6. Make notes on the blower shaft condition such as wear, grooving, rust, etc.

7. Roll the shaft on a flat surface to be sure it is straight and true.
8. Remove and inspect blower wheel bearings and replace if necessary.

Step 5 – Blower Reassembly
1. Install the original or a replacement blower shaft through the blower wheel bearings.
2. Install the original or replacement shaft pulley.
3. Align motor and blower pulleys with a straight edge. The straight edge should touch the outer edge of each pulley in at least two different places.
4. Lubricate the motor and shaft bearings.
5. Install new belt of the correct size.
6. Adjust the motor tension assembly for the correct belt tension. Consult the manufacturer recommendation for correct belt tension.

Step 6 – Blower Run Test
1. Spin the blower wheel slowly by hand and notice any drag noise or pulley wobble.
2. Install the clamp-on ammeter on one leg of power.
3. Place the blower door in the normal position for normal air flow and load on the motor.
4. Turn on the blower; read and record the motor amps in Table 90-1-1.
5. Compare measured amps with rated motor amps.
6. If the actual running amperage is incorrect the motor pulley will need to be adjusted or changed.
7. If the amperage is too high, the motor pulley diameter is too big.
   If the amperage is too low, the motor pulley is too small.
8. To adjust an adjustable pulley:
   a. loosen the motor belt tension and remove belt.
   b. Loosen the set screw on the outer sheave
   c. Turn the sheave in clockwise to increase pulley diameter
   d. Turn the sheave out counter-clockwise to decrease diameter
   e. Always adjust my ½ turns at a time.
9. Install the belt and adjust the belt tension.
10. Run the blower and once again measure the amperage.
11. Repeat steps 8 through 10 to obtain rated full load amps and rated CFM.

LAB 90.2 CLEANING EVAPORATOR COILS

LABORATORY OBJECTIVE
You will demonstrate your ability to safely clean an evaporator coil.

FUNDAMENTALS OF HVACR TEXT REFERENCE
Unit 16 Evaporators
Unit 90 Planned Maintenance

LABORATORY NOTES
You will clean an evaporator coil using both mechanical and chemical means. You will turn off power to the unit and remove any access panels in the way. Next, you will use a stiff brush and vacuum cleaner to remove debris on the surface of the coil. Then you will apply a chemical coil cleaner according to the manufacturer's instructions.

REQUIRED TOOLS, EQUIPMENT, AND MATERIALS
Safety glasses and gloves
Flat blade screwdriver
Phillips screwdriver
1/4" nut driver
5/16" nut driver
OR
6-in-1 with all the above
Adjustable jaw wrench
Units with dirty evaporator coils
Shop vacuum and brush
Bug sprayer with coil cleaner and bug sprayer with water

SAFETY REQUIREMENTS
A.   Make sure to wear safety glasses to protect against dust, debris, and chemicals.
B.   Make sure to wear gloves whenever handling cleaning solutions.

PROCEDURE
Safety glasses are required!

1.   Turn off and lock out power to the unit.

2.   You need access to the side of the evaporator coil where air enters.

   a.   For blower coils this usually is done by removing a panel to get to the coil.

   b.   Air Conditioning "A" coils mounted on furnaces require removal of the triangle side plate to gain access to the underside of the coil.

   c.   Occasionally air conditioning evaporator coils must be removed to gain access to the underside of the coil.

3. Use a stiff brush and/or shop vacuum to remove debris accumulated on the face of the coil.
    CAUTION: The coil fins are like thousands of razor blades ready to slice you up. Wear gloves and avoid contacting the fins while cleaning the coil.

4. Be careful not to bend the coil fins. They are very thin and easily damaged. The brush and/or vacuum should move in the same direction as the fins, not perpendicular to the fins.

5. Read the instructions for the coil cleaner and apply the coil cleaner according to the manufacturer's instructions.

LAB 90.3 CLEANING CONDENSER COILS

LABORATORY OBJECTIVE
You will demonstrate your ability to safely clean a condenser coil.

FUNDAMENTALS OF HVACR TEXT REFERENCE
Unit 14 Condensers
Unit 90 Planned Maintenance

LABORATORY NOTES
You will clean an condenser coil using both mechanical and chemical means. You will turn off power to the unit and remove any access panels in the way. Next, you will use a stiff brush and vacuum cleaner to remove debris on the surface of the coil. Then you will apply a chemical coil cleaner according to the manufacturer's instructions.

REQUIRED TOOLS, EQUIPMENT, AND MATERIALS
Safety glasses and gloves
Flat blade screwdriver
Phillips screwdriver
1/4" nut driver
5/16" nut driver
OR
6-in-1 with all the above
Adjustable jaw wrench
Units with dirty evaporator coils
Shop vacuum and brush
Bug sprayer with coil cleaner and bug sprayer with water

SAFETY REQUIREMENTS
  A. Make sure to wear safety glasses to protect against dust, debris, and chemicals.
  B. Make sure to wear gloves whenever handling cleaning solutions.

PROCEDURE
Safety glasses are required!

1. Turn off and lock out power to the unit.

2. You need access to the side of the condenser coil where air enters.

   You may need to remove an outer panel to gain access to the coil.

3. Use a stiff brush and/or shop vacuum to remove debris accumulated on the face of the coil.

   CAUTION: The coil fins are like thousands of razor blades ready to slice you up. Wear gloves and avoid contacting the fins while cleaning the coil.

4. Be careful not to bend the coil fins. They are very thin and easily damaged. The brush and/or vacuum should move in the same direction as the fins, not perpendicular to the fins.

   CAUTION: You should NOT brush spine-fin coils because brushing them will flatten the fins. You can vacuum them with a soft bristle brush if you hold the brush just off the surface of the fins and do not apply pressure.

5. Read the instructions for the coil cleaner and apply the coil cleaner according to the manufacturer's instructions.

   CAUTION: Most equipment manufacturers do NOT advise using chemicals on spine-fin coils.

LAB 90.4 TYPES OF MOTOR BEARINGS

LABORATORY OBJECTIVE
You will learn to identify the types of bearing used in electric motors.

FUNDAMENTALS OF HVACR TEXT REFERENCE
Unit 90 Planned Maintenance

LABORATORY NOTES

You will examine a selection of motors, identify the type of bearings in each motor, and identify the method of lubrication required for each motor.

REQUIRED TOOLS, EQUIPMENT, AND MATERIALS
Selection of electric motors

SAFETY REQUIREMENTS
Wear safety glasses

PROCEDURE
Select five motors in the shop and record their model number, type of bearing (ball or sleeve) and lubrication method used in Table 90-4-1.

Table 90-4-1 Motor Bearing Identification

Motor	Lubrication Method	Shaft Bearing	End Bell Bearing

LAB 91.1 MEASURING SYSTEM BASELINE PERFORMANCE

LABORATORY OBJECTIVE
You will measure and record the system baseline performance of an operating air conditioning system.

FUNDAMENTALS OF HVACR TEXT REFERENCE
Unit 91 Refrigeration System Cleanup

LABORATORY NOTES

Baseline performance shows what the system actually does in that particular installation, so it provides a benchmark to real operating conditions of that particular installation. Baseline data provides a benchmark for future service. Baseline performance data are often reference when replacing the original system refrigerant with a newer, climate-friendly refrigerant.

REQUIRED TOOLS, EQUIPMENT, AND MATERIALS
Safety glasses and gloves
Refrigeration gauges
Thermometer
Psychrometer
Multimeter
Clamp-on ammeter

SAFETY REQUIREMENTS
Wear safety glasses and gloves

PROCEDURE
1. Start the system and let it run until the temperatures and pressures stabilize, the compressor crankcase is warm, and the evaporator coil is wet (for air conditioning systems)
2. Take the measurements listed in Table 91-1-1 Compressor Data and complete that part of the table.
3. Take the measurements listed in Table 91-1-1 Condenser Data and complete that part of the table.
4. Take the measurements listed in Table 91-1-1 Evaporator Data and complete that part of the table.
5. Use the data collected to calculate the condenser heat of rejection.
6. Use the data collected to calculate the cooling capacity.

Table 91-1-1 Baseline Performance Data	
Compressor Data	
Suction pressure	
Suction saturation temperature	
Suction line temperature at compressor	
Suction superheat	
Discharge pressure	
Discharge saturation temperature	
Discharge line temperature at compressor	
Discharge superheat	
Amp draw of compressor	
Condenser Data	
Temperature of air or water entering condenser	
Temperature of air or water leaving condenser	
Condenser flow in CFM or GPM	
Air or water Delta T	
Condenser pressure	
Condenser saturation temperature	
Liquid-line temperature leaving condenser	
Subcooling	
Evaporator Data (Including Metering Device)	
Air Coils — Entering-air wet-bulb temperature	
Air Coils — Entering-air enthalpy from psychrometric chart	
Air Coils — Leaving-air wet bulb temperature	
Air Coils — Leaving-air enthalpy from psychrometric chart	
Temperature of air or water entering evaporator	
Temperature of air or water leaving evaporator	
Air or water Delta T	
Evaporator flow in CFM or GPM	
Liquid-line temperature before metering device	
Suction pressure	
Suction saturation temperature	
Suction-line temperature leaving evaporator	
Suction superheat at evaporator	
Performance Data	
Heat of Rejection BTUH for air cooled condensers CFM x Delta T x 1.08	
Heat of Rejection BTUH for water cooled condensers GPM x Delta T x 500	
Cooling Capacity BTUH for air conditioning (Leaving air enthalpy – entering air enthalpy) x CFM x 4.5	
Cooling Capacity BTUH for chillers GPM x Delta T x 500	

LAB 92.1 WATERFLOW EFFECT ON SYSTEM PERFORMANCE

LABORATORY OBJECTIVE
You will demonstrate the effect evaporator and condenser waterflow have on system performance.

FUNDAMENTALS OF HVACR TEXT REFERENCE
Unit 14 Condensers
Unit 16 Evaporators
Unit 92 Troubleshooting

LABORATORY NOTES
You will operate a water source heat pump in cooling, adjust the condenser waterflow and record the effect. You will them operate the unit in heating, adjust the evaporator waterflow, and record the effect.

REQUIRED TOOLS, EQUIPMENT, AND MATERIALS
Safety glasses and gloves
6-in-1
Temperature tester
Operating water source heat pump

SAFETY REQUIREMENTS
Wear safety glasses and gloves

PROCEDURE
Condenser Waterflow

1. Look up and record the system's high side test or design pressure on the system data plate.

2. Look up the manufacturer's specified water flow and inlet water temperature range for cooling.

3. Operate water-source system in cooling with the specified waterflow.

4. Check for proper operation, charge and adjust as necessary.

5. Record pressures and temperatures on Table 92-1-1.

6. Decrease the waterflow and observe for 5 minutes.

7. Record pressures/temperatures.

8. What effect does decreased condenser waterflow have on the system pressures/temperatures?

9. Return the waterflow to the manufacturer's specification and operate for 5 minutes until the pressures and temperatures return to normal.

10. Increase the waterflow and observe for 5 minutes.

11. Record pressures/temperatures.

12. What effect does increased condenser waterflow have on system pressures/temperatures?

13. Turn waterflow off and observe for 1 minute.

    Do not let the head pressure climb over the system's high side design pressure.

    Turn off the system if head pressure should start to climb close to the high side design pressure.

14. What effect does no condenser water flow have on system pressures/temperatures?

15. Summarize the effect of condenser water flow on total system operation.

Table 92-1-1 Condenser Waterflow Effects				
	Normal	**Decreased Waterflow**	**Increased Waterflow**	**No Waterflow**
Condenser Pressure				
Condenser Temperature				
Liquid Line Temperature				
Subcooling				
Evaporator Pressure				
Evaporator Temperature				
Suction Line Temperature				
Superheat				

Evaporator Waterflow

1. Look up the manufacturer's specified water flow and inlet water temperature range for heating.

2. Operate source system in heating with the specified waterflow.

3. Check for proper operation, charge and adjust as necessary.

4. Record pressures and temperatures on Table 92-1-2.

5. Increase the waterflow and observe for 5 minutes.

6. Record pressures/temperatures.

7. What effect does increased evaporator waterflow have on the system pressures/temperatures?

8. Return the water flow to the manufacturer's specification and operate for 5 minutes until the pressures and temperatures return to normal.

9. Decrease the water flow and observe for 5 minutes.

10. Record pressures/temperatures.

11. What effect does decreased condenser waterflow have on system pressures/temperatures?

12. Return the waterflow to the manufacturer's specification and operate for 5 minutes until the pressures and temperatures return to normal.

13. Turn waterflow off and observe for 1 minute.

    Do not let the low side saturation temperature remain below 32°F for more than a few minutes.

    Turn off the system if the low side saturation temperature falls below 32°F and remains there.

14. What effect does no evaporator water flow have on system pressures/temperatures?

15. Summarize the effect of evaporator water flow on total system operation.

Table 92-1-2 Evaporator Water flow Effects				
	Normal	Decreased Water flow	Increased Water flow	No Water flow
Condenser Pressure				
Condenser Temperature				
Liquid Line Temperature				
Subcooling				
Evaporator Pressure				
Evaporator Temperature				
Suction Line Temperature				
Superheat				

LAB 92.2 EFFECT OF AIRFLOW ON SYSTEM PERFORMANCE

LABORATORY OBJECTIVE
You will demonstrate the effect evaporator and condenser airflow have on system performance.

***FUNDAMENTALS OF HVACR* TEXT REFERENCE**
Unit 14 Condensers
Unit 16 Evaporators
Unit 92 Troubleshooting

LABORATORY NOTES
You will operate a refrigeration trainer, adjust the condenser airflow and record the effect. You will then adjust the evaporator airflow, and record the effect.

REQUIRED TOOLS, EQUIPMENT, AND MATERIALS
Safety glasses
Refrigeration trainer with variable speed blower controls

SAFETY REQUIREMENTS
Wear safety glasses

PROCEDURE
Condenser Airflow

1. Look in the instructions or on the unit for a maximum safe operating pressure.

2. Operate refrigeration trainer with both the evaporator and condenser fans set to normal airflow.

3. Check for proper operation, charge and adjust as necessary.

4. Record pressures and temperatures on Table 92-2-1.

5. Turn condenser fan speed to high and observe for 5 minutes.

6. Record pressures/temperatures.

   What effect does increased condenser airflow have on the system pressures/temperatures?

7. Return the condenser fan speed to normal and operate for 5 minutes or until the pressures and temperatures return to normal.

8. Turn the condenser fan speed to low and observe for 5 minutes.

9. Record pressures/temperatures on Table 92-2-1.

   What effect does decreased condenser airflow have on system pressures/temperatures?

10. Return the condenser fan speed to normal and operate for 5 minutes or until the pressures and temperatures return to normal.

11. Turn condenser fan off and observe for 1 minute. Do not let the head pressure climb over the maximum safe operating pressure that you looked up earlier. Turn the condenser fan to high and/or turn the compressor off if the high side pressure should rise to near the maximum high side operating pressure.

12. What effect does no condenser airflow have on system pressures/temperatures?

13. Summarize the effect of condenser airflow on total system operation.

Table 92-2-1 Condenser Airflow Effects				
	Normal	Decreased Airflow	Increased Airflow	No Airflow
Condenser Pressure				
Condenser Temperature				
Liquid Line Temperature				
Subcooling				
Evaporator Pressure				
Evaporator Temperature				
Suction Line Temperature				
Superheat				

Evaporator Airflow

1. Operate refrigeration trainer with both the evaporator and condenser fans set to normal airflow.

2. Check for proper operation, charge and adjust as necessary.

3. Record pressures and temperatures on Table 92-2-2.

4. Turn evaporator fan speed to high and observe for 5 minutes.

5. Record pressures/temperatures on Table 92-2-2.

   What effect did increased evaporator airflow have on the system operation?

6. Return the evaporator fan speed to normal and operate for 5 minutes or until the pressures and temperatures return to normal.

7. Turn evaporator fan speed to low and observe for 5 minutes.

8. Record the pressures and temperatures on Table 92-2-2.

What effect did decreased evaporator airflow have on system operation?

9. Return the evaporator fan speed to normal and operate for 5 minutes or until the pressures and temperatures return to normal.

10. Turn evaporator fan off and observe for 5 minutes.

11. Record pressures/temperatures on Table 92-2-2.

What effect did no evaporator airflow have on system operation?

12. Summarize the effect of evaporator airflow on total system operation. Be sure to include changes in superheat, subcooling, high side pressure and low side pressure.

Table 92-2-2 Evaporator Airflow Effects				
	Normal	Decreased Airflow	Increased Airflow	No Airflow
Condenser Pressure				
Condenser Temperature				
Liquid Line Temperature				
Subcooling				
Evaporator Pressure				
Evaporator Temperature				
Suction Line Temperature				
Superheat				

LAB 92.3 TROUBLESHOOTING SCENARIO

LABORATORY OBJECTIVE
Given a system with a problem, you will identify the problem, its root cause, and recommend corrective action.

***FUNDAMENTALS OF HVACR* TEXT REFERENCE**
Unit 43 Electrical Troubleshooting
Unit 53 Troubleshooting Air-Conditioning Systems
Unit 58 Troubleshooting Gas Furnaces
Unit 62 Troubleshooting Oil Heating Systems
Unit 73 Troubleshooting Heat Pump Systems
Unit 88 Troubleshooting Refrigeration Systems
Unit 92 Troubleshooting

LABORATORY NOTES
This lab is intended as a capstone lab. You are to use the knowledge and skills you have acquired to this point to solve a real-world problem.

REQUIRED TOOLS, EQUIPMENT, AND MATERIALS
Safety glasses and gloves
Hand tools
Manometer
Multimeter
Refrigeration gauges
System with a problem

SAFETY REQUIREMENTS
A. Wear safety glasses at all times
B. Check all circuits for voltage before beginning exercise.
C. Wear safety glasses and gloves when working with refrigerant.

PROCEDURE
Troubleshoot the system assigned by the instructor. Be sure to be complete in your description of the problem. You should include:
1. What is the unit is doing wrong?
2. What component or condition is causing this?
3. What tests did you perform that told you this?
4. How would you correct this?